The Rev Canon Robin Morrison writes with energy, enthusiasm and understanding as he tackles the fault lines between science and theology. The result is a thoughtful and thought-provoking argument about the origins of the universe which offers original insights into one of our deepest mysteries. *The Most Revd Dr Barry Morgan, Archbishop of Wales.*

Science and Theology melted together in a real page-turner. This book helps us harmonise religious beliefs and scientific observations throughout history in a multi-disciplinarian and thought provoking way. A must read for everybody who likes to reflect on these big questions. *Kristin Roaldseth. Formerly Director for Europe and Global Programs, Regional Representative for Central America, Director and Executive Officer for Latin America and the Caribbean in Norwegian Church Aid; Secretary for Human Resources Development in the Lutheran World Federation and Programme Director for Europe in the World YWCA.*

Love's Energy is a magisterial work, as thought-provoking in its content as it is impressive in its scope. It demonstrates a synthesis of cosmology, evolution and divinity in a refreshingly original and stimulating way. Here is an author steeped in theology who's also up to date with the sciences. If everything ultimately boils down to energy, Morrison's thesis is that love is both its genesis and its source. There's a sweep to this trilogy that is both breath-taking and absorbing. Buy all three, and begin the debate! *Rev Gethin Abraham-Williams, University Tutor and Author, Ecumenical Consultant & Former General Secretary, Churches Together in Wales.*

Each generation is challenged to rise above its perceived contradictions of religion and science. How do people living in a world of energy/matter and space/time as understood by scientists from Einstein to teams engaged in particle physics around the world understand those of the Abrahamic faiths who believe that God is love and created the universe? And vice versa. Rev. Canon Robin Morrison has given our generation a way to rise above our own perceived contradictions with a creative understanding of the interaction of love and energy. A must read in a time of religious fundamentalism and scientific preoccupation with weapons of mass destruction and new ways to monitor and control the human race. *Rev Dr John C. Moyer, formerly Director of Global Ecumenical Internships for Human Rights, Geneva Switzerland.*

Islam has always shown an interest in science and now this trilogy shows how important it is for Muslims and Christians today to engage with the latest and best scientific understanding of cosmology, quantum physics and evolution. I fully recommend this study to all Muslims and people of other faiths, believing that it contributes something special and perhaps unique to the wider debate about God and creation. *Dr Abdalla Yassin Mohamed OBE. Director of the 'Open Tent;' Director of Ihsaan Social Support Association Wales; Member of the Advisory Committee for the Centre for the Study of Islam in the UK, Cardiff University.*

Love's Energy is an ambitious and fascinating intervention in contemporary debates around creationism and science. Learned yet accessible and elegantly written, Morrison's work is both a synthesis of an extensive body of cosmological and evolutionary thought, and a sharply-considered meditation on the origins and complex indeterminacies of life. Whilst undertaking a thorough critique of reductive notions of creationism, *Love's Energy* offers to think creation differently – that is, through a concept of love grounded in contemporary theories of the nature of energy and matter. It will be of great interest to all readers, whether of faith or not, who wish to gain an informed perspective on the current, emotively-charged field of debate between the two cultures of theology and science.

Jeff Wallace, Professor of English,
Cardiff Metropolitan University

LOVE'S ENERGY
A Trilogy

'Beyond the entangled bank'
of cosmology, quantum and evolution

LOVE'S ENERGY   PART THREE

Pattern paradoxes in Evolution -
its origins and implications
in a universe that 'makes itself'

If creationism is dead,
what do we think we mean
when we say we believe in God,
as Creator or Source, in relation to a universe
that 'makes itself?'

DEDICATED
TO

Above all, Linda, my wife, for patiently supporting me through this project for the past twelve years and to Jane, my daughter and her family, Christian, Amelie and Theo, in the hope they will find something of love in it for them.

In memory of
the 'energy of love' in my much loved and much loving mother and father, Phyl and Len,

Also to the memory of
Dr. Hugh Montgomery, friend and physics lecturer at Edinburgh University with whom I had helpful, early discussions, a long time ago.

With thanks to
Michiu Kaku, Richard Feynman and Brian Cox for their inspiration and to all the great communicators of science through the ages – some of whom are featured in this work.

Lisa Martin who helped type the first drafts of the first chapters. The Rev John Moyer, until retirement, Geneva based Director of Frontier Intern Missions, Kristin Roaldseth, ex Director for Europe and Global Programs, Norwegian Church Aid; Rev Gethin Abraham-Williams, ex General Secretary of Churches Together in Wales; Avtarjeet S Dhanjal, an artist, philosopher and writer; Dr Lyndon Evans of CERN; Professor Colin Timms of Birmingham University; Sheikh Abdalla Yassin Mohammed; my wise Muslim friend. My thanks to them for looking at all, or parts of the manuscript and their advice and support. Also, to Terri Mackenzie SCHJ in the US for her continuing encouragement.

To Andy and Zoe Harcombe of Columba Publishing for their suggestions and to Mark Lane for his friendship and help in formatting

Love's Energy. The Trilogy Structure

The project is divided into three parts, each with its own book. Part One (LE1) develops the theme of Love's Energy in relation to cosmology and quantum science, within what Darwin called the *'entangled bank of creation.'* Certain sub themes emerged and these I call *pattern paradoxes* made up different kinds of possible, probable and potential relationships. These include, for example, *connectivity* and *separation; plentifulness* and *scarcity; predictability* and *indeterminacy; chance* and *plan; internal (process)* and *external (intervention); conditionality* and *autonomy.* These are not the same as the contradictions of opposites, because, as I learnt from Quantum physics, they are more like waves than fixed polarity positions. My analogy would be with the way different wave length frequencies oscillate rather than adopt set positions at different points along a continuum. In pattern paradoxes, there is an unpredictability in wave function, frequency and speed of movement. They are part of the fluctuating connectivity of physical reality. The pattern and the paradox relationship in each may differ and often this doesn't require further explanation in the text. These are pivotal to what I am claiming for Love's energy as ontological source and embedded processes in a universe that 'makes itself.' Part Two, LE2 looks at some theological dimensions of the argument, using particular early Christian mystics and other writers. It also explores something of the themes of Freedom, as part of the larger claim that the universe 'makes itself' and it ends with examples of Love's Energy in practice. This book, Part Three, looks at the theme in relation to evolutionary thinking.

In the introduction to LE1, I made the claim that creationism is untenable in a scientific age and is therefore dead, despite the fact that many, in all religions, still cling to it. This kind of literalism helps fuel a way of reading religious texts which may contribute to a culture of ideological and sectarian division, hatred and violence, with their own geo-political consequences. The trilogy as a whole is partly intended to challenge that kind of literalism, as well as suggest an alternative. Part Three, LE3, focuses on the key role of evolutionary thinking to move us forward beyond unscientific creationism. It offers to people from a non faith background, as well as thinking people in all faiths, a way of talking about the nature of God as Love and the Energy of that Love as the source of what we call 'creation.'

I have included some references to the theological development of Love's energy within LE1 and LE3 and not just in LE2 - theology should not be isolated from its context. I hope that readers more used to theological language will not turn away from the science and vice versa. For understandable reasons, they are treated as separate disciplines in our universities and schools (to say nothing of the attitude of publishers!) This, of course, was not always the case and, in that sense, this work is written in the tradition of what used to be called 'natural philosophy.' I hope it encourages more people, particularly those with a religious background, to cross boundaries and discover and make new connections. I have tried to inform the development of the argument by referring to notable individuals in science, theology and also in the arts. It has been a privilege to research something of the struggles and achievements of these individuals, who have added so much to our knowledge and understanding. I have, of course, only scratched the surface in most cases. Many readers will know more than I do about their lives and work and I am writing for the general reader who wants to know more. I hope what I have omitted, in terms of depth, is partly made up for in breadth. I have relished the opportunity to discover more about these thinkers and it has been difficult to restrict my comments. I was lucky enough to go to a good grammar school, but the way it taught science left me stranded and ignorant. When I came, near retirement, to write this trilogy, I had to start from scratch. If schools taught the big exciting science questions first, I suspect the motivation for the detail would then follow more naturally. I hope more specialist readers will forgive any mistakes I make in that detail. Inevitably, some of the science referred to will already be out of date as our knowledge is expanding so rapidly.

I have used photos of individuals where they are available and not restricted under copyright law. I have done my best to attribute and to comply with that law. Any failure to do so was unintentional. I have included chapter headings for LE1 and LE2, so that readers can see the territory covered and make their own connections.

CONTENTS - Part Three

Also available from Amazon as part of the Love's Energy Trilogy

LOVE'S ENERGY Part One. *Pattern Paradoxes in cosmology, physics and quantum.* **Contents** - From the Environment as Creation to the Creator as Love; The Hard Questions; The Beginning of Science to the Big Bang; The Big Bang; Big Bang Alternatives; Beyond Cause and Effect; The Physics of Energy; 'Does God Play Dice?' Certainty and Uncertainty' Entangled Separability and Connectivity.

LOVE'S ENERGY Part Two. *The Ecstasy in Energy. Pattern Paradoxes in the Theology and some thoughts on freedom as part of the autonomy of a universe that 'makes itself.'* **Contents** - Quantum Spirituality; Kenosis – the science of Potential and Propensity; Towards a Theology of Love and Energy; The Ek-stacy of Love; Love – the Dynamic Relationship between Energy and Matter; The Movement of Love as Energy; Love as Freedom Energy in Embedded Relationships; Some in between remarks about Freedom and Language; The Cost of Freedom as an Expression of Love's Energy; Love's Energy as Process; The Altruism and Mystery of Embedded Love – some case studies.

'It is interesting to contemplate an entangled bank, clothed with many plants of many kinds, with birds singing on the bushes, with various insects flitting about, and with worms crawling through the damp earth, and to reflect that these elaborately constructed forms, so different from each other, and dependent on each other in so complex a manner have all been produced by laws acting around us. These laws, taken in the largest sense, being Growth with Reproduction; Inheritance which is almost implied by reproduction; Variability from the indirect and direct action of the external conditions of life, and from use and disuse; Ratio of Increase so high as to lead to a Struggle for Life, and as a consequence to Natural Selection, entailing Divergence of Character and the Extinction of less improved forms. Thus, from the war of nature, from famine and death, the most exalted object which we are capable of conceiving, namely the production of the higher animals, directly follows. There is grandeur in this view of life, with its several powers, having been originally breathed into a few forms or into one; and that, whilst this planet has gone cycling on according the fixed law of gravity, from so simple a beginning endless forms most beautiful and most wonderful have been, and are being, evolved.'

Charles Darwin 'The Origin of Species,' The last words of the last chapter; 1979 reprint of 1976 issue of the 1968 edition Penguin Books Baltimore, Gramercy Books New York.

'I want to know how God created this world. I am not interested in this or that phenomenon, in the spectrum of this or that element. I want to know His thoughts, the rest are details.'

Albert Einstein from E. Salaman, 'A Talk With Einstein,' The Listener 54 (1955), pp. 370-371.

'Truth and untruth belong to the realm of signifance and values. I am not able to agree entirely with the assertion commonly made by scientific philosophers that science, being solely concerned with correct and colourless description has nothing to do with significances and values. If it were literally true, it would mean that, when the significance of our lives and of the universe around us is under discussion, science is altogether dumb.'

Sir Arthur Stanley Eddinton. 'Science and the unseen world.' Swarthmore lecture 1929. Quaker Books page 36.

LOVE'S ENERGY - PART THREE

Foreword

Sub headings

Admiration or understanding; Law and chance; The Energy of Profusion and Productivity; Connectivity and diversity; Difference and Division; Indeterminacy and Connection; Evo-physics and Aesthetics; Knowing and Not Knowing;

Few people have done more in our times to promote the science of evolution than Richard Dawkins. He is commonly perceived, by people of different faith communities, to be one of the highest profile enemies of their beliefs and what really matters to them in those beliefs. I do not believe this is the case, and even if it were, his challenge and contribution should be embraced as a critical friend, rather than as an enemy. Dawkins himself is haunted by the fact that in parts of America, with its constitutional separation of church and state, about forty per cent of people still believe that the earth is about 6000 years old, that all species are directly created by God and that the creation accounts in Genesis are literally true, according to at least one poll. While many accept such fundamentalist beliefs as being a private matter for the individuals concerned, or just irrelevant in modern societies, he is on a mission to challenge and change such attitudes. In this sense, he is doing what religious leaders seem to have struggled to do, ever since the work of Darwin, a century and a half before. While many, of course, have tried, it seems that adherents of such beliefs prefer to hold fast to their version of *creationism*, whatever the challenge of new knowledge. This holds as true for extremist Islam and its behaviours, as for fundamentalist Christianity.

Admiration or understanding

Dawkins accepts the place of wonder and admiration, as we contemplate the unlikely fact of human existence and the processes of evolution which brought us here. He is not asking religious people to give up this sense of wonder, but to transpose it into new levels of understanding. He is right. Wonder and admiration are not the enemy of understanding. If admiration stops short of curiosity, questioning and the search for knowledge, then it can enter a dangerous, ideological cul de sac, a protective edifice of defensive apologia. The move from admiration to curiosity and questioning does not remove the place of wonder. It enhances it. The paradigm shift he proposes is one that moves from an isolated statement of belief about creation, enhanced, as it must be, with an entirely appropriate sense of awe, to an exploration of the wonder of how nature itself evolved and 'made itself.' This exploration surely increases the sense of wonder for believers, even as it challenges their assumptions and paradigms. His own sense of wonder has been tested against nature's complexity and diversity; otherwise it would appear as a romantic sentiment which ignores many unpalatable sides of reality. This reality includes species behaviours which are abhorrently 'un-admirable' in their violence and destructiveness, especially, but not exclusively, in the human species. Some of these can be 'understood' as intrinsic to the processes of evolution and survival; others as dysfunctional expressions of human freedom which need understanding, challenging and transformation, as part of the way the universe evolves and 'makes itself.'

As part of the believer's search for a better understanding, she may have to tear down the protective defences of creationism to discover or rediscover the wonder of curiosity and connect it back with her underlying awe and belief at what is being considered and contemplated. In Love's Energy, we are invited to consider, contemplate and be in awe at the connection between Love, with its *emanating* energy, and the processes of natural evolution. In Part Three, we saw that the early mystical writers leaned heavily on the idea of incarnational *kenosis* as the moving towards and longing for the Other in the manifesting and emanating of God's Love. In this book, I am suggesting that this *emanating movement* of Love's Energy is embedded in the processes of evolution itself. This *embedded emanating* of Love's energy creates a connection with natural selection and its emanating movements of adaptation and change. It is this connectivity between the emanating movement of the Energy of Love and the processes of evolution which are at the heart of the idea in this book. I offer them as a framework for understanding the theological connectivity that is woven through the tapestry of this trilogy. It now comes into its own, in the way particular thinkers began to lay the foundations for Darwin's understanding of nature. The connection is, itself, an additional subject and object of admiration, if it allows for the movement from wonder to the search for greater understanding and back again. This is a lifelong and life embracing journey. As part of its searching and seeking, we come across the reality of *indeterminacy* within the connectivity, which, as we saw in LE1, is central to the way the universe 'makes itself.' Connectivity can mean many things in different disciplines but, in Love's Energy, I am following the findings of contemporary physics and biology that there is a certain indeterminacy within the potential of things to evolve, because nothing that exists is pre-planned or externally manipulated. In 'The Origin of Species' we shall see how Darwin struggled to identify connecting links 'between existing forms'[1] and to understand and articulate the nature of relationships across a huge variety of modifications taking place in species, dominant species,

---

[1] *The Origin* Gramercy books, Random House publishing 1979 p 440.

genera, varieties, groups, sub groups etc.[2] As he said *'variability is governed by many complex laws – by correlation of growth, by use and disuse and by the direct action of the physical conditions of life...Man does not actually produce variability; he only unintentially exposes organic beings to new conditions of life and then nature acts on the organisation and causes variability.'*[3] I see this as the biological equivalent of the way variability works within the connectivity of the physical universe. The idea of variability wasn't itself questioned when Darwin was writing. Aristotle himself had showed the importance of *particularity* in all its variety. Many had talked about the complexity and diversity of life. The problem was its explanation. As we shall see, he was 'standing on the shoulders of giants' – particularly Erasmus Darwin, Jean-Baptiste Lamarck and Charles Lyell[4] – when he courageously set out on his journey to explain how all things in the natural order changed without external intervention, although he still references a Creator, *necessity* and the secondary laws of nature along the way.

Law and chance

This was an age when various natural theologians were wrestling with the place of necessity and law in relation to the processes of change.[5] Many were building a case for God given, universal laws as the process through which God could relate to creation, if not by direct intervention. Law was a neat way out of the natural theologian's dilemma. Laws could be understood as part of a Creator's design and at the same time be observable and discussable in scientific endeavours. At the heart of Darwin's work was his bravery in talking about *chance* as well as *law*. It was one thing to replace natural theology with natural selection. It was quite another to imply that the later didn't work on the basis of a predictable and prescribed law in its processes of change from one state to another. Could chance be the cause of change? If so, was Darwin describing chance as a kind of causal law in the way *natural selection* operated? Was the implication of *'natural'* in tension with the word *'selection'* and was the latter a subtle reference to an 'agency' of direction within the former? Wallace pleaded with Darwin to drop the term, arguing that the metaphor of *selection* was inappropriately taken from *selective* breeding and implied a personification of nature's power.[6] If there was selection there must be a 'selector!' Darwin didn't want to use Spencer's *survival of the fittest* as it implied a hierarchy within the

---

[2] There is still dispute about the boundary lines between these and related terms which Darwin himself saw as ambiguous.

[3] Ibid p 441

[4] Darwin took his *Principles of Geology* with him on the Beagle and his beloved Milton.

[5] Many contributing to the Bridgwater Treatises, commissioned by the naturalist the Earl of Bridgewater, to explore *'the Power, Wisdom, and Goodness of God, as manifested in the Creation.'* (1833 to 1840).

[6] Darwin had of course used selective breeding as an analogy at the beginning of the Origin and disagreed with Wallace.

law,[7] but he couldn't escape the charge of inisipient design in the word 'selection' and maybe he didn't want to reject that implication completely. In his earlier work, he sometimes thought of Nature as having its own purposeful *being,* or as we might say 'agency.' For him, the metaphor *'natural selection'* had its own way of doing things, as if it were some kind of *vital force* weaving its way through the changes and chances of natural processes![8] We have to sympathise. It is one thing to include random chance in the process. It is quite another to produce a law that saw everything in general and in particular as happened solely by, or because of, chance. Clearly it didn't, because Darwin knew about the congruence and connection of living things and the influence of the past on the present, even if he could only glimpse what we now know about our historical and genetic inheritance. The rich ambiguity of his language positioned him away from the fixed certainties of Paley's perfect universe into the imperfections of chance contingencies within nature's congruence, connections and continuing constants. He was shifting our very perception of nature in order to sow the seeds of his theory. He was acutely aware of how entrenched traditional perceptions were, from which he sought to rescue us with his description of particular examples, case study by case study, in his long book. Just as subtely, he was using the *Origin* to shift our assumptions about nature in order to prepare the ground for a later shift in our understanding of the human within it. While Darwin hesitated to mention 'man' in his stories about pidgeons, beatles, finches and 'wild' life, Huxley took this on, in his 1863 *'Man's Place in Nature'* battle with Richard Owen – eight years before Darwin published his *'The Descent of Man and Selection in relation to sex'*[9] in 1871. The continuity with the *Origin* was symbolised by the repeated use of the word 'selection' in the title.

T.H.Huxley - an important figure in our story - found it very hard to accept that Darwin believed in chance or rather *spontaneity,*[10] rather than the regularity of laws. Clearly Darwin himself remained ambivalent about this dillemma.[11] He was not only including chance in his understanding of law, but making it a de facto law in itself. So, we must add *law* and *chance* to our list of *pattern paradoxes*, at least in the way it perplexed Darwin for most of his life. Struggling to replace design with *law* and *chance* was probably the hardest and most controversial

---

[7] But then used 'favoured' in the subtitle, as if someone/thing was doing the *favouring* of certain 'races,' and 'Struggle' which implies individuals competing or at war – but, he then explains, really meant something more like dependence. George Levine, in *Darwin and the Novelists Patterns of Science in Victorian Fiction* Chicago Pres 1991 p 100-103 identifies a metaphorical richness, fluidity and chosen ambiguity in his title which typifies his resistance to fixed

[8] In his 6th edition of the Origin he prefaces the para on natural selection with 'it may be metaphorically said..'

[9] Antoinnette Blackwell (1825-1921) criticised his view of sexual selection in 1875 for its gender bias in *'Sexes throughout nature.'* She was the first woman to be ordained (Congregationalist) in the US and a pioneer of women's rights.

[10] See T.H.Huxley *'On the Reception of the Origin of Species.'*

[11] *'If we consider the whole universe, the mind refuses to look at it as the outcome of chance – that is without design or purpose.'* Letter to Lord Farrer in Francis Darwin ed *More Letters of Charles Darwin* volume 11.p 395.

aspect of his theory, and not just for natural theologians. This was where he lost common ground with so many well wishers from that community. If all change, let alone *selection*, was caused by chaos and chance, this undermined the idea of meaningful rigour in scientific study as well as theology. Darwin was certainly rigourous in his collection of evidence, but remained open to the possibility that something that had happened, needn't always happen in the same way in the future. In divesting law of its metaphysical connotations, he was confronting and intuiting the presence of indeterminacy and uncertainty that would only appear decades later in the still controversial work of Werner Heisenberg. But are chance, chaos theory and indeterminacy the same thing, as they are bundled together in science? Clearly not, though they point in similar directions away from law, order and predictability! Just as design could mean many things, theologically, but points away from independent, natural processes, there are many dimensions of meaning hidden within the idea of chance. The moral philosopher, Mary Midgley certainly thinks so[12] and in many of her books she opposes 'scientism's' prententious claims or assumptions. If science moves from description to explanation of creation it encounters the challenge of agency, dependency and interdependency within the idea of autonomy. She dismisses what she calls the *casino model* of chance, arguing that this is well designed system that seduces gamesters with its claims to pure random chance. In our map making of how the world works, what does genuine chance mean, if there really is no clever method or system behind it, in our understanding of order or law? Is the casino analogy helpful? Probability theory can calculate the odds of dice being thrown in any one individual case, and then multiply this up. As we saw in LE1, chaos and swarm theories construct sophisticated maps of effects over distance and time in natural processes, so that what looks like chance may, from another perspective, be a dimension of cause and effect. Some mathematicians argue that there is no such thing as pure or random chance in reality and they talk of pseudorandom, number generators. Again, Mary Midgley is ueful here in her chapter on the '*Fascination of Chance.*'[13] In criticising Jacques Monod's *Chance and Necessity*[14] and Peter Atkin's insistence that Light discovers the briefest path by trying all other paths or that '*space time emerged by chance out of its own dust,*'[15] she says '*What sense this Just So Story can make when there is not supposed to be anything else in existence for these pot shot universes to fit with, or what meaning indeed the word 'chance' could have when nothing elses exists, must remain uncertain. I hate to be boring but the point must be made – the idea of Darwinian natural selection only makes sense inside a*

[12] *Science as Salvation*. A modern myth and its Meaning.Routledge 1992 p 36 onwards.
[13] Ibid p 43 ff
[14] London and Glasgow, Collins Fount, 1977
[15] See his *The Creation*. Oxford and San Franciso, W. H. Freeman 1987 p 51

relatively ordered system such as a biosphere. It presupposes an ongoing process, where there are specific conditions to meet and specific competitors who must also meet them. More widely, the idea of chance itself only makes sense within some specific existing order, an order to which it constitutes a partial or apparent exception.'[16] She warns against an 'inflationary' use of the ideas of chaos and chance. No wonder Darwin wrestled with the implications of 'natural selection!'

The philosopher and literary critic **G. H. Lewes** (1817-1878)[17] saw the law as *'merely the expression of the relations of co-existence and succession.'*[18] The word *merely* illustrates the contemporary project of removing all metaphysical implications. This was similar to **John Stuart Mill's** (1806-1873) view that laws referred only to uniformities among natural phenomena. The influential English Astronomer, **Sir John Herschel** (1792-1871), however, believed that a world without laws invited the danger of *mere* chance, given that the laws of nature *were not only permanent, but constant, intelligent and discoverable.*[19] He illustrates a widespread concern, if not sense of panic, when he wrote, *'We can no more accept the principle of arbitrary and casual variation and natural selection as a sufficient account per se of the past and present organic world than we can receive the Laputan method of composing books as a sufficient one of Shakespeare and the Principia. Equally in either case, an intelligence, guided by purpose must be continually in action to bias the directions of the steps of change – to regulate their amount – to limit their divergence – and to continue them in a definite course. We do not believe that Mr Darwin means to deny the necessity of such intelligent direction. But it does not, so far as we can see, enter into the formula of his law; and without it we are unable to conceive how the law can have led to the results.'*[20] This argument still resonates today[21] as people struggle to accept that the achievements of great writers (and for him it was Newton and Shakespeare), let alone of nature can be *merely arbitrary*. His view of *divergence* as needing some kind of limits was another mold that Darwin had to break! Meanwhile, Huxley[22] insisted that laws were not *agents of change in themselves but a mere record of experience.*[23]

---

[16] *Science as Salvation* p 44

[17] Who developed an interest in biology, encouraged Darwinism and religious scepticism and openly lived with George Eliot.

[18] *Compte's philosophy of the Sciences* London H Bohn 1853 p 52.

[19] In *Preliminary Discource on the study of natural philosophy* 1830 p 43

[20] *Physical Geography* Adam and Black 1862. p 12

[21] In some ways, Teilhard de Chardin, see chapter on Process theology in LE2, was still working in the tradition of Herschel rather than Huxley and Lewes.

[22] *Science and Christian Tradition* V 5 of Collected Essays Lond Macmillan 1893 p 77.

[23] See discussion in George Levine p 90 ff

The Anglican priest and scientist, **Rev Dr. William Whewell** (1794-1866), Master of Trinity Cambridge, refused to include the *Origin* in that college's library, mainly because it appeared to deny a moral distinction between animals and humans. In his *Plurality of Worlds* (1854), Whewell recognised the case for change and development and the congruence or consilience of a fact for other classes of facts or disciplines. Great theories could be confirmed by their relevance to, and agreement with, other areas of study. Whewell clearly saw development as teleological. Darwin saw that his own theory not only explained biological development but geological change as well – influenced of course by the work of the great **Sir Charles Lyell** (1797-1875). If Whewell's natural theology implied purpose and direction, if not final cause, Darwin also believed that nature was improving itself in the processes of chance selection. Whewell tried to keep final cause language out of his work, in the name of scientific enquiry. While Paley saw mechanistic evidence of design in all nature, Whewell was more cautious, even though, like so many in his era, he saw no conflict between a naturalistic study and theological belief. God had placed the design potential within the natural laws of the universe. This was certainly Chambers' view in his anonymous *Vestiges of Creation*. Law was the pivotal argument – it implied the regularity and consistency which natural theologians demanded of any divinely inspired or created design. It seemed to make good sound, English, common sense to think in this way! It was the half way house between theology and science. Looking back, we can see how an increasing use of *natural laws* as set between an extreme creationist view and emerging science had its own implications for theology. As a theological compromise, it drew natural theologians away from more dynamic ways of talking about the Creator - Creation relationship. If the purpose was to avoid the crudity of a deist designer, or interventionist God (whose shadow was always present) then one consequence was that Law became the dominant vehicle for understanding God's creative presence with its implication of 'fixed' and, therefore, inflexible categories in creation. The idea of law was appropriate to an age of mechanistic processes, but, as Darwin found, inadequate to explain the changing, dynamic and chance processes within nature. Darwin courteously disagreed with **Professor Asa Gray** that variation had been 'led' along definite lines in a revealing passage that includes many of his key ideas. *'If we assume that each particular variation was, from the beginning of all time, preordained, then that **plasticity of organisation** which leads to many **injurious deviations** of structure, as well as the redundant power of reproduction, which inevitably leads to a **struggle for existence**, and, as a consequence, to the **natural selection** or **survival of the fittest** must appear to us **superfluous laws of nature**. On the other hand, an **omnipotent** and **omniscient** Creator ordains everything and foresees everything. Thus we are brought face to face with a* *difficulty as insoluble as is that of free will and predestination.'*[24] It was the plasticity, struggles and failures of particular parts of nature that undermined or made superfluous any idea of a predetermined law. Of course, some theologians have argued that the fallibility and failures of nature could have been (albeit an inexplicable) part of the design; some have said that inconsistency and failure were part of the design in order to test our faith or for even greater purposes. Whewell believed they pointed to the *darkest and most tangled recesses of our knowledge*[25] in a science we haven't as yet understood. They certainly undermine the usual characteristics of a law and in that Darwin was surely and courageously right.

In this Trilogy, I replace the idea and functioning of Law with the idea of Love. Many have argued that a loving God would not allow the horrors of nature and this is a widespread and powerful argument. I offer the suggestion that Love allows the universe to become what it is becoming and work in its own ways of change and development because Love makes possible, and respects, the autonomy of its Otherness, even as it reaches out to make this possible and to be embedded within its consequences. Love, as we know it, longs to intervene to remove or prevent the suffering in those consequences, but it can only do so without contradicting its more basic longing for other *others* within this Otherness to find their own way of living out their creativity and survival within its autonomy; not least by evolving their own intellectual and emotional capacities and responses. Love therefore has an inbuilt *plasticity* which makes it more open ended, dynamic and creative than any fixed regularity of law. This coheres with the discoveries made by Darwin and Heisenberg and all they surmised about chance and indeterminacy within the order and disorder of natural processes – and this connects much of the thinking in LE1 and LE3.

Natural theologians could concur with any scientific study of natural processes if based on a law that pointed back to Divine intent or design. Whewell argued against 'intent' or purpose, even though it is implied by the idea of law – arguing we must remain open to discovery not to assumption. Darwin agreed, but went further, as to *assume* intent was to close down exploration. Both shared the eighteenth century inheritance of natural theology and the work of Paley in particular. A Law, by its very nature, implied repeatability and predictability. The past would be the same as the present on this basis. It would consist of the same fixed orders and processes of movement from one state to another. Here was the problem for Darwin. As he looked back from the present, he didn't see a fixed

---

[24] *The Variations of Animals and Plants under Domesticiation* (1868) Vol 11. New York 1900. p 415.
[25] *On Astronomy and General Physics considered with reference to Natural Theology* 1836 p 327-8

and ordered state, moving forward into a different fixed and ordered state, as if by divine intent through a naturalistic law. He discovered processes which were more gradual and dynamic, more unpredictable and random rather than fixed and prescribed. Could any causal law work like this? Darwin's struggle was truly heroic, given the shift he was making from Paley's understanding of adaption and also Whewell's scientific natural theology. From them, he inherited a belief that every particular detail of nature had its place and purpose in a wider and connected understanding of creation. He wrote in the *Descent,* when rethinking the arguments of the *Origin,* '*I was not able to annul the influence of my former belief then widely prevalent that each species had been purposely created; and this led to my tacitly assuming that every detail of structure, excepting rudiments was of some special though unrecognized service.*'[26] If every part of every form had its place and function, it was tempting to believe that everything was ordered according to a purpose or pattern that was fixed by a law, if not a Divine design! What if the idea of adaption he borrowed from Paley and others, could be turned to mean the absence rather than presence of design? What if every part of any form was in a process of change because of more random causes, and what if it could develop in any and different directions? Worse, what if change led to imperfection, aberrations, mistakes and extinctions?

The evidence began to stack up that some living things didn't develop or adapt but wilted away and failed. What did this say about the wisdom and order of any fixed law, let alone a divine intelligence? Darwin saw the particular, rather than the more abstract, if not mechanistic, laws of Paley and Whewell, as being crucial. Therefore any exception or 'failure' in the detail of the particular had to be taken seriously rather than assimilated into the general. Paley saw 'chance' aberrations as reflections of the observer's failures or ignorance. The particular didn't disprove the general theory. Darwin, like Heisenberg later, took the unpredictable and indeterminate as being inherent and not exceptional. The faith of the natural theologians could not be shaken by a particular observation of an aberration, but, for Darwin, each one was crucial and demanded an adjustment to his thinking. It was as if he were taking Aristotle's side against Plato's ideal forms – and the oscillating ideas about the general and the particular in that relationship have become, through time, their own *pattern paradox*! For Whewell, living things adapted to fit a connected organisation and arrangement based on fixed laws, and, in so doing, proved that every ordinary thing had its place in the universe.[27] Darwin found that as things adapted, some 'fitted' the connections and others didn't. For the natural theologians, the law said that all adaptions were perfectly fitted and connected (like Paley's mechanical model).

---

[26] Princeton university press edition 1981 p 153
[27] See his *Astronomy and General Physics*

Darwin observed the reality of disconnection and failure even within the *entanglement* of connection. He also observed the slowness of change and wrote in the Origin *'Why on the theory of Creation, should there be so much variety and so little real novelty? Why should all the parts and organs of many independent beings, each supposed to have been separately created for its proper place in nature, be so commonly linked together by graduated steps? Why should not Nature take a sudden leap from structure to structure? On the theory of natural selection, we can clearly understand why she should not; for natural selection acts only by taking advantage of slight successive variations; she can never take a great and sudden leap, but must advance by short and sure, though slow steps.'*[28]

Darwin was changing our myth making narratives, fiction, beliefs, perceptions and metaphors, as well as our science, when he described the present as only a moment in a much longer story to which it was naturally connected. Well before Einstein, he was feeling his way towards his own theory of time and relativity. In his model, unlike the more mechanistic images of some natural theologians, there was an organic and changing connection between time's different branches. The pivotal words he used to capture *change* – development, adaption, selection, mutation, metamorphosis and transmutation – were to lay the foundational language of *'evolution'* which became such an influential, narrative world map, not only in science but in literature and philosophy.[29] He combined the ontogeny of individual development with the phylogeny of species development in the organising idea of 'evolution' which was to influence how we have viewed all of nature ever since. This longer view of change and natural transformation owed much to Sir Charles Lyell's geology[30] and the pioneering work of the Scottish geologist **James Hutton's** (1726-1797) *'uniformitarianism'* – that all the 'land' of the globe had been produced by ongoing and slow moving, natural processes formed by the *'consolidation of loose or incoherent materials'* [31]which were then *elevated* from the bottom of the sea. At the time, it was common place to assume that the Flood was the only great catastrophre, and so must explain all drastic changes to natural phenomena. This was before the affect of glaciers was understood and at a time when it was mostly impossible to accept that some fossilised creatures were animals that had become extinct. This went

---

[28] 5th edition John Murray p 239

[29] See Professor Gillian Beer *Darwin's Plots –Evolutionary Narrative in Darwin, George Eiliot and Nineteenth –Century Fiction*. Cambridge 2009 now in its third edition. Also George Levine acknowledges his debt to Beer and works on the assumption that *the Victorian novel clearly joins with science in the pervasive **secularizing** of nature and society and ...the tradition of natural theology was threatened and largely dismantled by Darwin's science and in the process nature, society, narrative and language itself was **desacralised**. p viii* The words in bold (my emphasis) are, he admits, possiblly simplistic in his chapter on *Natural Theology; Whewell and Darwin p 24 ff*

[30] *'Principles of Geology: being an attempt to explain the former changes of the Earth's surface, by reference to causes now in operation'* 1830-33. A revealing title relating present processes to the past.

[31] *'Concerning the System of the Earth'* delivered to the Royal Society of Edinburgh July 1785.

completely against the assumption that God would never make such a mistake! Yet, the great **William Buckland** (1784-1856), geologist and Dean of Westminster had written *'No one who believes in the Bible to be the world of God has cause to fear any discrepancy between this, his word, and the results of any discoveries respecting the nature of his works.'* [32] Buckland is a key figure in our story, not only in his own right, but in his support for **Mary Anning**[33] (1779-1847) who discovered so many significant fossils in the Lyme Regis area, including Ichthyosaurs (which she described as crocodiles) and Plesiosaurus (later called a megalosaurus by Buckland) as well as countless ammonites, fish fossils and 'bezoar stones,' or fossilised faeces (Buckland named them coprolites in 1829). She helped keep her poor family solvent by selling such finds (this was perhaps the basis of *'She sells seashells on the sea shore'*[34]) in her fossil shop, and worked hard at cleaning and restructuring the fossilised skeletons of her larger finds. Her influence was far more extensive than her recognition,[35] and she was, in truth, treated abominally by most people who used her knowledge and profited from her finds. As a woman she was not even allowed into meetings, let alone membership, of the Geological Society[36] where the clergyman and geologist, **William Conybeare**, (1787-1857) Dean of Llandaff, never acknowledged her huge part in the story of a paradigm shift in science (and theology), even when lecturing on her finds based on her years of hard work which enabled such key skeletons to be studied and exhibited! She is of huge significance for our story of evolution, second only, perhaps, to Wallace and Darwin himself. Richard Owen and Charles Lyell both 'consulted' her in 1839![37] Mary also asked some very significant theological questions,[38] based on the implications of her finds - not least about extinction.

The Energy of Profusion and Productivity
The processes of change included extinction as well as development, destruction as well as creativity – *pattern paradoxes* which made Darwin's idea all the more

---

[32] Who took a *catastropist position and* like Lyell, believed in a much longer time frame which included countless extinctions. See his *Geology and Mineralogy Considered with Reference to Natural* 1837 p 19

[33] She survived a lightening strike in 1800 which killed 3 women with her, standing under a tree. She was only 15 months old and some locals considered this event may have contributed to her unusual intelligence!

[34] Terry Sullivan's 1908 tongue twister was allegedly based on her work, according to P.J. McCartney's 1978 work on *Henry de la Beche* whose painting *Duria Antiquior* – the first watercolour depiction of a pre-historic life forms – was probably based on her work.

[35] This was to change in the following century. See also the vivid novel *'Remarkable Creatures'* by Tracy Chevalier, Harper Collins 2010.

[36] Of which Buckland became President in 1840.

[37] Louis Agassiz, a Swiss palaeontologist visited her and Elizabeth Philpott in 1834 and benefited by their knowledge of fish fossils in the region and thanked them in his book 'Studies of fossil Fish' and named two species In the early 1840s he named two fossil fish species after her – Acrodus anningiae and *Belenostomus anningiae.*

[38] She began life as a Congregationalist before turing to the Church of England, but she proudly kept a volume of the *Dissenters' Theological Magazine and Review*, in which the family's pastor, the James Wheaton insisted God had created the world in six days and encouraged readers to study the new science of geology.

emotionally disturbing, as well as realistic. His was no simple linear teleology, despite its underlying optimism about the *beauty* and *grandeur* in this total system explanation about the evolving organisation of the natural world. The *endless forms most beautiful and wonderful,* in all their bewildering number and countless variety, depended upon an underlying fecundity for their profusion. This great productivity of nature – an idea the Victorians and natural theologians like Whewell would have applauded[39] – was essential to its capacity to adapt and it spoke of nature's boundless energy.[40] This *energy* would itself be adapted, by those who developed the myth for their own purposes, in the direction of an implied sense of teleological improvement and progress in industry, morals and Empire! For Whewell, the abundance, variety and complexity of laws in the universe is incalculable, but there is an adapting cohesion and coherence in their purpose and provision.[41] As we follow something of the constraints, temptations and dilemmas in this story, we are reminded of Darwin's courage in confronting such basic questions, in all their deceptive simplicity, about processes that were taken for granted politically and theologically, or, because of theology, only rarely explored. Darwin takes care, in his final chapter, to remind us again of artificial breeding processes as a gentle way into natural selection. '*It is certain that man can largely influence the character of a breed by selecting, in each successive generation individual **differences** so slight as to be quite inappreciable by an uneducated eye. This process of selection has been the great **agency** in the production of the most distinct and useful domestic breeds. That many of the breeds produced by man have to a large extent the character of natural species is shown by the inextricable doubts whether very many of them are varieties or aboriginal species. There is no obvious reason why the principles which have acted so efficiently under domestication should not have acted under nature.... If then we have under nature **variability** and a **powerful agent** always ready to act and select, why should we doubt that **variations**, in any way useful to beings under their excessively complex relations of life, would be preserved, accumulated and inherited?' Why if man can by patience select **variations** most useful to himself should nature fail in selecting variations useful under changing conditions of life to her living products? What limit can be put to this **power** acting during long ages and rigidly scrutinising the whole constitution structure and habits of each creature – favouring the good and rejecting the bad. I can see no limit to this power in slowly and beautifully adapting each form to the most **complex relations of life**.* [42] He then goes on to discuss the problem of demarcation lines

---

[39] The more numerous the more proof of intent and design.
[40] Darwin was sensitive to Malthus's concern about the effect of profusion on the vulnerability of populations unless they learnt to adapt.
[41] See *Astronomy and General Physics* p 11
[42] Ibid p 441 – 443. My emphasis in bold

between species and varieties on the basis of a commonly held view, in his day, that God created an *'original species by a special act of creation and varieties which are acknowledged to have been produced by* **secondary laws.***'*[43] But we note his use of *'power'* and *'agency'* in the lines above this. As we saw in LE1, and shall see later in this work, Darwin's predecessors all struggled with the question of the causes as well as nature of the links across diversity and variability. For some, it was a mechanistic succession of divine 'fiats' in the skeleton of time and history. For others, it was a more dynamic process and, as we saw in LE1, various terms were used to describe a *'living force'* or a *complexifying force'* within living things that explains the inner process of adaption and transformation. Darwin's challenge to the foundational myth of his contemporaries was to suggest that nature itself was capable of being the powerful agent of change and that this took place through graduated steps over a longer period of time, as evidenced by variation, but without external agency. Darwin's theory came to be perceived as providing a self-explaining and self-organising natural order that was constantly moving and changing, albeit more slowly than anything in the Biblical record, as it was then understood. What Copernicus had done for cosmology, Darwin did for evolution. Not only was the universe not 'created' around and for planet earth, but the endless forms of diverse species on this planet were not 'created' around and for humans, who themselves evolved from other species and only appeared very late in the timetable.

Connectivity and diversity

Since the work of Crick and Watson, leading up to the discovery of the double helix in 1953, we know more than Darwin could possibly have known about how the processes of evolution pass on coded information which ensures the reproduction of both *connectivity* and *diversity*. In LE1 and LE2, I attempted to hold *connectivity* and *diversity* together as part of the ontology of Love's Energy and the autonomy it creates in its longing for the Other to exist. We are all the same, or genetically similar,[44] *and* we are all different. We are also discovering how even our (genetic) nature can be influenced by our nurturing, or lack of it, in a complicated, but connected interaction between our brain chemistry and its wider environment. There is a vast literature on this subject. For example, Tom Leinster,(School of Mathematics and Statistics) and Christina A. Cobbold (Boyd Orr Centre for Population and Ecosystem Health), University of Glasgow, UK, argue that *'All the most commonly-used indices are based on a crude model in which distinct species are assumed to have nothing in common, contrary to what*

---

[43] Ibid 443 My emphasis in bold

[44] In *'Measuring diversity: the importance of species Similarity.'*

every biologist knows. Non-specialists are amazed to learn that a community of six dramatically different species is said to be no more diverse than a community of six species of barnacle. There is a mismatch between the general understanding of biodiversity as the variety of life, and the diversity indices used by biologists every day... 'Diversity' is one of those words that is used freely in both scientific and nonscientific contexts, often with different meanings (Adams et al. 1997)..The OECD guide to biodiversity for policy makers states that 'associated with the idea of diversity is the concept of "distance", i.e. some measure of the dissimilarity of the resources in question' (OECD 2002). But the conventional measures of diversity ignore this aspect altogether...The best-known similarity-sensitive diversity measure is the quadratic entropy of Rao (1982a,b). This is receiving increasing attention, but is still a minor player. Perhaps theoretical ecologists have been hesitant to introduce new diversity indices when the profusion of similarity-insensitive indices is already perceived to form an impenetrable jungle (Ricotta 2005)—although work of Jost (2006, 2007, 2009) dispels that myth. We present a new family of similarity-sensitive diversity measures.'

Difference and Division

In LE2, we also saw that the early mystical theologians insisted that *difference* needn't lead to *division* or *hostility*. I believe this is relevant to Darwin's genius, as he discerned how intermediate stages of varieties of species were connected. One of the remarkable statements in the New Testament is that in Christ the *wall of hostility or division* has been broken down in the very nature of the potentiality of things. In Ephesians 2,14 we find *'For he himself is our peace, who has made us both one and has broken down in his flesh the dividing wall of hostility.'* The context for this was certainly not contemporary evolutionary biology or mathematics, but rather the divisions caused by Jew/Gentile relationships and the way the law separated people out based on the exclusions emanating from many of its commandments. This early insight has been applied to many international and local disputes ever since, particularly where difference has led to tragic divisions, as we saw very acutely in Christian schisms of the past and in sectarian Islam today. I like to think that, in addition to these racial, religious and political walls of hostility, this verse has a more general meaning as well, and, therefore, can be applied to the debate about genetic and physical similarity and dissimilarity.

Indeterminacy and Connection

In LE2, we saw that some of the early mystical writers had intuited the theological and physical *indeterminacy* within the *connectivity* between what we have called 'A' and 'B,' and their related differences. They were not, of course, aware of

Heisenberg or modern quantum physics, but they have something to contribute to the thesis of Love's Energy and the contemporary debate. Darwin's challenge to any theology, or science of development, is to take chance and indeterminacy as seriously as any fixed laws. In LE1, we also claimed there is a connection between the processes of cosmological evolution and species evolution on this planet. There are certainly differences, but also many similarities, particularly in the physics involved. One of these is the effect of gravity and this is worth recalling, briefly, at this point, [45] not least to honour its part in the final paragraph of *The Origin* –'*and that whilst this planet has gone cycling on according to the **fixed law of gravity**, from so simple a beginning endless forms most beautiful and most wonderful have been, and are being, evolved.*' While evolution sculpts the nature of physical reality through the chance processes of natural selection, it doesn't have a completely free, or independent hand. Biology is constrained by the laws of physics. The mass of things is governed by its own universal laws and the law of gravity. For example, the physics of how fluids behave is universal and unaffected by biology, but it affects the processes of natural selection. The density and viscosity of fluids are related to their inter-molecular forces. Water is eight hundred times denser than air. It provides support for marine animals, but they use huge amounts of energy to counteract drag and friction. This influences the way evolution has shaped larger animals such as whales, dolphins and sharks. They have a similar streamlined 'aircraft' shape, where length is proportionate to width, to aid movement through the water.[46] If weight is the force exerted by gravity on a particular mass, then in water, as we learnt long ago, from Archimedes,[47] weight is reduced, because another force of buoyancy is operating proportional to the weight of water displaced. Animals in water depend upon this – much of their mass is made up of water, so it approximates to the balance of equal and opposite forces, making water supportive, because it releases marine animals from the force of gravity. Meanwhile, on land, physics presents evolution with a different challenge.

Evo-physics and aesthetics

Gravity limits the size and shape of things cosmologically, and in all evolved species – so, the height of trees is limited by physics as they find the best way of supporting their own mass against the pressure of gravity. Each species has to find its own way of balancing size, weight, mass and surface areas within the laws of physics. Gravity limits what evolution can do in selecting the best solutions over time. In the *otherness* of nature, there is space for a plurality of solutions to

---

[45] I am grateful to Brian Cox for this idea which appeared in his 17.2.13. BBC 2 Programme, *Wonders of life*. See also his book with Andrew Cohen under the same title Harper Collins 2013. P 124 ff.

[46] And skin texture where 'scales' called dermal denticles assist with lowering the 'Reynolds' number (quantifying efficiency of a shape moving through a fluid)

[47] Scyracus c 287 B.C

the shape and type of each species' evolving otherness. The great diversity of species demonstrates nature's different ways of coping with the laws of gravity. The shapes of this evo-physics evoke in humans an aesthetic sensitivity. We have evolved to appreciate and enjoy the beauty of a tree, or the shape of a dolphin. However, many humans squirm at the 'sight' of some insects, because we have not evolved the same visual capacity to observe the details of their shapes, colours and textures. So, we perceive them inhabiting a different world from our own, where different rules seem to apply – but of course they don't. At this much smaller scale, gravity doesn't seem to operate in the same way. Insects can often lift more than their own weight and walk up vertical structures. They can fall from a great height, without harm, because of the low gravity effect. While all things fall at the same rate under gravity, their surface area, in relationship to volume, changes their speed of fall because of resistance of air to size.[48] The cohesive force of electro magnetism works for smaller animals in a way it can't for larger ones and some insects can literally walk on water. Paradoxically, as animals get smaller, they seem to become stronger. They have very low mass in relation to the forces of gravity and size greatly influences energy use within all living things. There are also so many of them (10 quintillion?) and so many different species (between two and thirty million?), many of which are always hidden from our view. These facts, in our human perception, may combine to make us feel very insecure. It is one thing to admire the beauty of an insect under a microscope, when our visual aesthetics comes into play. It is quite another thing to observe them from the 'distance' of our comparative sizes when they appear so far away and alien to us. At this point, many find that their evolved aesthetic appreciation breaks down at the sight! So, evo-physics and aesthetics seem to be related in human experience. As we reflect theologically on this aesthetic of evo-physics, using the simple code A and B (used in LE1 and LE2) we might say that A learns to perceive B with a sense of wonder at its otherness in a world of diverse otherness. It can do this partly because A shares with B the same ontology of evo-physics. A and B are both part of the same Otherness, despite the differences which have evolved in their own sense of otherness, by contrast with the otherness of others. They are energetically related, as part of Love's longing and, in that connectivity, we sense the presence of embedded energy, flowing away from itself to make the existence and diversity of other *others* possible.

Within this overall thesis, we can say that it is possible for believers to stretch their *admiration* into a search for greater *understanding* of what they think they mean when they say *they believe that God created the world*. This is a crucial

---

[48] All things fall at the same rate under gravity, but the surface area relationship to volume changes the speed of their fall because of resistance of air to size. Size affects Kinetic Energy. The larger the object falling, the more dissipation of energy takes place when it hits the ground. Kinetic energy =1/2 mass x velocity^2

question for all religions and for many in science. In asking it, they will be extending their admiration to encounter the world of curiosity represented by Dawkins, particularly in his claims for the importance of Darwin, which is the focal point of this last book in the trilogy. So, for believers, as well as for those who follow the a-theism of Dawkins, there is a way of connecting *admiration* and *understanding*, even though the scope of connectivity is different for each. For Dawkins, there is an *a priori* rejection of a Creator as the source of the processes, however understood. For believers, there is an openness to their encounter and relationship with this source in the very processes of both *admiration* and *understanding* in their changing and symbiotic inter relationships. This relational openness is implicit within the development of spirituality within its *pattern paradox* of contemplation and action.

Knowing and Not Knowing

At the heart of this connection, between the sense of wonder in any belief and its continuing search for a better understanding of natural processes, we find a theological indeterminacy. This asks believers to remain open to the truth that our understanding of the source will always be partial and risky, because it is, in and of itself, an indeterminate knowledge, even as it is relationally connected. To some extent, this parallels the way science itself (but not all scientists) remains open to what it does not, as yet, know and certainly to what Darwin wrestled with in relation to law and chance. In the case of our theological knowledge, there is an intrinsic 'not knowing' at the heart of our knowing. This will always be the case, whatever the state of our intellectual knowledge. In this context, admiration and understanding are different, although not opposites. Each can include and needs the other. A spiritual admiration of the wonder of what we understand scientifically can be part of a relational understanding of what fills us with wonder, when we contemplate the science through our beliefs. This is not to imply that scientists are not filled with wonder as they contemplate new questions about natural processes. This is not to say that believers aren't filled with wonder when they, too, discover something more about their understanding of nature and intuit or make a theological connection. Believers will have different, often conflictual ways of making theological connections, but the belief behind Love's Energy is that those connections are no longer credible when they deny or conflict with advances in scientific understanding.

As science changes, theology must continue to be in a flexible and critical relationship with any one stage of its development. That critical relationship should never again turn into a dogmatic resistance to new understandings, nor to those dimensions of existing scientific knowledge which are evidentially true – such as the principle of evolution in our understanding of how the universe can and does 'make itself.' Within such broad brush principles, there is room for asking more questions of the detail to increase our understanding. For example, in LE1, we saw how early natural philosophers, as they were then known, asked questions about the basic forces in nature; how physicists and engineers gradually increased their understanding of how these forces behaved, for example electricity. We saw how the early pioneers experimented with the effects of electricity, at the physical and chemical levels, leading eventually to our knowledge of the connection between magnetism and electricity and the makeup of charges in the atom. Identifying the way positive and negative charges 'behave' in the nucleus, or the way atoms combine and bond through ions and the operation of 'spare' electrons, does not solve the more basic question – *what causes the presence of this charge and in any case 'what is electricity.'* So, better understanding about the detail can lead us to ask even more basic questions about the nature of things in themselves, as well as how they behave.

It has been tempting for people of faith to retreat to this area of *'things in themselves'* as a last ditch, theological defence. So, for example, some believers might argue that because we do not, as yet, fully know what electricity really *is in itself* – despite all our equations and experiments – then this can only be explained by the mystery of God's creative presence. Locating mystery only in these ever reducing areas of the unknown is, as we saw in LE1, a mistake, however tempting it might be. It is tempting because these questions are in the true sense, fundamental questions. However, there must be a connection between what is fundamental or ontological about nature, and its forms, behaviours and practical operation. All of these things are expressions of its nature and therefore part of the whole, made possible by its existence and meaning. Therefore, they have theological significance too, because of their connections within this whole made up of many and diverse 'other *others'* in the essential evolutionary plurality of the *Otherness* of 'creation.' It is theologically important to ask ever more fundamental questions about the nature of things, as well as their behaviour, context and the 'external' relationships which affect them. Despite the expanding specialisation of science into increasing numbers of different disciplines, there is a growing recognition of the connection between the fundamental questions about what we do, or don't understand. Indeed, as we saw in LE1, many physicists spend their lives working on how we might understand and then describe a *'theory of everything.'* So, in many different ways, humans use their sense of wonder to inspire their search for greater understanding and use this to increase their sense of wonder. Religion, itself, surely exists to serve and encourage those dynamic connections. Sometimes the 'myths' of religion do indeed achieve just that. At other times, they, or the uses to which they are put, close down those kinds of connections.

The cultural vehicles of all religions carry their connected associations, but the content of one may seem strange to the followers of another, and even to some of its own adherents! For example, as a Christian, I find the plurality of Gods in Hinduism very culturally and theologically alien. Yet, I have to accept that Muslims for example challenge the Christian idea of the Trinity for the same reason and many point to the 'worship' of Mary, (particularly in Southern European countries), as a kind of semi-divine goddess in the same connection. It is interesting that Indian origin narratives produce and are, to some extent, produced by an underlying attitude that doesn't appear to struggle with either the cosmology of modern physics or with Darwin. This is particularly so in those Hindu traditions which focus on Shakti – as a divine energy force, or life principle. These latter ideas are ones that Dawkins would reject and replace with the science of DNA coding itself. Of course, these early origin narratives didn't know anything about modern physics or genetics, but, in their plurality of metaphors and myths, viewed the origin of things in a comparatively, open ended way. As we saw in LE1, there are different examples of this. The Rig Veda hymns, for example, focus on the idea of the void of 'nothing,' which allowed for the non-being of things in creation. There are also myths of the egg which divides and the idea that a creator God, sitting just below the higher God (a similar idea to the demi-urge of neo-Gnosticism) brings material things out of pre-existing things. Hinduism also has the idea of creation as a manifestation or emanation out of Brahma and, in LE2, we saw how the early Christian mystics valued and used the idea of creation as an emanation and manifestation from the very essence of God. The other God in Hindu origin stories is Vishnu, who seems to be viewed as both a creator and a destroyer at the same time. A Hindu friend, Avtarjeet S Dhanjal, claims that, with such a variety of (sometimes conflictual) origin narratives, Hindus find it much easier to accept the findings of modern physics and evolutionary biology than many Christians do.

So, we embark now on the third and final book in this trilogy to explore something of the development of evolutionary thinking and its implications for people of any faith as well as those fascinated with the subject for its own sake. Before we set off on that compelling and developing journey, it might be helpful to remind readers of the basic thesis of Love's Energy, as it has appeared in Parts One and Two. This is followed by a chapter on 'Life's Longing for itself' and various creation myths before we come to the central figure of Darwin, those who influenced him and those he has so universally influenced.

Chapter 1
Summary of the argument in LE1 and LE2

Sub Headings
Love and the place of Jesus; Summary; The energy of Life's Longing for Life.

Science tells us that we live in an expanding universe which came from a point of singularity, ironically called the *Big Bang* by the theory's greatest critic Fred Hoyle. Science is now studying that moment of incredible density, energy and heat, not least through the work at CERN and its focus on the dynamic relationship between energy and matter, including dark energy and dark matter. As scientists at CERN speed up small particles to bounce off other particles, they are hoping to see what energy and/or matter is created in the hope that this will increase our knowledge of what happened in the moment of 'creation' or soon after it. I believe this research and much of the physics after Einstein is relevant to our understanding of what has been called the Big Birth. But to fully understand the Big Bang we are forced to ask questions about its provenance. The only possible scientific, as opposed to theological, approach to this question is to keep projecting backwards towards the first seconds or parts of a second, *after* the Big Bang. Even the attempt to get closer to the point of singularity itself hits a wall of un-knowability, as illustrated by the Big Bang diagram from CERN shown in LE1. All language in maths, physics and chemistry breaks down at this point. It seems that science will go as far as the smallest part of second after the Big Bang, but not to Point 0 itself. To address Point 0 is to raise impossible and awkward questions about what happened 'before' Point 0. 'Before' is the language of Einstein's time/space which we know began with Point 0, so how can we ask anything about 'before' when there was no time-space? We may never know, but we need to ask the question – 'what *caused* this or indeed other big bangs, given the possibility of multi-verses, in string and related theories?' As we saw in the section on cosmology in LE1, the response that our Big Bang may be just one of many, in a multi verse, only postpones the question of 'before' and of causation (itself the subject of a chapter in LE1).

Different Religions have, in different ways, places and times attempted to celebrate, engage with and even 'explain' how the universe was created through their belief in a creator God or source. The context of such beliefs has always influenced these different attempts, as we shall see in relation to ancient near eastern creation myths. The 'how' in the creation myths is often influenced by the 'why' – the need to find purpose in what observably exists, and the need to explain what is difficult to understand or interpret. So, the writers of the first account of creation in Genesis may be as much concerned with explaining the *continuation* of creation as its *origin*. Some Hindu origin narratives specifically say that the gods came after creation, presumably inferring something similar. This is the function of the mythos in most religions and cultures. Mythos is a way of sharing the telos or 'purpose and end' type questions of life. The sharing often involves telling stories and repeating rituals to help sustain the structure and values of participatory, community life in its different, en-cultured forms of what the Greeks called the *polis,* as set in the wider *oikumene* or known world in the *cosmos* itself. This helps those of us living at the level of the local community and household, family, or *oikos* to fit ourselves into something not just geographically larger, but larger in meaning and purpose. This Trilogy focuses on the *mythos* of creation and how to connect a possible understanding of what we mean when we say God *created* the cosmic *oikumene* to the experience of being in relationship with each other, with non human species and with nature. In LE2, I developed a theology of what I called the 'ek-stasis' of Love's Energy and this lies behind my observations of the science in LE1 and in this book. It may help to briefly restate the theological approach now, rather than repeat it frequently throughout the rest of the book, where I will attempt to integrate it at different points. The argument went as follows

> ➤ For Christians and many in other religions, the defining nature of God is Love. Therefore, the being of God, which is unknowable, can best be described as having the being and nature of Love. There are many 'names' for God's being and attributes in all religions. Islam in particular has many names for God. Some might argue that all are equally important. Others might say that all are subsumed in the primary nature of God as Love. I am proposing that God's nature is Love. That is as close as we can come to contemplating the nature of the being of God.

> ➤ For Christians, this 'nature' of God is unveiled by and through the life of Jesus and in other, different ways, through the lives of people from all backgrounds and religions when they display something of the ways of love. In this sense, Christianity recognises the value of loving behaviours, wherever they are found, whether or not they are motivated by religious belief.

> ➤ The nature of Love is most itself when it moves away from itself in the direction of the Other (the universe). This *emanating* is the way Love fulfils, as well as expresses, its own nature and being. In Part Three, we considered, in some detail, what some early mystical writers said about the emanating and kenotic nature of God as a kind of ekstasis – a 'standing outside of itself.' In this sense its external purpose or Telos is part of its nature.

> When our ways of Love move in the opposite direction, they destroy their own nature. Love cannot be love without moving in the direction of the Other, to make possible the autonomous *potential* of the Other and, within that Otherness, the *possibility* of other *others* to make themselves and adapt in a diverse, but relational connectivity. By relating this proposition to the first order belief that God is the creator of all things, we can then say that Love is the source, meaning and purpose of the act of *reaching out* which is known as 'creating' and makes possible the embedded creativity of physical and evolutionary processes in their *reaching out*.

> If this is so, then the physics as well as the psychology, physiology and social science of Love are important ways of understanding its meaning. If this has no reference points in contemporary science, then it may still have its own appeal, but only in a self referential, separate and isolating, religious sense. Mythos of that kind compels little credibility in today's world.

So, central to the thesis of Love's Energy is the claim that Love always moves away from itself in the direction of the Other. This *'dunamis'* or dynamic energy of movement is part of the ontology of Love and comes from the desire, longing or yearning of Love for the Other to exist. Love's longing and yearning constitutes the 'energy' or physics of Love. The 'longing' is part of the nature of God which is inseparable from the emanating, kenotic and ecstatic movement away from that nature in the direction of the Other, therefore making possible the conditions for the Other to exist and develop in its own way. Because the society of God is constituted by Love, its 'internal' nature, although unknowable, is constituted and expressed by its external expression in creation. The *perichoresis* of the society of God, as Love, is about the internal momentum of communion. Because the nature of the being of this God is Love, it achieves the completeness and consistency of its *communion* through its kenotic ekstasis outside of itself. Some key theologians in the early church, though working, of course, within a very different cosmology, understood this as being central to their belief in God. This made it possible for them to speak of the incarnation as reflecting exactly the same movement and longing of Love's energy in kenotic ekstasis, as in the creating of the conditions for creation. Energy is universal and implicit in all natural processes. The physical movement of light or sound based waves/particles, the movement of larger, day to day objects, the processes of production and creativity, the cosmic scale movements of planetary or asteroid size objects, the unobservable movement within particles themselves, and the processes of evolution – these all require a movement and transference of energy. The forces of energy transfers, involve movements of different kinds from entropy to potential and kinetic energy. The movement of Love's longing for the Other to exist can only be metaphorically related to such movements. But it is posited as being its own kind of energy – albeit beyond our easy articulation – and one that is embedded in the processes of all energy. It is a way of linking what faith communities mean by *metaphysics* back with *physics*. Love's longing is the expression and source of Love's energy. This energy is the momentum from which the movement in the direction of the Other begins. Love's energy, whatever its pre-physical properties, creates the physical conditions for energy matter to come into being. There are equations for 'rest energy' which imply or require a mass to begin with, so these won't do. There are energy equations which demand a 'frame of reference,' but as yet there is no equation to measure the effects of Love's longing, pre Big Bang, in ways that link with the equations, post Big Bang. The equations that exist for measuring or explaining energy-matter relationships after the Big Bang, have developed significantly following the work of Planck, Einstein and others referred to in LE1. These too are relevant to what is taking place in the physics of evolution. At the end of this book, I offer a simple metaphorical equation for its thesis.

In classical Christian theology, it is argued that the universe is created by the omnipotent, omniscient will, wisdom or Logos of God. The idea of 'Logos' was popular in stoic philosophy. It included the idea of rationality and wisdom, at almost divinised levels of abstraction, and led to the idea of other logoi or beings. It is understandable why such 'rational' and powerful terms were used to describe the work of creation and indeed God. These terms reflect a political and social context of all-powerful rulers whose command and control systems were incontrovertible, and produced their own unquestioning responses. If the *Almighty* One spoke, then his (and it was mostly a 'his') *will* was done, because the 'Almighty' was perceived as all powerful and all knowing. The proposition I am offering, by contrast to many dimensions of this classical approach to God as 'Almighty,' is that

1   God is Love and it is the Longing of Love for the Other to exist which produces the energy for that to happen.

2   This Love - energy is embedded relationally in the way the Other becomes other *others* (cosmological and evolutionary development) in the diversity of nature and in human experience.

3   This happens in ways that affirm the radical freedom of the Other making possible the coming into being of other *others* within that Otherness.[49] The other *'others'* participate in the same ontological freedom of the Other so that the cosmological and evolutionary processes which follow are *internal* to the Otherness of a universe that 'makes itself.' It is therefore a very different kind of process from that implied by the intervening of an omnipotent, all powerful, All-mighty, Creator God.

In each of the above 3 statements the nature of God as Love is taken from

> ➢ the words and actions of Jesus,
> ➢ the values and behaviours described in many of the books of the New Testament (see appendix of references to love at the end of this book)
> ➢ and from particular, early church writers.

The 'longing energy' of Love is its own kind of 'being in expression,' movement and fulfilment, away from a source which is unknowable, in the direction of some thing that is knowable and tangible within the existence of the Other as the matter-energy of the connected particles, waves, forces, relationships which make up the universe and its evolutionary processes. At this point, Love is embedded in both the physical ontology of the universe and in the relational experience of connectivity within its structure. This assumes that there is a need for something or some energy or force to make possible the conditions for the energy in the Big Bang to exist. There is no pre-existing matter out of which this energy matter appears. I believe this *something* is the energy of Love as it moves in kenotic and ecstatic ways away from its own source to make possible the primal existence of the physical force of energy in the Big Bang. In this sense, this is the *something* constituting the *nothing* normal posited (or rejected) by physicists when they ask *how can something* (Big Bang) *come out of nothing*. This *something*, in turn, makes possible the primal relationship of energy - matter to exist in the process so that the Otherness of the Universe can come into being with all its potential propensities. This 'point' of intense energy singularity, and what happens a few micro seconds afterwards, is beyond our imagining, even as we try to calculate the equations and figures involved. I believe that the energy of Love's longing is the source of this singularity or the conditions which made it possible as a spontaneously generating process/event – out of which flows the

---

49 In Genesis 1 God says 'let there be' as a way of creating in the list of things which follow. This *'let there be'* implies a giving of permission, allowing, facilitating, encouraging the things mentioned to come into existence. There is no forced imposition or intervention. They come into being almost of their own accord once the divine word has spoken. This 'let there be' seems to give space for things to come into existence in their own way and, therefore, by implication, develop in their own way.

energy - matter forces and movements which constitute the other *others* of a universe that 'makes itself.' Love is the energetic 'longing' which creates the possibility of energy to become energy-matter in different kinds of syn-ergeia. This syn-ergeia is echoed in what scientists have discovered about the way heat, light, gravity, different kinds of electromagnetism and other quanta packets of wave/particle energy behave and relate. This energy-matter synergeia, which constitutes the Otherness of the universe, continues to make possible a pluralism of other *others* to evolve within this connected Otherness. Therefore, they share an underlying *connectedness*, whatever their *diversity*, and this will take many (Darwin's *endless*) forms in the *pattern paradoxes* of nature. This connectedness has genetic, chemical, biological and physical, as well as social and other kinds of dimensions. This is the 'original,' underlying and ultimate source of communion, connectivity and cohesion, whatever cultural and other kinds of diversity fracture or deny it. This pluralism and diversity follows from the freedom created by Love's energy for evolution to take place from within, rather than from any kind of imposed external intervention. In this sense, believers can hold their theological heads high when they proclaim that the universe 'makes itself,' as an expression of the freedom at the heart of the processes of the Energy of Love. This is a radical freedom which displays itself in connected and apparently unpredictable, evolutionary randomness, operating, for example, alongside or within natural selection. It includes all the observable risks of suffering, species extinctions or evolutionary cul de sacs, natural disasters and negative human behaviours and choices. This is why the work of Darwin is so crucial for Christians, as they rethink their understanding of the reality of evolutionary processes.

In LE2, we also saw the relationship between this kind of science with the social and relational experience of Love as the energy which enables and requires moral autonomy at the ontological heart of the Other. Love's longing for the Other to exist continues through embedded relationships within the Other and between different kinds of others. Love continues to move and motivate a reaching out in the direction of other *others*. In the case of human and some non-human species, this reaching out takes the form of altruistic, sym-pathetic and em-pathetic relationships which operate across, as well as within, many species. Love is therefore present within the experience of reaching out that takes place within the genetic and social conditioning of human society and many non human groups. This will be discussed at the end of this book in relation to the higher primates and in the chapter on altruism which draws on the work of Frans de Waal in particular. Love, therefore, risks and fulfils itself by being in relationship to the complexity and potential of otherness, without compromising the latter's freedom to chose, adapt, respond, act with either negative or positive effects on other *others*. This is true cosmologically in the diversity of different galaxies, solar systems and even life forms where or if they exist. It is also true in the processes of evolution on this planet. Love remains true to its own nature and being, even when it is being denied or undermined in the way humans behave. It's creative 'let be' allows for and helps to *explain* the problem of theodicy, as we saw in LE1. It is tangibly present whenever people respond to their own or to other *others'* capacity and need for love. Love, as we know from its expression in many people, and as it is revealed in Jesus, chooses to suffer rather than impose or force its purpose and presence, its ways of being and doing on others. It chooses to take the cost of ensuring and protecting the freedom of the Other, as part of its own nature and purpose. The relationship between an evolutionary neuroscience of the brain and questions of human freedom and moral agency is complicated and still a matter of debate. We understand how culture and morality are explicable in evolutionary terms, and how a human being, with a unique personality, is, at the same time, an animal with an innate evolutionary nature, laid down over millennia. The drive to genetic survival is one of the great discoveries of contemporary post-Darwinian science, but can become, on its own, a reduction of the complexity of the human condition and individual personhood. Our capacity for and exercise of freedom is conditioned by genetically innate and learnt culture (including for example good or bad parenting and peer group pressures), but it is still legitimate to talk of human responsibility for a range of decisions and choices.

The uniqueness of each separate *other* can be experienced as part of the freedom of Love and its connectivity within the Otherness of the universe. The human response to this possibility always includes mixed motives, choices and outcomes. The suffering which often follows should not be seen as either the failure of God to intervene or a failure in *design*, but as radical evidence of the risk and cost of Love as the higher value of all that is involved in human fallibility and natural processes. Even within our fallibility and mortality, we can move closer to the possibility and source of Love within ourselves and others, thus making it possible for other *others* to exist in freedom, but also in conscious and committed connection and communion. The Love who longs for the Other to exist, within which other *others* can develop in their own ways, is more than a prime mover, cause, or designer of physical laws. This Love does not design all the detail, as a divine architect or engineer might, nor intervene to create or destroy any particular part of creation, as a divine game keeper might. Rather this Love provides the energy for energy matter to exist within its own freedom and limits. This Love is very different from the sadism and masochism imputed to God in the way some have interpreted natural and human disasters and diseases. This Love is embedded in relationship with all things through their own processes. The longing of this Love continues as a form of energy which takes many different forms in human and non human species. It is an energy that can be used well or badly, as humans choose different ways to develop their relationship within all the diverse and different opportunities for their reaching out in the direction of others. This Love is embedded in evolutionary processes, as in cosmological processes, without undermining its own essential longing for the other to exist within its own radical freedom.

Love and the place of Jesus

'Meaning and purpose' language is now ubiquitous in the study of spirituality, whether in mental and physical health, or in organisations and work places.[50] These are higher level words that can be abstracted from or embedded in human experience, often to pin point the fractured or dysfunctional culture of an organisation, or fragile, disconnected and disjointed relationships. In LE2, we saw how the early church wrestled with this language in relationship to the idea of creation, and the relationship between divinity and humanity in the person of Christ. This consumed the energy of many, early, doctrinal arguments. They used Greek philosophical ideas to explore the meaning of the story of Jesus, as the incarnation or kenosis of God. Greek thinking had already influenced parts of the Hebrew Scriptures with their different worldviews. Greek philosophy was also incorporated into dimensions of the Roman Empire's cultural life and it was within this context that early Christianity developed and fought for survival. Ironically, this new religion looked very unlike any religion the Romans could recognise. They were comfortable with the idea of powerful Gods intervening to influence events, fate and destiny. To them, early Christianity seemed almost atheistic by comparison. Its God 'behaved' in very strange ways.

The early church constructed its own variants of a Jewish - Greek synthesis about the divine nature, meaning and purpose of Jesus. They believed that in him the transcendent was immanent; the invisible visible, as all that could be said about God was accessible within the words and actions of his humanity. Many early church theologians feared the spiritual implications of theological dualisms which were circulating at the time - separating spirit and matter, soul and body, the religious and the profane. Their concerns are still relevant to the way we do theology in our modern, scientific world, with its plethora of spirituality cults and different religions. In the Hebrew Scriptures we come across the inspiring word for spirituality - 'Ruach,' with its inclusive and physical reference points. For those writers, the Spirit of God was breath, life, spirit and inspiration. Spirituality was part of our physical make up. It was the very stuff of life itself, for without

---

[50] These were pivotal words in many of the papers given at the first international conference of the British Association for the study of Spirituality which met in Cumberland Lodge, Windsor Great Park, May 4-6th 2010.

breath we cannot live. The Greek word 'pneuma,' as used in the New Testament, is similar and takes spirituality into the anatomy and biology of human nature and daily living and makes possible its transformation.

At the heart of the early theological opposition to some of the Gnostics cults, to Manichaeism and Docetism, was the conviction that, in Jesus, we see the old dualistic kinds of religion abolished, as the nature and meaning of God is revealed and embodied in his humanity. In Jesus they saw the unveiling of the ontology of God's nature and being - the Energeia of God moving in a self emptying and kenotic way in the direction of creation and redemption. This energeia was at the heart of the early Byzantine understanding of *leitourgeia* as *'that act of public works which benefits the common good of all citizens.'* They freely applied this word to what we call, with so much reduction, a church service. Liturgy was the act by which we participate in the creative and redeeming act of God in the continuing tasks of creation and transformation. It comes from God as the source of Love and depends on the energy of love working in *perichoresis* through our humanity, created in the image and likeness of God, for the benefit and common good of all people, in our relationship with each other and with all things. For them, this idea of Liturgy enabled the perichoresis within the love of the society of God to be cascaded through all created beings in the energy of love which created them and the basic conditions of life itself in the Big Bang and the Big Birth. There is an urgent need to rediscover the non dualistic basis of Christian spirituality and to relate this to post Copernican and post Darwinian science and worldviews.

Summary

The main argument of this Trilogy is that religious people can talk positively about creator and creation based on the idea of God as Love, and Love's Energy as the context source for the physical ontology behind and within creation. This is embedded in natural processes and in the freedom of human action within them. This view of God is defendable in terms of biblical witness and human experience. It avoids the traps of deism and creationism. It opens doors for further debate and learning from contemporary science. It invites us to explore physics and biology alongside each other as well as the social sciences of human behaviour. It provides inquiring minds with a space for creativity and new connections. It invites scientists and theologians to discover new synergy in their different thinking and approach; a synergy which will create a new common language in the middle ground between their world maps. It is offered as a way of enlarging the middle ground space within which religion can enter the exploration of new scientific thinking, and scientists and other contemporary thinkers, can understand, if not engage with the place of religious enquiry and belief. At its heart, is its own *pattern paradox* which holds the thesis together across cosmology and evolution. The paradox is that of an *embedded source*. We are used to thinking of *transcendence* and *immanence* as distinct, different and distant from one another. So also, we are used to the language of God as creator or source of a process that is then separated off in many kinds of creation theology. The *pattern* in the *paradox* of an embedded source is the way Love longs for the Other. This pattern is found in the movement of the longing and reaching out and this is the same in and as *source* and as *embedded process.* The fact that we tend to distinguish source and process through the filter of our space-time understanding may be confusing our perception of the *paradox*. In both cases, Love emanates away from itself in ways that fulfil its nature by being outside of itself, without any kind of outside interference in the process in which it is, itself, embedded. It is hard enough to imagine something embedded in a process without interfering with it, but this *something*, as source, can only emanate from itself in this way. It reproduces this same truth about its nature as it embeds itself in relationship to the Otherness it longs to make possible in both cosmology and evolution. Its emanating nature takes it incarnationally away from itself as source into an embedded relationship in ways that don't contradict the freedom of the otherness that this emanating makes possible - as implied in a creating and creative 'letting be.'

The energy of Life's Longing for Life

Life loves to live. We can see this modelled in the way viruses and bacteria move and mutate, using other, including human, bodies to their great advantage and often to our disadvantage.[51] In this microcosm of a universe or universes, viruses and bacteria are omnipresent and have been persent since soon after the Big Birth of organic life. Many bugs are an essential, as well as unavoidable, part of the functioning of the human body, but many are a serious threat and we live in an age when they can adapt faster than we can produce new anti-biotics. As they move and mutate, they naturally search for new environments in which to live and pass on their experience of living. We are being used on a huge scale of unimaginable statistics, but that is part of their purpose and it is built into their structure and its biological - physical mechanisms. Life longs for life and will act to fulfil this longing in ways that are appropriate to its different species, variety and make up. This has always been the case, right back to our common ancestors or even ancestor, if there was such a thing, in those dark mysteries of the first transition from inorganic to organic life, from singularity to pluralism, from simplicity to complexity. This is part of the inheritance of who we are and the makeup of all species, though much has changed and is changing still. In our

---

[51] The micro-biologist, Anne Glover who has been the science adviser for Scotland and the EU says there are 10 times more microbes than human cells in our bodies.

human experience of *life longing for life,* we acknowledge both our 'dependency' on this primal movement and the sophisticated ways in which we have reacted to it, learning a myriad ways to transform or adjust the behaviour of this relational impulse. There is an ontology behind these processes in what Darwin called *'the entangled bank of creation.'* We share this with our common ancestor in that first period of transition from one state to another and everything that has happened since – in all its mutating, developing and transforming of diverse and different shapes and forms *'most wonderful'.*[52] This ontology, as we saw in LE1, is about how the universe 'makes itself,' as it moves from comparative simplicity into pluralism and diversity, from one state to another in the embedded energy of Love's energy. This movement happens in two ways at the same time. Some might argue they are mutually exclusive. I believe it is possible to view them differently - as being mutually inclusive and that is part of the thesis of this trilogy.

➢ The first way, much studied in the history of physics, is to do with the laws of forces, energy and movement. In this way of viewing and experiencing physical reality, there is a predictable relationship between A and B. If we do a certain thing to A, then it will affect B in a way that is consistent and predictable. So A and B can be understood as being in relationship with each other, according to these laws which seem to apply everywhere in the universe. Of course, the relationship isn't always a simple linear one, but it conforms to laws that have been worked out over the centuries. If there is a cause A, then there will be an effect B. Even Einstein accepted this had to be the case and, as we saw in LE1, he stood his ground against a new generation of physicists, and to the surprise of many was seen as taking up a conservative position in this respect. It may not always be possible to compute or calculate the connections between A and B, but they have to be there, however complicated the example chosen.

➢ According to the second way, it is argued, as we saw in the case of Heisenberg against Einstein, that there is an indeterminate unpredictability in the relationship between A and B and if this is the case, it must be the case everywhere in the universe. Yes, society managed very well without knowing this, thank you very much, until quantum mechanics pointed it out. For centuries of scientific enquiry and cosmological assumptions and myths, religious or not, we experienced

---

[52] *There is grandeur in this view of life, with its several powers, having been originally **breathed** into a **few forms or into one**; and that, whilst this planet has gone cycling on according the **fixed law of gravity**, from so simple a beginning **endless forms most beautiful and most wonderful** have been, and are being, evolved.* Final words of *The Origin*

the first way as functioning perfectly well, and we had no need of another. Indeed, it was practical and fitted with common sense. We observed it taking place in everyday things, as forces where applied to them and energy used and transformed from one state to another. The industrial revolution depended on such laws. Indeed, this way continues to work and we have, for example, flown a heavy craft to the moon on this basis and calculated the orbits which will sling shot it most effectively and efficiently around the moon on its return journey. So there can be no doubt that this first way is an accurate description of reality. But is it complete? It seemed to Heisenberg and those who followed him, and most physicists did, that there was another, accurate description of reality at the quantum level that said something apparently contradictory; that at the smallest, particle level, forces of energy or energy mass behave in ways that are random and unpredictable and that this is part of the ontology of things and always will be, regardless of our future capacity to observe and measure, even in the development of super computers' faster and more accurate calculations of the complexity involved.

So, any ontology has to be big enough to include both ways of understanding the physical reality of things in the processes used by the universe to make itself. Initially, this was a matter of the physics of cosmology and quantum, in the first decades of the twentieth century, but then people looked back to an era, sixty or more years before and thought of Darwin. It is part of the thesis of this trilogy that what we see in the ontology of the *predictability* and *unpredictability* of all things is true in the way that species mutate ever since the era of the Big Birth. It is also the case that there is more to discover about these processes at the micro-biological and genetic levels, where the physics of electro-chemical and quantum processes is again relevant. This is, or certainly was, just as challenging to theology as to science and remains so in the former case for those people whom we have described as creationists.[53] There is a complicated *pattern paradox* between ontology and epistemology which is as relevant to science as to theology. As *life longs for itself*, we, who are inheritors of many of its early evolutionary processes, find ourselves interrogating the very processes of which we are a part. As we saw in LE1, if we are to be true to Heisenberg's challenge to Einstein, we will accept the indeterminacy of our observations about evolutionary processes, not in order to protect the legitimacy of creationist views, but to challenge all kinds of fundamentalism. So, at the heart of this

---

[53] We shall also see that there is a fundamentalism in the way that Richard Dawkins reacts to religious fundamentalism.

trilogy, I have developed the argument that the *pattern paradox,* sub themes of connectivity and isolation, predictability and unpredictability, difference and division, diversity and similarity are part of the observable ontology of creation, at least in our epistemologies. It is how things are, as well as how they work and how they are perceived. Therefore, when religious people are asked the question *'what do you think you mean when you say you believe that God created the world'* it is important that they engage with these parts of our epistemology of the ontology of things - that is with what we can, at any one point, say about their basic physics, chemistry and biology and their interrelationship. Any ontology which ignores what we understand about the reality of these things is untenable. Any religion that continues to articulate its central beliefs in denial of our knowledge of these things is also untenable. We also argued, in LE1, that any particular stage of knowledge and our different experiences of knowledge have to be revisable in the light of new evidence. Therefore belief in the ontology of things changes in relation to this changing state of knowledge and this process is unavoidable. We, therefore, need to build into our understanding of reality a certain open mindedness and flexibility. But, for the present, at least, it is possible to argue that the two ways of understanding reality, described above, both hold and therefore any believer, in any religion, needs to engage with both as part of the same ontology.

To be clear, I am not arguing that every believer has to have a physics or biology degree, or understand the detail. I certainly don't. I am saying that every believer should do their best, within their different situations, to think about these things and to avoid constructing, or defending their belief on the basis of something which contradicts the science. When given an opportunity to upgrade and rethink their beliefs, in relationship to the broad challenges of science, they should embrace such opportunities wholeheartedly, not fearing them, or viewing them as any kind of threat. I am fortunate to have found my own opportunity to engage with new scientific learning in the writing of this trilogy, and I hope that readers will find it stimulates new interests and motivation in their own explorations. I am arguing that Love and Love's energy does fit with some of the science and with the hard question about theodicy, as we saw in LE1. Love creates the conditions for the indeterminacy and random processes to take place as the universe mutatesmaking itself, cosmologically and in evolution. At the heart of this process of making itself, there is a continuing movement in the direction of the other which in turn creates further *possibilities* and *propensities*, however indeterminate. We saw that the process known as *proton gradients* seemed to be present in this movement from one state to another. This takes place in the chemical, electrical and biological processes, as a physical expression of life's longing for life to exist. In this longing, life continues to exist and develop in its movement across difference and distance through electrical/particle membranes, as we shall see in the chapter on Cells, Genes and DNA. In this movement, we see the story of life on our planet, and any other life carrying planet in the universe, going back to the beginning of the Big Birth and behind that the Big Bang. The processes of evolution are a well known and well told story, but it is a living, changing narrative, as the latest science writes new chapters in the book that began with Wallace and Darwin, who in turn were building on and confronting the work and assumptions of their predecessors. If Einstein had to rethink Newton's comparatively fixed laws of the universe and introduce the notion of relativity, so Darwin had to rethink most of the geological and biological assumptions he had inherited from the natural philosophers, in order to explain how species adapted, mutated and developed, relative to their environments and each other. Life moves on in its longing for life to exist, in all its *forms most beautiful and wonderful,'* and often extremely damaging and destructive. So, what was the background to Darwin's words and work and why were they and why are they still so controversial for many who believe God created the world? Before we explore the nature and implications of this powerful and formative science, we take a brief look at the relationship between science and myths.

Chapter 3
Science, History and Myths

Sub Headings

Science and Rationality; Random and Complex; Origin narratives; Reaching out beyond the paradigm

'As regards religion, on the other hand, one is generally agreed that it deals with goals and evaluations and, in general, with the emotional foundation of human thinking and acting, as far as these are not predetermined by the inalterable hereditary disposition of the human species. Religion is concerned with man's attitude toward nature at large, with the establishing of ideals for the individual and communal life, and with mutual human relationship. These ideals religion attempts to attain by exerting an educational influence on tradition and through the development and promulgation of certain easily accessible thoughts and narratives (epics and myths) which are apt to influence evaluation and action along the lines of the accepted ideals. It is this mythical, or rather this symbolic, content of the religious traditions which is likely to come into conflict with science. This occurs whenever this religious stock of ideas contains dogmatically fixed statements on subjects which belong in the domain of science. Thus, it is of vital importance for the preservation of true religion that such conflicts be avoided when they arise from subjects which, in fact, are not really essential for the pursuance of the religious aims.'

Albert Einstein 'Science, Philosophy and Religion,' A Symposium, published by the Conference on Science, Philosophy and Religion in Their Relation to the Democratic Way of Life, Inc., New York, 1941.

Science and Rationality

Sherlock Holmes was an interesting figure in late, 19th century literature, not only for his exotic personality with all its 'sociopath' vulnerabilities, but as an expression of a cultural and religious worldview. This was an age struggling to combine reason, and its different methods, with the claims of the imagination in Romanticism's attitudes to nature, art and spirituality. At the end of that century, many, like the inventor of Sherlock Holmes himself, turned to spiritualism to compensate for the body blow given to religion by the implications of rationality, not least in the word of Darwin. **Sir Arthur Conan Doyle** (1859-1930) was born one year after the birth of **Max Planck** (1858-1947), the father of quantum physics, and in the same year Darwin's *The Origin of Species* was finally published for the first time. He lived through the era of its debate and development. He gives Holmes a famous method, sometimes called abduction, rather than induction or deduction. Certainly, he made Holmes remind his readers of the importance of a scrupulous and scientific attention to detail. He also relied upon the imagination to face the impossible conclusion, sometimes going beyond all reason, when all other explanations had been excluded. Reason alone was apparently not enough to satisfy all human instincts and situations. The ideological analysis of Marxism and Fascism were based on their own kind of 'rationality' and science, but they brought new levels of chaos and horror to the twentieth century, when taken to extremes by powerful dictators. The 'Age of Reason' and the Enlightenment seemed to reach their nemesis in these distortions. Reason was no defence against the worst instincts of humanity and despite its impetus to different kinds of industrial progress, it didn't satisfy the deeper longings in our human condition and its place in the universe. Some say that Darwin gave permission to those who took, and still take, a negative view of human behaviours. We remained myth making creatures, including the narratives of our own individual and social power, potential, identity and entitlements in the face of changing geo political and personal need and wants. Reason is needed to understand Myths and to uncover their provenance and reference points, and to challenge and update their utility. However, reason needs to stretch itself and go much deeper to touch the human spirit and nourish its search for a sense of greater meaning and purpose.

Science and rationality did not begin with Darwin or with Britain.[54] It would be hubris and foolishness to make such a claim, ignoring the foundations of science established much earlier and in other parts of the world, not least in early Islam and before that in ancient China, Mesopotamia, Greece, Egypt and other parts of Asia. Darwin and his contemporaries inherited much from the thinking and practical experiments that took place in the previous two centuries. Nor was Britain isolated from the continent. On a November evening in 1660, something rather special happened. A few people had gathered at Gresham College in London to listen to an unknown speaker talk about astronomy. His name was **Sir Christopher Wren** (1632-1723). When he finished his talk, the philosopher **Francis Bacon** (1561-1626) inspired the group to form some kind of association to promote the acquisition of knowledge. The called it *'The Society'* and it met weekly to discuss scientific issues. By 1662 it had become the Royal Society with a charter from Charles II.[55] Significantly, it was not the *British* Royal Society and it committed itself to be inclusive of scientists from whatever national background. Its name described an activity not a location. In 1665, **Henry Oldenburg,** a German, became editor of its first journal and so the Royal Society had a foreign secretary. It became the least nationalist of all institutions. When **Benjamin Franklin** was supporting revolution against Britain, he still remained a member.

---

[54] although many of the scientists profiled in this Trilogy are from Britain
[55] See *'Seeing Further' The Story of Science and the Royal Society'* edited by Bill Bryson. Harper Press 2010

When **James Cooke** (1728-1779) circumnavigated the globe in the search of new knowledge, he was assured that he would not be attacked by American ships. The Society refused to expel Fellows who came from countries at war with Britain, and **Napoleon** (1769-1821) gave **Humphrey Davey** (1778-1829) permission to travel across Europe on scientific business. The extensive correspondence between **Rev Samuel Clarke** (1675-1729), who was acting as a surrogate for Newton[56] and **Wilhelm Leibniz** (1646-1716), illustrates the connections being made over a range of philosophical, theological and scientific areas. As we shall see, there were many different kinds of provincial, philosophical and scientific associations and societies during this and the Georgian period - not least the Lunar society in the Midlands - all exploring the Enlightenment belief in reason as the basis for invention, innovation, discovery and industrial progress. *Natural Philosophy was frequently justified because of its apparent utility in the arts and sciences in addition to its fashionableness in politice culture.*[57] Within this atmosphere, theology was implicitly woven into the liberal arts and the idea of experimental and industrial progress flowed naturally out of a belief in the development of knowledge of how things worked. The Enlightenment excitement over electricity was only part of a range of discoveries which stretched from a growing interest in astronomy to rocks and plants. As we shall see, Erasmus Darwin, Charles's Grandfather, was central to the Derby Philosophers' Society and his view of development was highly influential in industrial thinking as well as evolution, but so too was the Derby based astronomer, geologist and clock maker, **John Whitehurst** (1713-1788). *Discoveries in natural philosophy were likened to Masonic stages, with increased knowledge of the 'sublime mysteries' as initiates ascended the degrees revealing the laws of nature both at a universal level and in terms of the self.*[58]

There was already something special happening in the seventeenth century and much of it took place in Oxford and Cambridge. Great individuals were inspiring each other to satisfy their hunger for more knowledge about the natural world - **Sir Christopher Wren's** courageous experiments with his own eye and puzzling about a comet's orbits; **Isaac Newton's** (1643-1727) mathematics and natural theology; the natural philosopher and polymath, **Robert Hooke's** (1635-1703) influential drawings of his early microscope's observations of fleas and flies in '*Micrographia*;' the natural philosopher and chemist, **Robert Boyle's** (1627-1691) experiments with bell jars and oxygen and the astronomer **Edmond Halley's** (1656-1742) study of the stars from the southern ocean and reaction to the

---

[56] Famous for his Boyle lectures and his position in philosophy between John Locke and George Berkeley

[57] See '*the Derby Philosophers, Science and Culture in British Urban Society 1700-1850*' Paul A Elliott Manchester University Press 2009. p 39.

[58] Ibid p 63

strange orbits of a comet. In fact, the comet was one of the significant events of their age. The different personalities of these 'natural philosophers' was an apparently random factor, working for and against intellectual cooperation in the early days of the Royal Society. Newton's apparently introverted and difficult personality, his jealousy of and paranoia about Hooke – if that is what it was[59] - isolated him from other colleagues. It was Halley who found a tactful way of bringing him back into the discussions, seducing him away from his obsession with alchemy to the problem of the comet's 'ellipses' and their possible relationship with the invisible forces of gravity. Only Newton could work out the maths, and this, in part, lead to the publication of one of the most formative books of those times - *'Principia Mathematica'* - creating a new science of mathematics in the process.

Hooke's work created the new science of biology and Halley's observations led to even more excitements in the observation of the stars. It was Wren and Hooke, working together, who responded quickly to the great fire of London (1666) and used maths to create the geometry of architecture in the great Dome of St Paul's and the Greenwich observatory. To commemorate the fire, they build the Doric shaped, tallest monument in London, designed to house the tallest and largest telescope in the world. This is the kind of world that Darwin inherited and into which he developed his own distinctive contribution. In the 17th and 18th centuries, most people in the Western world believed God was the creator and divine architect of everything. This was the intellectual context within which individuals who were studying the natural world – not as yet called scientists – did their work and Newton is only one example. Many were Anglican or Dissenting clergy. Bill Bryson begins his section of 'Seeing Further' in this way *'I can tell you at once that my favourite Fellow of the Royal Society was the Rev Thomas Bayes from Tunbridge Wells in Kent who lived from about 1701-1761. He was by all accounts a hopeless preacher, but a brilliant mathematician. At some point – it is not certain when – he devised the complex mathematical equation that has come to be known as the Bayes Theorem.. to work out various probability distributions or inverse probabilities.'[60]* This references the important idea of *probability* which we saw as a sub-theme of LE1, and because it well illustrates the connection between the clergy, theology and natural science at the times. All of this was part of Darwin's inheritance.

As we shall see, Darwin himself went up to Cambridge in 1828, expecting to become a clergyman and was inspired by the many ordained academics at Cambridge studying the natural sciences. One of the more influential was **Rev**

---

[59] Believing that Hooke had stolen many of his ideas in the former's Micrographica.
[60] *Seeing Further* p2

John Stevens Henslow (1796-1861) who was to commend the young Darwin to the captain of the Beagle as a 'gentleman's companion' in 1831.

J. S. Henslow. Lithograph Thomas Herbert Maguire. Public Domain Copyright expired

Henslow well illustrated the spirit of the age, particularly at Cambridge. He was a Professor of Mineralogy and was influenced in his early years by the geology teaching of Rev Adam Sedgwick. Henslow later turned from geology to become a Professor of Botany. He was an academic and country parson, who ran his parish at a distance.[61] He was productive and influential. He was a teacher and a mentor, particularly of Darwin. Cambridge University, at the time, was very influenced by the church in its organisation, beliefs and standards. St Mary's Church was a seat of clerical governance. **Rev William Paley** (1743-1805) was one of the great intellectual giants of these years and, as we shall see later, certainly influenced the way Darwin's contemporaries thought of God as Creator. His watch maker analogy became widely known and used, although we have to be careful when applying scientific metaphors to particular scientists. Newton, for example, disliked mechanistic metaphors for the universe, including Descartes' billiard balls. He saw heavenly objects more as ideas in the mind of God. Paley's argument went something like this. When we look at a watch and admire the intricacies of its construction, it is natural to think of the watchmaker as the cause or creator of the watch. By extension, using this mechanical age image, Paley looked at objects in the natural world in the same way. Like the watch, every living thing must have had a divine creator. If one starts with the premise of an *a priori* belief in God, it was a natural assumption to observe everything else in nature from within that premise. The Scottish Enlightenment philosopher, **David Hume** (1711-1776), however, pointed out the difference between an empirically observed *watchmaker* and the *watch* and the case of *God* and *creation* as follows.[62] We can observe the existence of both the watchmaker and the watch. We cannot observe God (A) as creator of all things in creation (B), even though we can observe B. We can derive the connection between watch and watchmaker in a way that we cannot do for A and B. To begin with a belief in A and then make a connection with B is a legitimate approach, but not the other way around. Whewell put it like this – '*the principle that a design must have had a designer can be of no avail to one whom the contemplation or the description of the world does not impress with the perception of design. It is not therefore at the end, but at the beginning of our syllogisms, not among remote conclusions, but among original principles, that we must place the truth, that such*

---

[61] In 1833 he was made vicar of Cholsey-cum-Moulsford but continued to live and work in Cambridge, only visiting the parish in vacations. In 1837 he was appointed to the richer Crown living as Rector of Hitcham in Suffolk where he moved to the parish until the end of his life. He kept his chair at Cambridge and gave lectures, set exams and contributed to university affairs.

[62] In his *Dialogues Concerning Natural Religion* and *An Enquiry Concerning Human Understanding*.

arrangements, manifestations, and proceedings as we behold about us imply a Being endowed with consciousness, design, and will, from whom they proceed.'[63]

Creationists, broadly speaking, believe that everything that exists in the natural world, and indeed in the universe, has been created directly by God. What does it mean to say 'everything that exists in the natural world'? If we look at the diversity of species, as Darwin did, can we similarly assert that God has created every species and variety of species as we see them now? When a new species appears, is that as a result of God's direct creating? If so, some questions arise including:

➤ Does that mean that when species become extinct,[64] that is also the result of God's intention and intervention?
➤ Does divine intervention make possible the appearance of a new species in a way that bypasses the previous development of species or modifications within them?
➤ If so, does that same divine intervention cause the extinction of a previous species or the mutation that produces a new one?
➤ If so, does that same intervention work through environmental changes that influence mutations and adaptations in different climates and conditions?
➤ What is the process of this intervention and what is the relationship between it and what takes place in the material substance of a species - how much of the change comes from within the substance itself and how much from 'outside it?'

---

[63] On *Astronomy and Gneral Physics considered in relation to Natural Theology*. 1836. London Bohn 1852 p 296. The model in his mind was not so much a watch but the regularity (essential to any law) of the movements of the planets as described by Herschel.

[64] In a fascinating paper on 'Energetics and genetics across the prokaryote-eukaryote divide, 2011,' Dr Nick Lane of UCL argues *'However, a single ancestor is perfectly consistent with multiple origins if all 'protoeukaryotic' lines arising later were driven to extinction by fully-fledged eukaryotes already occupying every niche, and if all earlier protoeukaryotes were displaced by modern eukaryotes (or fell extinct for some other reason). This cannot be addressed phylogenetically, as any phylogenetic evidence for their existence is lost. Nor is the fossil record any help. It is hard to distinguish between eukaryotic and prokaryotic microfossils let alone prove the existence of extinct lines of protoeukaryotes. While asserting the unprovable existence of extinct lines of eukaryotes is unsatisfying, if not unscientific, extinction is commonplace, and the argument seems, on the face of it, irrefutable. But there are several reasons to doubt that prokaryotes have repeatedly given rise to more complex 'protoeukaryotes', which were ultimately all driven to extinction by modern eukaryotes that came to occupy every niche. The periodic mass extinctions of plants and animals, followed by evolutionary radiations of hitherto suppressed groups, are not characteristic of microbial evolution-such radiations explore morphological, not metabolic, space. Moreover, large animals and plants generally have tiny populations in comparison with microbes, and cannot acquire life-saving genes by lateral gene transfer, making animals and plants much more vulnerable to extinction. The continuity of global geochemical cycles over three billion years shows that no major prokaryotic group has been driven to extinction, not even methanogens and acetogens, the most energetically tenuous forms of life. The abundance of apparently parallel niches suggests that extinction is not the rule.'*

As we begin to ask the latter question, it is as if we are taken back into the middle ages and the Aristotelian understanding of accidents and substance in the 'transubstantiation' of bread and wine in the Eucharist. If we continue with the argument that God does create every new or existing species, does God create every single organism within that species, or is there some other process by which individual growth takes place? Clearly, we now know enough about the germination and reproduction of natural life to explain what happens within those natural processes. We also realise how much more there is to discover about their physics, chemistry and biology. We understand how an egg is fertilised by a sperm, or not. We understand how cells grow and reproduce, or not (diseases etc). There seems to be, within the natural order, a process that functions independently of any external or divine intervention. If God is the creator, then this implies an indirect rather than direct relationship with the development and natural processes of living things. If we reject external (divine) intervention, what options are left for a theological argument? Either

1   God is mysteriously present within and alongside the natural process. On this basis, those processes do not *require* the presence of God to *explain* their origin or development and mutations, but believers in A may legitimately look at B and see a relationship in this way. How they understand that relationship will vary widely.

2   God was originally *directly* present in the processes which began life on earth and through that is *indirectly* involved in the emergence, development and mutation of all living things ever since. This implies that God's creative presence at the 'beginning' was different from God's presence subsequently. The difficulty is that from what we know of the 'beginning,' the cosmological processes involved seemed to have evolved from a universe that was *making itself* in a similar, if not the same way that species evolved on this planet. If that is the case, at what point did *direct* intervention stop and *indirect* intervention or presence continue from it?

There may be other possibilities and even harder questions to face, but these are challenging enough and they have many subtle variations. So, it is worth considering them, for a moment, in relation to what has been said in LE1 and LE2 and for what follows in this book. In the Love's Energy thesis, I am exploring a way of talking about Love in relationship to the *longing of life for itself* – the latter having some ontological significance. I am also trying to combine a pre-Big Bang *longing and reaching out type,* Energy of Love with an *embedded* post Big Bang and Big Birth presence of Love in the *free* or *autonomous processes* of a universe that *makes itself.* I cohere with some dimensions of position 1 above when I propose that Love is mysteriously present and does function within the *energy* of natural processes as an expression of its longing for connectivity, as well as the *indeterminate* way in which that happens. This has to be stressed in order to do justice not only to the Heisenberg argument (with its implications of random potential and probabilities), but to allow for genuine autonomy in the theological sense. The latter allows for the full scientific truth of a universe that *makes itself* to be taken seriously and to become a sine qua non of any theological argument.

Random and Complex
It also relates to a *complexity* which goes to the heart of the contemporary debate about the meaning and nature of biodiversity. This, in its own way, parallels what we saw about *indeterminacy* in quantum physics in LE1. Not only is that present at the quantum level of physics, chemistry and biology, but it also operates in its own way in evolution. If Darwin's task on the Beagle was to describe rather than explain biodiversity, in *The Origin* he began to work out just how this *complexity* worked and why. The debate between laws and randomness in the process has continued ever since, and has become a subject of widespread, specialised discussion, in itself. I am not simplistically equating randomness with either *indeterminacy* or *complexity*. I am saying that any theology of creation has to take very seriously the operation of indeterminacy in its complexity, because science has shown us that biodiversity isn't produced by any simple rules or linear processes. Professor Steve Jones puts it like this *'The tension between necessity and chance still pervades that science (evolution) and its handmaidens, genetics and ecology. Shared disagreements in each of those fields have appeared and have been (at least temporarily) resolved again and again. The early twentieth century saw a disjunction between genetics and evolution for it seemed that sudden leaps —the origin of new species – could be explained by the chance appearance of major mutations. Then population genetics claimed to show that natural selection on variants of minor effect could explain the origin of novel forms of life. We now know that under certain circumstances large mutations can indeed give rise to new species...The perceived importance of selection versus genetic drift – the accidental change of gene frequencies through sampling errors – in maintaining variation has also oscillated. From snail shell patterns to blood groups and to protein variation it was once assumed that most inherited diversity had no influence on fitness; a claim often followed by a belated realisation that in fact the opposite is true. The discovery of extraordinary levels of individual variation in human DNA has caused the pendulum to swing again and most molecular geneticists assume that most such diversity – and perhaps much of the genome – is adaptively irrelevant. ..The genome is now seen as a system filled with non linear interactions and speciation*

as a side effect on an incompatibility between intricate organisations. Geneticists sometimes need to remind themselves of the stark simplicity of Mendel for reassurance that their subject has any laws at all. [65] So, it is scientifically true to say, however unpalatably, that biological life, in all its remarkable diversity, is less simple that any rules imply, and even more complex than the natural laws of selection worked out by Darwin. It is important to remind ourselves of this as part of the context for considering the contribution of Darwin and its central role in the question about how species evolve in a universe that *makes itself*. We have to include random chance in the process and this may point to an implicit indeterminacy. As Steve Jones puts it *'Ecology which once saw ordered communities moving through predictable stages to a more or less stable climax, their structure determined by energy flow or predator pressure, now accepts that many may be little more than a random bunch of functionally equivalent creatures and that changes in space or time may often result from accident.'* [66]

Darwin himself accepted some measure of random change when observing species communities on island life driven by accidents of migration and extinction. We know how the 'accidents' of geological events such as continental drift, ice ages, meteor strikes, volcanoes etc affected changes in species diversity. Each individual 'accident' may have its own external and internal explanations or causation but their affect on biodiversity has to be seen as part of a wider process of chance effects. It seems that in ecology, evolution and genetics even a small disturbance, let alone an extinction event, can lead to unpredictable change elsewhere. The word 'complexity' appears in *The Origin* many times (at least forty). If we are to be true to the science, any theological understanding of nature also has to accept the presence of *indeterminacy* and *complexity*. Compared with this kind of science, the watchmaker design argument seems to be a crude, mechanistic reduction, scientifically as well as theologically. Nature doesn't operate like a watch and the complex and often random interactions of its connectivity don't follow any simple laws in their processes. However, I do not follow the Deist[67] implications of position 2 above which distinguishes an initial, direct, divine involvement from a subsequent, indirect presence. I did propose in LE1 that the Energy needed to make possible the emergence of energy-matter in the Big Bang did come from the emanating nature of God as Love, and that this ek-static nature continues in the complicated way the energy in natural processes enables them to 'make themselves.' When religious believers try to answer the difficult question *'what do we think we mean when we say we believe*

[65] *Seeing Further* p 278/9

[66] Ibid p 279

[67] An extreme form of this approach would be that the 'Watchmaker' designs a universe's laws and functions, winds up the machine and then lets it run. There are many variations of this argument still heard today which challenge us to find a way of relating creative source with ongoing process.

that God created the universe' they often refer to their origin stories or creation myths. These differ from religion to religion and even within religions. We pause at this point to have a brief look at the formative place of the Genesis texts which have been the subject of much discussion through the ages, not least during the Enlightenment period and certainly in the context of Darwin's work. A quick look at blogs on the subject of creation or creationism shows how such discussion of these texts seem to be increasing rather than decreasing.

Origin narratives

Genesis 1 begins with these words *'In the beginning God created the heaven and the earth.[2] And the earth was without form, and void; and darkness was upon the face of the deep. And the Spirit of God moved upon the face of the waters.[3] And God said, Let there be light: and there was light.[4] And God saw the light, that it was good: and God divided the light from the darkness.'* This is the King James translation and of course there are many other versions and variants. For the Hebrews, as for the Mesopotamian people in general, but not the Egyptians, the ultimate threat was the void which is formless. It is not only dark, it is uncontrollable and chaotic. This experience was based, as we shall see, on the geography and hydrology of the two key rivers of the area – the Tigris and the Euphrates, with their irregular, unpredictable flooding. This physical fear of the chaos of the deep in unpredictable waters became a metaphor for their 'cosmology.' This is what the creation story had to 'solve,' in addition to other aetiological challenges – why are things as they are? The word *'beginning'* implies some kind of time space reference point. At the beginning of all time and space, there is presumably a moment when the beginning had not happened and this 'moment' or state cannot have had time or space attached to it, even though they are implied. Nothing is present that we can observe or imagine. The idea of *beginning* already contains its own transition implications, referring to a state of affairs before the *beginning* has happened in order to achieve the *beginning* in ways that make it possible to talk about all that follows after the *beginning*. The *beginning* in this text is part of and makes possible what happens next – the *creating* of something. To make a beginning is to *create* and vice versa. The Hebrew word used here is 'bara.' It was only applied to the action of God. A distinction was made between the kind of creating humans do, in many different situations and the way God acts. It is true that, in Genesis 1 and 2, God also *makes* (asah) *fashions* (banar) and *forms* (yatsar), so this is a complicated textual area. The implication of *bara* is that there was a beginning and everything that follows - making fashioning and forming - comes from this kind of creating. Some creation myths are far more circular, as if there wasn't a beginning, just as some physicists do not think there was one Big Bang or that the universe and perhaps multi-verses have *always* been there. Some creation myths imply that one creation ends and another begins, that life repeats itself in different kinds of cycle. Other creation myths position the Gods in a particular physical (heavenly) place from which they first discuss and then create new things. Other creation myths imagine an abstract, ideal world which has nothing to do with this, from which passes some kind of divine initiative or spark which sets this universe into being. *And the earth was without form, and void; and darkness was upon the face of the deep.* In the Genesis account, the earth is created and then described as being without form and void. What follows is a way of dealing with this, although no explicit connection is made between the void and the effect of creating light and dividing light from darkness into day and night. We are told that it was the Spirit of God who *moved* on the face of the waters which is the place of the void and the chaos. It is as if the Spirit acts as an agent of creation, either as the voice that speaks 'let there be light,' or through its 'movement' over the face of the waters. The beginning - for the writers of this myth - is an ordering of something that already exists after God creates the heavens and the earth, even if it is not anything like the universe we now know. Is it more like the early universe when planets were formed out of the debris of collapsing suns, or, even earlier, when the first gas molecules combined? The point of the myth is not so much to give a scientific explanation, but to talk about God as creator and sustainer in difficult situations – bringing order out of chaos, something out of nothing, bringing light our of darkness, or separating the two. If we locate it within its original context of being conquered and exiled, where displacement and suffering forces questions of helplessness, frustration and bewilderment in a people who are still trying to believe, then Genesis I makes a new kind of sense. There is a God who made a beginning, separated the light from the darkness, brought order out of chaos and who is a giver of good gifts in the things we see around us. There is a God who wants us to exist and find our own way of being human in relationship to these things, both our use of them and our understanding of their place in the wider context. Of course, the writers and editors were also struggling with particular aetiological questions, particularly in the different creation account found in Genesis 2. They asked questions such as why is child birth so hard, why is work so hard, why do we fear the snake, a good part of God's creation? Why are the good things which God made also a threat to others? What is the relationship between man and woman? How do we talk about them as both, organically connected yet separate, each with their own nature, perceptions and biological roles? How, out of the best of motives, including a curiosity for more knowledge, do we so hurt and damage each other? What is the part of each man and woman in this process? How is it that light and darkness return on a reliable basis? How did light appear in the first place out of darkness? This is a pre scientific age. To answer these kinds of questions, they created stories, myths and rituals rather than scientific experiments and equations.

The existence of light from the sun was of course central to all life and the Egyptians made this part of their religion and origin narratives. For them, the water of the Nile flooded *predictably*, unlike the Tigris and the Euphrates. It produced the silt and the water essential to the fertility of the Nile Delta and beyond, and so was also part of their creation myths. The rituals of many ancient religions showed an anxiety about the daily return of the Sun and the fertility of their crops. Their cosmology meant that these things depended upon the intervention of the Gods. The writers of the two main creation myths in Genesis had their own picture or model of the cosmos, as we saw in LE1. It was virtually the same as those found throughout the ancient Middle East, but different from the cosmology of Egypt. It was from out of that Mesopotamian/Sumerian worldview that they developed their stories of creation and we need to understand this when reading their texts – not just Genesis but the Psalms and Wisdom literature. There is no point in using their model as a vehicle for scientific truth about the early universe - that is the creationist way. There is no point in judging them for being pre- Copernican and pre-Darwinian. We have to look closer within the myth to discover insights which can still be of benefit. The writers were reaching out to metaphysical questions which transcended, but helped, their daily experiences of captivity in Babylon, when their sense of identity, previously focused on the temple cult in Jerusalem, had been destroyed. The covenant promises of God had gone. They were facing the awful possibility that either they had offended God – thus causing their suffering in exile – or that God had deserted them for some inexplicable reasons. That was the context for Genesis I - an act of faith, showing that God was the creator of the structures of everything, even the void of darkness and its unpredictability. To do this, they constructed a paradigm shift away from the narrow, tribal history of the people of Israel. They were living in a more 'multi-cultural society.' They looked around for new ways to talk about God as a something bigger than a narrow, tribal God. The Persians conquered Babylon and brought with them a more inclusive approach to different religions. Edicts of toleration were proclaimed. Now, the word the writers of editors of Genesis 1 used for God was not the nationalistic YHWH, but Elohim, a Hebrew version of a Mesopotamian name for God, used in the plural as well as the singular. They were reaching for a larger vision which transcended their particular problems and included everything that existed and all men and women, not just their own Davidic kings or heroes. The inspiration could well have come from the different cultural, political and religious context of their exile. God must want the best for all people and therefore, would create the best of conditions in the goodness of everything that was created. Despite evertying, *bara* had been done well. This set the context for the possibility of positive human experience within that goodness. Creation could, after all, continue, despite the chaos monsters of the deep, the flooding, the constant risk of unpredictability just under the surface. There was a God who created all things despite the uncertainty, unpredictability (indeterminacy) of exile and suffering. This was not a scientific statement. It was one of hope forged out of the cries of agony and disorientation in the face of random chance. The stories they told, and eventually wrote down, helped them adapt to their changing environment in ways that Darwin might have understood.

Reaching out beyond the paradigm

Out of the same experience, other writers focused their hopes on the restoration of the Temple and the Davidic kinship in a restored and powerful Israel - what migrant or refugee group hasn't longed for something similar. The Genesis writers could have become more narrowly religious, as many did – creating and policing religious boundaries and producing increasing regulations to protect the community from the impurity of all that was alien - eventually much in post exilic Judaism turned in that direction. What we find in Genesis One is, therefore, surprising and breaks the mould. These writers thought on a larger canvas making connections between their own situation and all people and all things, rather than just their own situation. When it came to speaking of the creation of human beings, they could have reverted to a story about the history of the Israelite tribes. Instead they chose to say that all men and all women are made in the image and likeness of God – an inclusive vision which would later become diluted and corrupted. In Babylon, they would have witnessed two other kinds of *image and likeness* of God. Firstly, the stone or wooden statues carried annually round the temple area at the time of the New Year festival. Secondly, the king priest, vice-regent who represented the likeness of God on earth and took his political role and power from the religious cult. This was transformed into something entirely knew – the claim that all men and women are created in this 'image and likeness.' In modern terms, this is one of the most equality driven, politically correct, religious statements found in the ancient Near East, created in the context of political and cultural diversity against the pressures of a nationalistic, narrow minded understanding of God. It challenged much in the functioning of the temple cult of the past. The implication must have been radically new and clear – connectivity with God can now be seen in all men and all women, everywhere across the known world. Reaching out to include their diversity in the creation story was the opposite of what was later implied by some other parts of the Hebrew Scriptures and Zionism's interpretation of their significance for ersatz Israel.[68]

---

[68] I have not read Israel Shahak's *Jewish History, Jewish Religion* , *the weight of 3000 years,* Pluto Press, 1994 but it is often quoted in this context.He was a Belsen survivor and Professor of Chemistry at the Hebrew University of Jerusalem (1933-2001).

Darwin teaches us, in his own way, and from his own experience, how species reach out to others and react in relationships that are both competitive and cooperative. In this process, many species have suffered and disappeared, tragically and traumatically, through a primal and formative reaching out for life which is responsible for the free growth of the almost uncountable number and diversity of life forms on this one planet, in the solar system in the Milky Way. This *reaching out* includes suffering and harm, selfish and more enlightened self-interest[69] and the perpetuation of what we now understand as genetic survival. According to the thesis of this book, this evolutionary act of reaching out reflects the way Love creates new opportunities for life to exist, in all its conflictual and harmonious variety. Just a few decades before the work of Einstein on relativity, Darwin knew that all life is connected in time as well as space. All that Darwin observed, of different species, in his international travels, and his own back garden, comes from what life was, millions of years ago in a simple form that had none of the features of a higher primate, except its potential for life and mutability into other forms of life. This potential is theologically part of the capacity of Love, longing for life to exist, if not for the other *other*, then at least in ways that lead to the development of life for the other, as species differentiation takes life into ever new directions.

There is energy in this process of Life's longing for the other to exist. It is the energy of Love, longing to reproduce itself, longing to find different and new ways to exist, in new environments and in relation to other *others*. This is part of the ekstactic emanation of God, ontologically and physically present in the processes of creation and Love's relational energy embedded in creation. Darwin taught us that living forms aren't separate and immutable. We owe who we are to what others have been. This constantly changing variety of life forms is evidence of our capacity to survive, to change and to do this in the conditioned freedom of internal, rather than externally imposed processes. There are always limits to any kind of freedom and Darwin knew this, as he gazed at structures under his microscope and at different habitats and their affects. Species are mutable and therefore participate in a freedom to be mutable. This freedom requires its own energy and it motivates us to develop our capacity to be something different and something 'more,' in relationship to other *others.* In the historical story of Love's energy, longing for life to exist, Darwin's understanding of how this happens is formative and, at least for now, normative. He was surely right – we do live in an *entangled bank* with all its economic, personal, social and political interrelationships. In such an entangled space, one result of freedom is

---

[69] see the influence of Adam Smith in the Chapter on Darwin

not only the diversity of species, finding their own way to survive and develop, but the realities of suffering and death, of which he was so personally aware.[70]

Like the physicist working backward to the Big Bang, he used the present to work out the past. He made connections and saw new relationships. In this, he went back further than many had gone in the story of living things. In rejecting the independent creation of species, he had paradoxically discovered an *internal* connection between all species, which takes us back to the heart of our questions about the meaning of creation itself. If these connections don't exist, then we can only talk about things in isolation from each other. The writers of the Genesis One account had seen the relationship between the goodness of all things. A separately created view of each species was always at risk of missing the whole, and undermining its connectivity. Darwin assumed his theological critics drew their view of separate creation out of the Genesis accounts we have been considering. It is easy to see how they did this - as do indeed most creationists, but it is hard to understand. Paradoxically, they describe a God who creates separate categories of animals plants and things, but from within the same process of *bara* and according to a sense of accumulating ordering – even if it makes no sense scientifically.

The chronological six day – evening and morning - framework in the story may reflect a complicated mixture of liturgical pattern and Persian court structures and rituals. The acts of creation were split up to reflect the cycle of the week and its pinnacle in the creation of man and woman. The structure of the six days emphasises the purpose of the holiness of the Sabbath day – not so much to be separate, as to pull all things together by seeing their connected goodness. Taking this six day pattern literally - which is hard, given we are reading it across centuries of different translations and interpretations – what does the text say that supports a creationist or a Darwinian view? Even asking such a question shows how tempting it is to project our conscious, as well as unconscious assumptions onto the text, and I pause now to make some brief comments on this process. This is a well know phenomena in our use of any significant texts and applies to Darwin's writings as well. For example, we assume the *Origin* is about the Origin because of its title and impact. It preceded, of course, Gregor Mendel's 1863 discoveries and the genetics of the next century.[71] Darwin replaced direct creation with Laws which determined processes through 'secondary causes.' Just as cosmology could only go back so far, so he takes us back not to *The* Origin of Species, but a time just after. He does say that '*all organic beings.. have descended from some primordial form into which life was*

---

[70] As we shall see in the chapter on Love and Loss
[71] See chapter on Cells and Genes.

first breathed,' but prefaces this with the word *'probably.'*[72] He is tentative about any claims for a primordial original, and creationist critics such as **Thomas Wollaston** (1822-1878) made the point that the wonder of creation was in *the* Original, not the number of times the creative act was repeated. If God could intervene at the beginning, God could do so continuously in species and what Darwin called their *varieties* (as a more *fluctuating* form of species.)[73] So, the *Origin* was understandably interpreted as a text about the bginning because of later reactions to it.[74] As with *Genesis,* or any significant text, the words both veil and reveal a literal meaning.[75] Our perception of its meaning oscillates through our knowledge of its original purpose and impact and our changing cultural interpretations of these things over time. It is hard to consider or *observe,* in the sense used by Heisenberg,[76] either the *Origin* or *Genesis* separately from our participation in these oscillations. Both texts have embedded themselves in our wider narratives about much later and contemporary debates about the universe and life on this planet. As **Mary Midgely**[77] has so vividly pointed out, in several of her books, Evolution and Cosmology have become contemporary meta-narratives or myths that now frame our story telling and beliefs about ourselves, in relation to what we know about the physical world. We accommodate our understanding of identity, meaning and purpose to these myths, because they come from a science that has overthrown what is so obviously unsatisfactory or unscientific in previous, dominant myths – the authoritative narratives of religion. This is not to argue that the previous myth is better than the new one, especially where the former has now been proved to be untenable. It is to warn us that the science itself can take on the status of myth in ways that give permission for its development into a dominant narrative which itself needs continuing questioning. If the contemporary high *priests* and myth makers of evolution are the biologists, then *prophets* will be needed (from within as well as outside their communities) to challenge their assumptions when they drift into fundamentalist territory. Most scientists, of course, ask prophetically critical questions of themselves and were trained so to do. Some turn the authorative narrative of evolution into a campaign to oppose religion - on the dogmatic assumption that all religion, whatever its attitudes, is untenable because many

---

[72] Penguin English Library Harmondsworth version 1968 ed by J Burrow. p 454-5. We find ' *with its several powers having been originally **breathed** into a **few forms** or into **one*** ' in the final paragraph of the *Origin.* My emphasis in bold

[73] He admitted the distinctions were somewhat arbitrary. Ibid p 108

[74] I am grateful to Professor Jeff Wallace for this insight. See his chapter in 'Charles Darwin's The Origin of Species,' New interdisciplinary studies, Manchester University Press, 1995. p 3 ff.

[75] Because of its title and position, we assume, wrongly, it was the oldest book in the Bible.

[76] See LE1 of this trilogy

[77] See her Science as Salvation: A Modern Myth and its Meaning, Routledge 1994 where she traces Science as a Myth presenting itself as fact in a crusade against superstition; Evolution as a Religion and The Solitary Self Routledge 2002, where she unpacks evolution as a creation myth for our times; Darwin and the Selfish Gene (Heretics). Acumen Publishing 2010, where she takes on Dawkin's reductionism as a Hobbesian approach and not true to Darwin himself.

adherents use it to defend myths that have been proved to be unscientific.[78] The point is that myths need followers and converts, who then become believers, who then become champions, consciously or unconsciously. When we become believers in a myth, our ongoing participation in the narrative affects our reading of the 'original' text, not least from the particularity of our own 'time' and 'place' within that ongoing process of interpretation. As these cultural contexts change, so does our perception of the myth. We may think, as believers, that we can pick up the original text and read it, independently of our participation in the narrative of interpretation which it subsequently created. This is a temptation for all believers in a myth, and not just in religion.[79] True, in the latter case it is more prevalent, as believers will argue that the Holy Spirit, or another divine being who wrote (Islam) or inspired (Christianity) the original text, enables us to go directly to it now, bypassing all that has happened ever since. This is ironic because during everything that has happened since, the claim is also being made that the Holy Spirit is involved in the inspiration process. So we are looking back through the 'inspired' processes of accumulating interpretation in the way the myth has embedded itself in culture, time and place.[80] The following diagrammes illustrate only something of the complexity involved.[81] The first shows the movement from previous world views to the impact of a new text's myth making, within changing and adapting interpretations and cultural assumptions.

'GETTING BACK' TO THE 'LITERAL' TEXT?

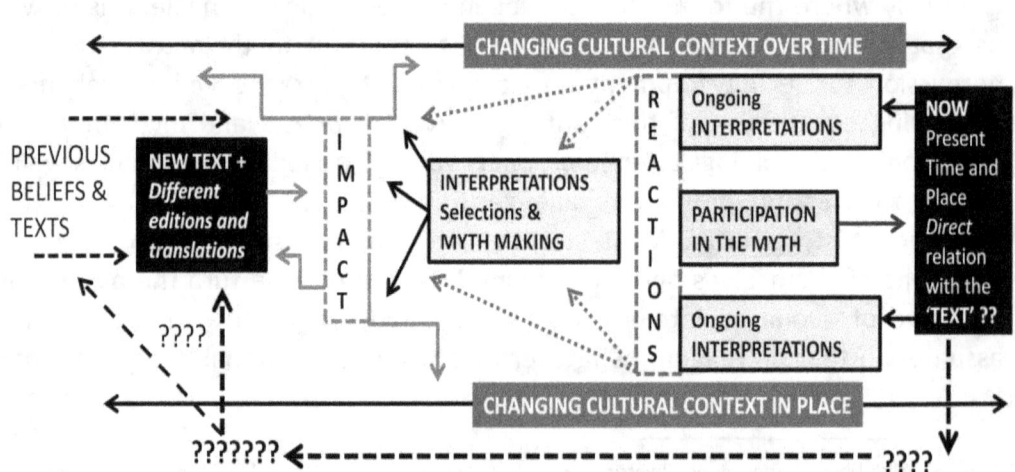

---

[78] To be clear, I am not commenting on the comparative truth of different myths here, nor their different purposes, although I clearly believe that the 'myth' of evolution works scientifically in a way that Genesis or rather its interpretation by creationists cannot or does not.

[79] To be clear, I am not saying that scientific myths and religious myths are the same kind of myths, although both can function as myths.

[80] I take the view that the inspiration of the Holy Spirit can be found in any book, art work, or person, inspiring and changing lives for the better, otherwise we may be forcing a boundary limit on the processes of inspiration adn truth seeking.

[81] Each interpretation stage could be an individual, group or era and this is accumulated over time.

The second illustrates the way different understandings of myths function as part of this process and how they are used as a pivot point in the *present* to interpret the *past* in order to create *future* world views.

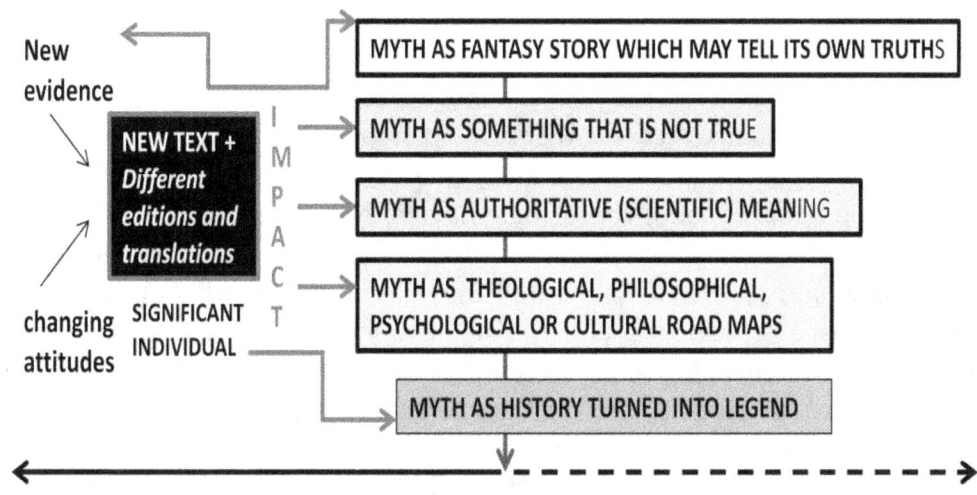

At the point of appearance of a new text —whether Genesis or Darwin — its oral and textual prehistory is also complicated, and will affect how people react to it, depending on what is at sake for their own interests and positions. That too is part of the dynamic and sheds light on its impact and later interpretation. If we believe Darwin's text was revolutionary and controversial for the theology of his times, we may be surprised when we read it from the perspective of our own science. The obvious hesitancy, humility and ambiguity may reflect his own anxiety about its impact, or his own respect for influential figures such as Sedgwick and his own wife, Emma, but this reminds us it is hardly a campaigning, revolutionary text — although that is how some interpreted it. The hesitancy didn't dilute its impact on more general, philosophical, cultural and religious views. **George Levine**, setting the Origin alongside other literature of the time, sees it as participating in the *'decentering'* of humanity *'radically undercutting the tradition that invested language, definition, and idea with some ultimate and absolute authority,'* setting its *'most important points in the conditional mode,'* and *'usng rhetoric that reflects the conditions of the dysteleological world it invokes; various, democratic, multitudinous, constantly transforming, intricately entangled.'*[82] He sees its style as the *'perfect compliment to uniformitarianism.'*[83] It contains inherited and implied theological, philosophical and economic assumptions from his times and before. On the surface, it shows few overt

---

[82] *Darwin and the Novelists. Patterns of science in Victorian fiction.* University of Chicago Press 1988 p 86
[83] Ibid p 87

references to the political economy, but its assumptions inform his perception of the natural order, within which natural selection takes place. As a Victorian liberal, his view of adaption is hardly compatible with a revolutionary view of top down, or bottom up social change. This was an age when science wasn't as siloed in fixed disciplinary boundaries as ours. He was drawing on wider patterns of intellectual thought, *modifying* and *adapting* them to his observations, however tentatively. The dominant paradigms, in those patterns of thought, included metaphysical assumptions that scientists would rarely, if ever, use today. That adds another layer of complexity as we observe their commentary on his text. We have to dig deep beneath these many layers, not least those of different kinds of Social Darwinism, as we shall see. The further we move historically from his time of writing, the less likely it is that the general reader will have actually read the text because the title and the name Darwin already conjures later assumptions about meaning and an implied major, intellectual revolution. Similarly, it is hard for Christians to read the text of Genesis separately from the assumptions we bring to it from the New Testament, and later centuries of theological dispute, for example about the Fall.[84] There is a process of *natural selection* at work in the Darwin industry and its studies of the history of *modifications,* reformulations and *varieties* of the original *species* know as *The Origin of Species.* Even within its different editions, Darwin made many changes as he reacted to other peoples' reactions to it. Mutations have already happened within these editions of the original species of his idea, let alone what followed later. With the Genesis texts, a similar process has been taking place, but over a longer time frame of complex fracturing and the development of attitudes, assumptions and interpretations in different times and places.[85] Genesis uses its own patterns of language and association. Within the rythym of the six day pattern of creation, we find different kinds of compelling repetition e.g. *'then God said, "Let the earth put forth vegetation; plants yielding seed and fruit trees of every kind on earth that bear fruit with the seed in it" And it was so, the earth brought forth vegetation.'* There is something surprising here for those who assume Genesis automatically supports a creationist view. The phrase *'let the earth put forth'* doesn't prescribe any species detail, nor imply direct hands on intervention in the creation of each separate plant, sea creature, or animal. It is the earth that does the work, once the divine *'let the earth put forth'* is uttered. The result is *'And it was so,'* but again without any detailed lists of separate species. This is a good example of an ontological freedom given to the internal processes of creation itself. The Creator creates the possibility and conditions for

---

[84] A simple text is to ask people who tempted Eve with the apple – many will reply Satan or the Devil! It didn't help that in the Latin translations, 'malum' can mean either apple or evil!

[85] See also the search for the 'historical' Jesus in the work of e.g. Albert Schweitzer (1875-1965) as an attempt to get back to the 'essence' of the original through the accumulated layers around the text.

the earth to bring forth these things – within the generic term 'vegetation' or 'plants,' and the 'living things' which 'teem' in the sea (v 21). We note the language of the next verse – *'Be fruitful and increase in number and fill the water in the seas, and let the birds increase on the earth.'* Again, we note with surprise to the creationist position that the *increasing in number* is achieved by things, in themselves, being 'fruitful' rather than any direct intervention in the process. The order of creation is, of course, pre-Darwinian in its understanding of the common ancestors of all species - the creatures in the sea follow the plants on the earth not vice versa. Also, the two great lights of the moon and sun (v. 16), as distinct from 'the Light' already mentioned in verse 3, were created *after* the plants appeared – so much for photosynthesis and our usual understanding of the link between light and plant growth. There is more than a hint of what we might call mutability in the *'trees of every kind that bear fruit with seed in it.'* (v 12) The next tree will therefore grow from the seed of the previous one. There is a sense of diversity freedom in the phrase *'every kind.'* The extraordinary challenge to any creationist assumptions brought to the text is that growth and development emerge or 'evolve' from within the earth or thing in itself. The idea of being *fruitful* and *bringing forth* allows for variety, change and chance and indeed more than hints at processes which are complex and independent of the original *'let there be…'* Meanwhile, Darwin's *Origin* became a catalyst for a new interest in archaeology as a way of testing the historicity of myths – for example the evidence for the existence of Troy[86] and the Trojan wars which were so formative of the ideology of Greek unity (as was Delphi) and indirectly to the early formation of Roman and Greek ideals which in turn influenced the thinking of Western societies. In 1871, just twelve years after *The Origin*, George Smith discovered the 'flood tablets' from the eighth or seventh century B.C. A 19th century passion for geology and archaeology now led to new questions about the dates of the Bible and its historicity. Smith's flood tablet is one of the many tablets of the *Epic of Gilgamesh*[87] and refers back to Mesopotamian stories about the great Flood which were at least a thousand years earlier than the Hebrew Scriptures. They reflected on its moral and theological significance in relation to the whole of the creation story. We now know this catastrophic event was probably historical – it appeared in different sources.[88] We know how floods, like other natural disasters, happen and mostly what caused them. We do not need to explain them as a divine intervention, but that would have been a natural assumption at the time. The early writers were not just recording events but learning to reflect on their implications for moral behaviour and belief.

---

[86] Schliemann found what he thought was Homer's Troy in 1871

[87] One of the earliest origin narratives from Uruk and part of teh many tablets that make up *Shūtur eli sharrī* (*Surpassing all other Kings*) or later the *Sha naqba īmuru* (He who saw the deep or the unknown) from perhaps 13th century B.C.. The tablets covered many different stories concerning Gilgamesh, Enkidu, Istar, Humbaba, Siduri, Utnaspistim…

[88] Until the science of geology, natural philosophers had problems explaining natural disasters given a 'good' creation.

Chapter 4
The Big Birth

Sub Headings
Creationism is dead; intervention and healing; the big birth story; the physics of the big birth; the heavens declare the glory of God; energy in the death and birth of stars; energy capture; the energy and physics of water; the conservation of energy; cells; carbon; the conditions for the origin of life; spontaneous generation; energy exchange; variables, catastrophes and adaptions.

'The kingdom of God is as if someone would scatter seed on the ground and would sleep and rise by night and day and the seed would sprout and grow, he does not know how. The earth produces of itself, first the stalk, then the head, then the full grain in the head...Jesus also said with what can we compare the kingdom of god or what parable shall we use for it? It is like a mustard seed which when sown upon the ground is the smallest of all the seeds on earth, yet when it grows up becomes the greatest of shrubs and puts forth large branches so that the birds of the air can make nests in its shade.'   *Mark 4. 26-32*

Creationism is dead
Nearly two millennia before the work of Richard Chamberlain, Wallace or Darwin, the writer of Mark's Gospel has Jesus saying these words *'the earth produces of itself.'* If we are to be true to the New Testament, we have to say that in this parable - painting a picture of the Kingdom of God - we are given a glimpse of how Jesus saw the processes of creation. It comes as something of a surprise to those who assume every part of the Bible takes a creationist position. Here Jesus seems to be implying that Creation *'makes itself'* from within its own processes with the small growing into something larger and more complex. The issues behind *'he does not know how'* can now, at least to some degree, be resolved by evolutionary science. Yet, despite this gospel idea of natural processes, emanating from other natural processes, all producing of and from themselves, too many theologies of direct, divine intervention, in separate species creation, have continued through the centuries and still discredit the belief they ironically seek to proclaim in the name of this same Jesus. Creationism cannot be justified by a reading of the New Testament, if this passage is to be understood as reflecting the words and views of Jesus. In his 1920 *'Outline of History,'* H.G Wells wrote with horror and concern about the effects of creationism, and he did this well before the work of Richard Dawkins. The passage is worth quoting at length and we will return to his comments later in the Chapter on Darwin. *'We have already noted that this story of the special creation of the world and of Adam and Eve and the serpent was also*

an ancient Babylonian story and probably a still more ancient Sumerian story and that the Jewish sacred books were the medium by which this very ancient and primitive 'heliolithic' serpent legend entered Christianity. Wherever Christianity has gone it has taken this story with it. …. Until a century and less ago the whole Christianised world felt bound to believe and did believe that the universe had been specially created in the course of six days by the word of God a few thousand years before..Upon this historical assumption rested the religious fabric of the Western and Westernised civilization and yet the whole world was littered, the hills, mountains, deltas and seas were bursting with evidence of its utter absurdity. The religious life of the leading nations, still a very intense and sincere religious life, was going on in a house of history built upon sand…If all the animals and man had been evolved in this ascendant manner then there had been no first parents, no Eden and no Fall. And if there hand been no fall then the entire historical fabric of Christianity the story of the first sin and the reason for an atonement .. collapsed like a house of cards…Many men and women are still living who can remember the dismay and distress among ordinary intelligent people …as the invincible case of the biologist and geologists against orthodox Christian cosmogony unfolded itself….their whole moral edifice was built upon false history…they believed that to assent to it would be to prepare a moral collapse for the world. And so they produced a moral collapse by not assenting to it.'[89] Creationism as an explanation for the appearance of stars, planets and biological life[90] is now dead, although of course many still cling to it. It is time to let it go and take a leap of faith into God's calling to seek and be lead into all truth, and to pursue it with whoever and wherever it takes us - for the truth can be no enemy of God. It is time for leaders in all religions to be clear about this. It matters to their credibility that they pronounce the death of creationism and point out its dangers. To be silent is to be seen to collude when militant creationist voices go on the attack. It matters that people of faith engage with the latest science and locate their beliefs alongside it. It matters to what some call their mission and their evangelism that they demonstrate a new, open ended interest in what they can learn from others and from science.

Intervention and healing

As we have seen, Creationism also depends upon a theology of divine intervention, some of which is patently crude and undermining of natural laws and processes.

---

[89] *The Outline of History*, H.G Wells, Waverely, 1920 page 522

[90] E.g. that the age of the universe is only a few thousand years and that God directly and separately creates '*all creatures great and small*' so denying the reality of natural selection. It was the Calvinist Archbishop of Ireland James Ussher (1581-1656), a reputable academic, who famously calculated, after much correspondence and scholarship, that the earth was created in October 4004 BC. For centuries it had been assumed that the age of the earth - which was of course seen as at the centre of the whole universe - corresponded with the age of human life on the earth – a double pre-Copernican whammy! The study of geology had not developed and it was some time before science could break through religious dogmas about the age of creation. He vigorously rejected all forms of 'Papism,' but clearly shared their view of creation!

Such a view adds hugely to the problems of theodicy. This was discussed in LE1 and will be picked up again in the Chapter on Darwin. Of course, Christianity holds to a belief in the incarnation – which can be seen as a form of divine intervention, and likewise the guiding presence of the Holy Spirit in peoples' lives, bringing transformation, inspiration, strength and change. Within the idea of incarnation itself, however, we see a contradiction to an interventionist theology. In the desert temptations of Jesus, the Garden of Gethsemane, and on the cross Jesus rejects the use of such an approach, however tempted he is by it. In Jesus, we do see healings and 'miracles,' which seem to imply a direct intervention to change natural processes. There is no doubt that most people in his audience believed these acts were divine interventions. This coheres with a pre-modern view that suffering is the result of sin and therefore a kind of divine punishment, or the work of devils through human intermediaries. Jesus was part of this culture, but asked people not to tell others about the healings, nor did he heal everyone that came to him. Of course, the presence of Jesus was a healing to many, but mostly he put the emphasis on their action and responsibility, saying, for example in Luke 17.19, *'Rise up, your faith has healed you.'* The psychosomatic effects of positive (faith) attitudes are well documented in contemporary medicine.

All religions have oscillated between spirit and matter and their effects on each other. As we saw in LE2, early Christianity sought to avoid such dualisms, even though many now talk about their beliefs in just this way. A spirituality that is embedded, related to, or involved in material things compels us to ask how God is present in natural processes. Christians who take Darwin seriously have to find a way of believing in God's presence in and through natural processes, if they are to avoid the charge of Deism. This presence, however, must not be thought of in ways that deny or undermine those processes. It cannot be an *intervention* that bypasses them. If we see God's presence through the ways of Love's energy, then we would have to critique the idea of *intervention* from the perspective of the ways of Love. The way love *intervenes* is very particular in the Christian tradition. It works transformatively from within, as an embedded, self-giving process. If it did not, it would have destroyed its own nature as Love. The incarnation reflects the same presence of Love as in the processes of the Big Bang and the Big Birth. What happened, must have happened from within those processes, not from outside them. We need to continually enlarge our understanding of those processes and be open to what Darwin saw as the random parts of selection, however conditioned, and what Heisenberg saw as the unpredictable in the Indeterminacy Principle. If Jesus stands for anything, he stands for this idea of transformation from within. Because we claim that Jesus was Love's energy, unveiled in our humanity as Love incarnate, the workings of this Love - from within his own humanity - had healing powers that look as if they came, miraculously, from outside – at least to those who thought in that way.[91] The energy of healing and of life changing events and experiences comes from within the movement of Love from A to B, not from outside of it – this is the message of Jesus and his temptations, his agony in Gethsemane and the cross. This is how Love works and moves among us and within us. The church interprets Jesus as God *with us* for this reason. Today people of faith still ask for divine healing and other kinds of intervention, sometimes over very small matters indeed. Faced with tragedy or terrible suffering, particularly involving loved ones, most people can be tempted to ask for divine intervention. But, if healing had been the point of the incarnation, no doubt we wouldn't have needed a health service and very few medicines would have been discovered and developed. This is clearly a large and controversial area and not the main subject of this book.

The Big Birth story

The story of the appearance of the Big Birth of organic life on this planet is constantly developing and being revised. It is one of the most exciting stories in science and remains a compelling one, as new discoveries are made. Those who stick to a creationist view are missing out on learning that can be taken from this story. In LE1, we saw the usefulness of the ideas of *propensity, potential, probability* and *possibility* in understanding the 'randomness' of natural processes in cosmology and quantum thinking. We saw these ideas were essential to the thesis that the universe 'makes itself.' This is, for many, a controversial claim and one that is only very recently recognised in the long history of scientific speculation and discovery. It is one thing to make the claim and quite another to understand how we arrived at this point and to reflect on its implications. Some will accept that the universe makes itself, but only from a certain point, before which it was made or created – to set it going as it were (very much along the lines of the Deists). Perhaps, most people of faith end up with a version of this conclusion. But then we have to ask – were there only two points of original intervention i.e. the Big Bang and the Big Birth and did these explain what followed in each of them? Was the first intervention in the Big Bang enough to explain how the Big Birth could make itself, or was the latter only possible because of another, separate, divine intervention? In fact, we still don't know for sure what started the Big Birth or its earliest processes, but we believe there is a connection between them and the conditions which preceded them.

The Physics of the Big Birth

Before we go any further, there is something that needs to be said to connect this book with LE1 - which is mainly about Big Bang cosmology and quantum physics. It

---

[91] In medieval times, women healers who used natural products and their own knowledge to heal, were accused of witchcraft or devilry by the ignorant and those who had most to lose.

is a simple but profound connection with many implications that are not always taught in our schools. It is the bridge between everything we observe in our space time and the energy-matter of the Big Bang. The physics of the big bang and its energy must be the basis of any new theology of creation, and certainly one that links us with what we believe about the energy before, in and after the Big Bang itself, including the processes leading up to the Big Birth. We need to understand the latter within the wider and connected context of the former. The processes observed by Darwin and Wallace started in this wider context.

The heavens declare the glory of God

We know that early humans saw the stars and constellations as Gods or the home of the Gods. This was a formative imprint on the human imagination and one that has been hard to dislodge in myths and stories about the human condition. Some religions still see heaven as a divine space, somewhere up there, in the far reaches of the sky. 'Up there' has become a special metaphor for the Gods, heaven and the place we pass to after death. Astrologers and wise men looked to the stars, if not to see Gods, then to find the will of the Gods - the fates and omens that guide human destiny. In our times, many people still look up their 'stars,' or star 'signs.' We now know much more about how stars are formed – how clouds of gases and dust are pulled closer together, becoming denser and hotter. At least fifteen million degrees are needed to create the conditions for star formation or fusion. The plasma builds up and atoms fuse together creating star light. At the centre of each hydrogen atom is a proton. With enough heat and pressure these become positively charged and stick together - helium is created, with energy given off. Mass disappears in the process as energy increases, so that only a small amount of a star's mass is used to create huge energy. Thanks to Copernicus and his 1543 intellectual revolution in our knowledge of the heavenly spheres, we learnt that our sun like all stars is at the centre of the movement of planets around them.[92] But Copernicus didn't realise that a star isn't just the centre, but the creator of a solar system. Within the sun, there are two forces - the gravity which pulls it into itself and the fusion creating energy which pushes energy away from itself. We might see this as the forces of order and disorder in a constantly changing, dynamic tension which creates the energy of the sun. Galileo observed some of the effect of this and now we can see more clearly how the sun emits charged particles of energy thousands of miles into space; the magnetic field fluctuations causing flares of energy outbursts. The solar 'wind' is full of energised particles passing the earth and the other solar planets to eventually form a protective boundary with deep space. The energy at the centre of the sun is transformed into the heat and light which provide the energy for life, in all its forms, including Darwin's processes of evolution on earth. No wonder ancient religions worshiped the Sun.

---

[92] John Donne said this new philosophy calls all into doubt.

Energy in the death and birth of stars

Our Sun has been burning for about 5 billion years and it will burn for at least that much in the future. The implications of this truth are greater even than the Copernican revolution. Our sun will die and take with it our whole solar system of planets. Not only are we not at the centre of the universe, but 'the end of the world' will come from our star's death when hydrogen is exhausted within it. It will become a red giant with a catastrophic expansion which will be both hugely destructive and creative. As its energy spreads over a greater area, the red giant will become cooler. Its core will collapse under the force of gravity. The fusion that used to take place in its core will now happen in its outer atmosphere. It will become at least a thousand times bigger and more chaotic – an expanding ball of fiery gases that will destroy the planets around it. In the last stages of the great battle between gravity and fusion, helium fusion will take place in its core and this will produce carbon and oxygen which will then be spread outwards in its destructive expansion. It will have collapsed into a white dwarf,[93] which is denser than anything known. Fusion has now stopped. Its fuel is exhausted, but it is still shining. All that is left is carbon and oxygen. It is now so dense that its small mass has huge weight. It doesn't collapse because Quantum mechanics has shown that electrons like to be different from each other, so tend to move or push away from each other. This is enough apparently to resist the force of gravity. Supernovae, on the other hand, occur when the most massive stars explode with energy approximate to ten million suns. They are very rare as far as we can see – perhaps only two or three in each galaxy per year. Their remnants give off different colours (violet from potassium for example) according to their different elements (e.g. sodium, hydrogen, iron, calcium). These elements are produced in their fight against gravity. With fusion at an end, gravity wins. When the core of iron gets to critical mass, electron pressure is no longer able to push particles apart and gravity collapse causes an implosion, and then an explosion of energy to produce and then expel the heaviest elements which wouldn't otherwise be created (including iron and gold). The energy remains, however, and new stars are created out of the expelled dust, elements and gases of a supernova.

Neutron stars only appear to us as feint signals from deep space, although they were predicted by the Swiss astronomer, **Fritz Zwicki** (1898-1974),[94] before they were detected. His idea was dismissed – which is a common story in our trilogy - until radio astronomy picked them up in 1967 as a pattern of predictable pulses

---

[93] In 1922 the name 'white dwarf' was given to Sirius A, a companion to Sirius (one of the brightest of stars), but hidden by it. White dwarfs are hard to trace.

[94] Who spent most of his working life at the California Institute of Technology. He was known for his lateral thinking and is responsible for many big ideas. He coined the word 'supernova' with Walter Baade while working on Neutron stars, five years before Oppenheimer's paper on the latter. He 'located' 120 supernova over fifty years. In 1933 he inferred the existence of dark matter.

which gave them their nickname - pulsars. They appeared to be only the size of a planet, but with huge energy. They had to be denser than other stars. The Crab pulsar was located deep within the remnants of a supernova. As Zwicki had predicted, the only thing capable of creating a pulsar is a supernova. The pattern of pulses proved to be coming from the intense, magnetic field created by a neutron star. As it spins, it sends out radio pulses. Black holes could be described as the last stage of a star's life. Einstein's relativity' taken to its extremes' implied their existence, even though he didn't believe they existed. They remain perplexing and complex, but seem to be at the centre of galaxies as well as elsewhere. They appear as a spot in space where gravity is so intense that neither light, nor anything else can escape from it. They seem to be created by the death of the very largest stars – those which don't turn into pulsars but keep on collapsing. The matter of these stars seems to shrinks to nothing, so all their mass appears to be contained in zero volume. Some now speculate that their zero volume is analogous to the infinite density of a singularity but it is hard to see how this can explain the big bang, if black holes only result from the previous collapse of existing energy and matter.

So, what is there in the physics of cosmology that relates us back to the physics of life in evolution? Are organic molecules, as well as all the elements, created in the nebulae coming from supernova explosions? **Dr Scott Sanford** of NASA believes that as nebulae create the conditions for a new star, they also make the building blocks of organic life. The molecules created are delivered to earth on meteorites. Amino acids originated in space, were scattered by nebulae and not just onto our planet. Whether this 'extra terrestrial' origin story explains the emergence of all life on our planet is another matter. As we shall soon see, there are other explanations which seem to command more widespread support. However, all energy comes from cosmological energy. Energy is the source of what we understand as Life but how does energy capture and adaption work? This is a huge question in the history of biology and chemistry, as well as physics. Darwin's contribution is only part of the story, although a crucial one.

Energy Capture
Animals and plants have evolved in ways that increase their efficiency at harnessing the energy of the Big Bang as it filters through the energy available from the sun. There are more bacteria on our planet than stars in the universe. We remind ourselves they have been around longer than most other life forms and they know how to adapt. This is why Dame Sally Davies, the Chief Medical Officer in the UK said recently 'we may to not even get to climate change if we

are killed off by infections.[95] She equated the seriousness of the risk with terrorism, as well as climate change. Many bugs have learnt to protect themselves from the UV energy of the sun, notably cyanobacteria. Bacterial genes use the energy of the Sun to create sugar or ATP. Without the energy of the sun, photosynthesis would not be possible. The equation describes an incredible and complex process of energy transformation. *6CO$_2$ (carbon dioxide) + 6H$_2$O (water) + photons (light from the sun) = C$_6$H$_{12}$O (sugar) + 6O$_2$ (oxygen).* This extraordinary, beautiful and very complex process – which we mostly take for granted - allows energy capture to take place, as well as the manufacture and release of sugars. The 'self interest' of the plant is to make sugars and ATP. To do this it needs sunlight, carbon dioxide and some electrons. The sugars are made by taking electrons and using the energy from the sun which has been collected, as chlorophyll forces the electrons onto the carbon dioxide. The ATP is made slightly differently, using another form of chlorophyll. Instead of 'forcing' the electrons into carbon dioxide, it cycles the energised electrons around a circuit which then stores the ATP. The plant has no interest in where the electrons come from, but they come from water – although getting the electrons out of water is a difficult task. In the process, they split this highly bonded molecule, with its recycling processes of precipitation, infiltration, evaporation, atmospheric storage and condensation. The plants split the water molecule as they extract the electrons they need, and this releases the waste (in their terms not needed) gas – oxygen, which may not be essential to plants but is essential to animal life including our own. Plant life turns water into oxygen and light energy into a transformed kind of energy, which in turn is used in the food chain as accessible energy for all animal life. In the chemistry laboratories of plant cells, chloroplasts (once bacteria) are crucial. So is the proton gradient in the membranes of the cells. So, we discovered that light, water and energy are all related in the processes which create the possibility of life. The most basic of all energy is Big Bang energy converted into Big Birth and continuing, evolutionary energy. Until the time of Priestley[96] much of this was not only taken for granted, but unknown. In his work at Bowden House in England, Priestley put a sprig of mint in a bell jar and discovered the presence of oxygen and the role of 'vegetation.' It was 1772; the first travellers cheques went on sale in London; slavery was outlawed in England; Poland was being partitioned into Austria, Russia and Prussia; the anti English Committee of Correspondence[97] was formed in Boston.

---

[95] Radio 4 ,26.05.13 Programme on the ability of bacteria and viruses to adapt to antibiotics and on their wrong or over use. It seems this is reaching catastrophic proportions and that the NHS in England do not mandate enough data to track the more sinister developments in bug resistance, despite warnings by previous CMOs and many clinicians and experts.

[96] And of course the experimental work of Boyle in 1659 working with Hooke on vacuums showing that when air has been removed, a candle goes out and animals die.

[97] These were shadow governments organised by the Patriot leaders of the thirteen colonies on the eve of the American Revolution.

The simple truth is that all biology depends on an underlying chemistry and all chemistry depends on an underlying geo-physics and, in that physics, the story of energy and its effects remains crucially important. Every living creature on this planet and presumably on others too, if such exist, contain basic atoms and molecules and in particular, combinations of Hydrogen, Oxygen and Carbon. Where did these come from? We now know that the water which covers 71% of the surface of this planet came from the Hydrogen and Oxygen which in turn came from the effects of the Big Bang. *'Hydrogen forms 74% of all the elemental mass. The second lightest element, helium, comprises 24 per cent. These two elements dominate because they were formed in the first few minutes after the Big Bang. Oxygen is the third most abundant element in the cosmos at around 1 per cent by mass. Most of the rest is carbon.. All the oxygen and carbon atoms in the Universe today were produced in the cores of stars by nuclear fusion and scattered out into space as the stars died.'*[98]

The Energy and physics of Water

Something as apparently simple as the chemistry of water and its electrolysis takes us back to the way electric charges in sub-particle physics connect things together. Water is a most extraordinary illustration of how this happens. It takes a considerable amount of energy to split water into hydrogen and oxygen because the latter 'longs' to acquire the two extra electrons necessary to complete its outer shell - and it turns out that hydrogen is a convenient source of those electrons. This longing and bonding connectivity means that water is very stable and, therefore, in its liquid and solid forms, is hard to split apart into its constituent parts. In the science of water bubbles, we know that water forms a sphere because in the physics of water tension that is the most efficient, minimal shape possible. In this science, the point of transition moment in surface tension change is called a singularity! Water and air don't want to mix so water tends to try to squeeze air into the smallest possible area. Singularity is a term we have already used in LE1 in relation to the Big Bang. Here it means any rapid or radical change in shape. We know that nature tends to ensure that such changes or singularities occur with the minimum use of energy; that the very process of minimisation sometimes generates its own transition point, in the energy state transfer. In the case of bubbles, this transition point might be the pop of energy transferred into a sound. Bubbles reduce the density of water and so increase the speed of things moving through water (penguins, fish, boats, swimmers) when travelling through the area of bubbles. The wave movement of water or any fluid (even in the plasma flowing from the sun) creates wonderful shapes. Water, in its three different states, is often quoted as an example of how energy is transformed but conserved. Energy is crucial to the thesis of this book as well

---

[98] *Wonders of Life,* Brian Cox and Andrew Cohen. HarperCollins 2013. Page 28.

as the whole Trilogy, because it is crucial to the description of any physical process. So, the energy we see now, in its different forms and stages of transformation, is the same energy that appeared in, and immediately after the Big Bang itself. As we look at the processes of the Big Birth, we are studying the same energy in the same, but changing universe. This means that the universe depends upon things changing, but, at the same time, energy staying the same. However much some find it difficult to accept that change happens, without the processes of change there would be no Big Birth, evolution, or life on earth. Underneath all changes in the state of all things is this ubiquitous and universal thing we call energy. This may take different forms, but somehow stays the same. So, whether or not we contemplate its science, or its theology, we know that in every possible state, in the changing form of B, as matter in space-time, there is the presence of A as energy. A makes possible the pluralism of B, in all its different shapes and sizes. We could spend our whole lives counting, identifying, living with and probably taking for granted all these different types of B, without realizing that the invisible A of energy is what they have in common.

The relationship then between A and the forms of B is varied and generated in different ways, but is itself constant or conserved. A stays the same whatever the changing shape of B, in all its changing diversity. This diversity in all its forms over time and space cannot escape from this basic ontological truth and we humans need to be far more creative in the way we perceive diversity, in all its forms, in relationship to the energy that is its source and embedded nature. If A is the energy of Love, then Love stays the same, even though its particular expression, in the changing shape of B, is so varied as to make A invisible to each kind of B. At the point that one kind of B transforms into another, as with liquid water boiling off into gases or freezing down into ice, there is a singularity transition, or exchange of energy, which ensures that A is present in the next form of B. Scientists, in our times, have been clear that this is the case, but we should not underestimate how strangely mysterious and unobvious this was, for a very long time and how counter intuitive it would have seemed in pre-modern science. So in similar ways, pre-modern theologians would have thought of God (A) as relating equally to all things (B) as a direct line creator, and not understood that the infinite variety and wonder of B could possibly be related in any other way to A. So, theology also had to change in the light of new science, particularly the idea that energy is universal and ubiquitous and the source of many different and diverse forms of the transformation of matter. As we saw in LE2, there were early mystic theologians who understood this, and there are those who now resist it. Before we turn properly to Darwin, it is worth reminding ourselves of the contribution of earlier scientists to our growing understanding of how energy is present in everything. Many of these scientists were working on theories of energy from within an overall cultural and theological assumption that the driving force behind all life was some kind of unifying, primal *vis viva*. This theological dimension is part of our story. It comes to the surface in particular ways when we consider Darwin's struggle with his own contemporary, and highly dominant, church personalities and their beliefs.

The Conservation of Energy

It is probably the case that several scientists came upon the truth about the conservation of energy in different ways, even before the work of **James Prescott Joule** (1818-1889).

James Joules 1906 taken by Henry Roscoe. Copyright expired

He was an English brewer and physicist, born in Salford who became obsessed with the nature of heat and its relationship to energy or mechanical work done. He worked with Lord Kelvin to develop a scale of temperature, and he did pioneering work on the relationship between the current that passed through a resistor and the heat dissipated in what is now called Joules' First Law. He is famous for measuring energy in *'foot pounds force of work'* - 772.55 to raise the heat of one pound of water, by one degree Fahrenheit! When he announced these results to the British Association for the Advancement of Science in Cork in 1843, he was met by an eerie silence of incomprehension or misunderstanding. He continued with different experiments to show how work could be diverted or transformed into heat. His studies on the mechanical equivalent of heat showed that heat and work are interchangeable. The caloric theory of *Carnot* and *Clapeyron*[99] asserted that heat could neither be created nor destroyed and this he refuted in a paper which was rejected by the Royal Society, but published in the Philosophical Magazine. *'I conceive that this theory ... is opposed to the recognised principles of philosophy because it leads to the conclusion that vis viva may be destroyed by an improper disposition of the apparatus: Thus Mr Clapeyron draws the inference that 'the temperature of the fire being 1000 °C to 2000 °C higher than that of the boiler there is an enormous loss of vis viva in the passage of the heat from the furnace to the boiler.' Believing that the power to destroy belongs to the Creator alone I affirm ... that any theory which, when carried out, demands the annihilation of force, is necessarily erroneous.'* His adoption of the (theological) language of *vis viva*[100] for energy was not surprising for the times, but was probably influenced by **Peter Ewart's** 1813 paper *On the measure of moving force* read to the Literary and

---

[99] In 1798, Benjamin Thompson (Count Rumford), had measured the frictional heat generated in boring cannons and so developed the idea that heat is a form of kinetic energy. These measurements challenged caloric theory, but were inevitably too imprecise to disprove it completely.

[100] It is interesting to see how widely and flexibly the Chinese word Ch'i is used for energy force in Tai Ch'i and parts of Chinese medicine. Practitioners and teachers of Tai Ch'i believe that energy in the body can be located, centred, moved, stabilised and nourished by its gentle movements, breathing and meditation.

Philosophical Society in April 1844. In 1845, he read his paper *On the mechanical equivalent of heat* to the British Association meeting in Cambridge describing his most famous experiment using a falling weight to do the 'mechanical work' spinning a paddle-wheel in an insulated barrel of water. He demonstrated how this work, or *vis viva*, increased the temperature of the water. Joule, a pupil of **John Dalton, (**1766 –1844)[101] began to see a relationship between his experiments and the kinetic theory of heat which he believed to be a rotational motion. He also saw himself as developing the work of **Francis Bacon, Isaac Newton, Benjamin Thompson** and **Humphry Davy.** At the same time, in 1844, the Welsh born **William Robert Grove** (1811-1896) developed this work by asserting a relationship between mechanics, heat, light, electricity and magnetism, seeing them all as manifestations of a single 'force.' In 1846, he published his theory in *The Correlation of Physical Forces*. In 1847, drawing on the earlier work of Joule, Sadi Carnot and Emile Clapeyron, **Hermann von Helmholtz** came to similar conclusions in his book *Über die Erhaltung der Kraft (On the Conservation of Force)*. As we saw in LE1, this work proved to be formative for our later understanding of energy. It was only later in the twentieth century that we would understand mass energy equivalence and different forms of energy – kinetic, potential and electromagnetic radiant energy.[102]

Sir Benjamin Thomson (1753-1814), later made a Count of the Holy Roman

Empire – Count Rumford of Bavaria, was an American born, British physicist, inventor and designer of warships.

Benjamin Thomson. Author and date unknown. Public Domain, Copyright expired.

His experiments on gunnery and explosives led to an interest in heat. He developed a method for measuring specific heat, but the Swedish son of a clergyman, **Johan Wilcke** (1732-96)[103] beat him to being the first to publish. He worked on the insulating properties of natural materials such as fur, wool and feathers, realising they inhibit the convection of air. This led him to believe, wrongly as it happened, that air and all gases were non conductors of heat. In fact he saw this as evidence of the theological argument from design – God had obviously designed fur on animals to provide for their comfort! He then moved to what he concluded was the non conductivity of liquids in 1797 – again seeing God's design in water as helping to regulate the temperature of the human body. It was while working on cannon for the Duke of Bavaria, that he discovered the effect of energy conservation/transformation. He

---

[101] Best known for his 'atomic theory' and work on colour blindness

[102] An electron and a positron have rest mass and when this 'perishes' their combined rest energy is converted into photons with electromagnetic radiant energy but no rest mass. If this happens in a closed system with no release of the photons or their energy then the mass and total energy of the system remains unchanged. Non material forms of energy can likewise be converted into rest mass.

[103] In 1762 he invented an electrostatic generator a devise that was named and made popular in 1775 by the work of Alessandro Volta.

noticed the heat generated by friction when boring a cannon barrel. When immersed in a barrel of water, the heated barrel would then boil the water in about two and half hours. He noticed that no physical change had taken place in the material of the cannon. Many people through history had, of course, found a way of transferring heat from one substance to another in this way, not least in domestic usage, but he was one of the first to reflect on his observations. He argued that the apparently inexhaustible generation of heat was incompatible with the caloric theory, and that the only additional factor was the friction motion energy given to the barrel. More calculations could have been done, measuring the mechanical equivalent of the heat generated by friction. Perhaps he gave up on further exploration because this work was not taken seriously until later in the next century. **Julius Von Mayer** (1814-1878) was a German physicist working on thermodynamics. He is best known for his work on the conservation of energy (first law of thermodynamics) and gave us its early articulation *'energy*

can be neither created nor destroyed.'
Julius Von Mayer, Public Domain Copyright expired

The tragedy of his life was that **von Helmholtz** and **Joules**, like many others, looked down on his abilities. He had worked as a doctor on a Dutch, merchant ship sailing for Java in 1840, and is said to have learnt from the ship's navigator that the ocean seemed warmer at times of a storm and cooler when calmed.

This may have inspired him to turn from medicine to physics and particularly the idea that motion can be converted into heat, as he described in a paper to the Annalen der Physik in 1841. This was rejected, but he continued to work on the mechanical equivalent or measurement of heat and in 1848 – a year of revolutions - calculated that the sun would cool without some outside force to keep it hot which came from the impact of meteors and so heat energy was passed from one body to another. Despite his calculations on the numerical value of the mechanical equivalent of heat, published in a booklet in 1842 – *'the Organic movement in connection with the metabolism,'* his work was little recognised and the fact that **Joules** won public recognition for similar work led to

Mayer's breakdown and admission to a mental hospital. His work was finally recognised and affirmed by the Irish physicist **John Tyndall** (1820-1893) [104] in a Royal Institution lecture given in 1862.

Tyndall probably 1872 Cropped version of engraving based on earlier photo. Smithsonian Institution's digital collection from the Dibner Library of the History of Science and Technology. P-D. Copyright Expired.

---

[104] Professor of Physics at the Royal Institution with a prolific career and life; known for his work on magnetism, crystals, the role of water vapour as the strongest absorber of radiant heat in the atmosphere, the proof of a green house effect in the atmosphere, the building of many experimental apparatus, the writer of many physics books. With Huxley and others he joined the club that supported Darwin and advocated the separation of science and religion.

Surely Meyer had done more than anyone, apart from Helmholtz and Joules, in showing the relationship between work and heat as transferable forms of the same energy (definitively stated by **Helmholtz** in 1847). Meyer had also described the chemical process of oxidation as the primary source of energy in living creatures, and that plants convert light into energy. What an extraordinary life these late eighteenth and nineteenth century scientists lived, working without the technology of modern science or the advantage of advanced theories. They gave us the intellectual basis for understanding how energy is that flowing, connecting, animating, generating force of and in all things, not just organic life. Energy is the reaching out which facilitates the movement of life and change in the universe. If the thesis of this Trilogy is a helpful way of envisioning what is true theologically about all physical reality, then it tells its own analogous story of how Love's energy reaches out to facilitate life itself and is present within it.

We now know, although it is sometimes hard to visualise, that all of Love's energy is conserved and whatever its form, shape or appearance it is the same energy. The energy that invigorates life now, in the entangled bank of creation, is the same energy that performed the same tasks, albeit in different forms, 13.7 billion years ago. Amazingly, there is no exception to this law. This energy underlies all the changes of organic and inorganic matter through space time. So, the energy in the hot gases of the early seas or sea vents of sulphur minerals and structures is the same energy as in everything else we see. It is from the biochemistry and physics of those early, superheated vents and under water gases - the black smokers of the Sea of Cortez, or the very deep and different samples discovered from the Atlantic Massif, or the fossils of Marble Bar in Australia - that the first, single celled organisms found appeared 3.45 billion years ago. As we saw in the chapter on the physics of energy, it might well have been the physics of proton gradients that made possible this appearance of early life forms (from volcanically released molecules such as hydrogen sulphide). The black smokers produce an acidic environment in the sea, while in the Atlantic Massif the sea water becomes alkaline. The physics of these naturally occurring proton gradients with their non-equilibrium conditions may well have triggered the early bio-chemistry of early life forms. So, perhaps before the first cells appeared, our common ancestor was the chemical reaction taking place in the rocks powered by the energy of the inner heat of the Earth. Their common ancestor was the nuclear radioactive reactions in the inner earth, coming from clouds of dust from supernovae, carrying the heavier elements of uranium and thorirum dating back to the energy transfers of the Big Bang. It was the decay of those elements, along with its gravitational collapse that heated the inner earth. This heat energy is then released, not least by volcanoes and superheated vents, into boiling water and steam and the chemical energy trapped or transformed in the water of the seas and lakes evaporates into climate energy. As we saw in LE1, we know much about the processes which appeared in the miniscule time eras soon after teh Big Bang happened, but not much about the processes during, let alone 'before' it. The same is true with the Big Birth and its processes *before* the appearance of the first single cell and then multi cellular life. Many scientists who study the Big Birth claim that it comes from the same laws of physics and of nature that we see in everything, ever since the moment of the Big Bang. But, as some argue, there was nothing in the Big Bang that pre-determined that the Big Birth was inevitable. Its nature and appearance was entirely random and unlikely. But, what is the Big Birth and what is the nature of organic life?

Cells

As we shall see, the history of how we understand the emergence of life is a long one and cells are centre stage in that early story. Before we look at that history, we pause to look at where we are now. A good example might be the developmental biologist and noble prize winner, **Sir John Bertrand Gurdon FRS,** working with **Shinya Yamanaka**. John Gurdon is famous for his work on stem cells which are already proving invaluable therapeutically and will probably become an even more crucial part of medicine in the future.

John Gurdon. Photo Deryck Chan taken at Annual Scholars Dinner Magdalene College, Cambridge 2012

He discovered that a single cell egg had the potential to develop all the specialities of a developing animal. A could become B – to use the simple code from the theme of this book. Gourdon led an extraordinary life. He came last in biology exams at Eton but went to Christ Church Oxford to read classics, and then changed to zoology. His most significant work at Oxford was perhaps done on frogs, discovering that mature cells can be converted to stem cells. As early as 1958, well before Dolly the sheep, he hypothesised that a single cell egg had the propensity to develop all the specialities of a total organism. In his classic experiment, published in 1962, he showed that the genome of a mature cell contains all the information needed to drive its development into all the different cell types of an organism. He replaced the immature cell nucleus in an egg cell of a frog with the nucleus from a mature intestinal cell. This modified egg cell developed into a normal tadpole. The DNA of the mature cell did have all the information needed to develop all the cells in the frog. This discovery goes to the heart of what we mean when we say that life *'makes itself'* from within its own resources – in its DNA coding. The fact that this is twentieth century science shows how Darwin's big idea was so significant. Despite all the evidence he collected, there was much that he couldn't have known and that makes the significance of his innovation and conversion even greater. As we shall see later, he made a transition that was difficult even for him.

When on the Beagle, his Journal gives little sign of this conversion, though he is asking many of the right questions of the evidence before him, during that five year expedition. In 1832, aged only 23, he arrived in Argentina. In that empty landscape, he discovered gigantic, fossilised bones on the coast and thought they might be allied to the rhinoceros. He hoped this would be an ancient, extinct species. The normal religious assumption was that extinctions had been caused by the great Biblical Flood, but the evidence showed the animal had not been killed by a catastrophe.[105] He began to think that extinction might be a regular feature and he found fossils of sea creatures which supported this. He lived rough and lived off the land during these times. When picking through the bones of an animal on his dinner plate, he saw similarities with the giant fossilised bones he had found. He thought it might be an armadillo and realised there must be a link between ancient fossils and animals still living. He asked himself why modern species seemed so similar to long dead ones. Was extinction a necessary part of evolution, and how did species originate? He sent the fossils to his former tutor, **Henslow** at Cambridge who put them on show. Even **Sedgewick** was enthusiastic and said so to Darwin's father. When he returned in 1836, he was already something of a celebrity, because of his fossils,[106] but retreated from public view, knowing he had the beginnings of a dangerous idea – that species weren't fixed and unchanging, otherwise more extinctions would follow. He knew that only a fraction of the creatures born survived and wondered if the same forces drove species to change. Was it only those species which adapted well to their environment that survived and passed on their adaptions? Was all life connected through these adaptions? He confided in **Hooker** that suggesting species weren't immutable felt like confessing a murder. It was a year later, in July 1837, that he wrote in his notebooks '*In July opened first notebook on Transmutation of Species. Had been greatly struck from about the previous March on character of South American fossils and species of Galapagos*

---

[105] The Flood was extensively seen as the second divine intervention after the Creation. Theologians and scientists spoke of pre diluvian and post diluvian. Theologically, it was very hard to accept the idea of extinctions. This implied that a mistake had been made in creation. Even the great geologist and Dean of Westminster, **William Buckland** found this question very troubling and many of his contempory clergymen rejected the possibility altogether. This is part of thetheological and scientific background in which **Mary Anning** and **Elizabeth Philpott** struggled to gain acceptance for their fossil finds in the early part of the nineteenth century. They wrestled with the growing realisation that the fossils they were finding were not representative of any living creature anywhere in the world, and, therefore, must be extinct. Even the great *George Cuvier* rejected Mary Anning's drawing of a plesiosaur as representing a fake fossil. They pointed to a much older era well before humans appeared and so contradicted the Bible. Mary seems to have accepted an interpretation given her by Buckland that 'one day' in the Genesis account represented a much longer epoch of time.
[106] Ironic in view of the disputes with Owen that were to follow and the problem of the lack of evidence in the pre-Cambrian.

Archipelago. These facts (especially later) origin of all my views.'[107] From the immutability of species to their transmutation is a big intellectual jump even though, as we shall see, the idea had its precursors. Why are plants and animals sufficiently alike to be grouped into species and sufficiently different to vary amongst themselves within a species? Are species related to each other, and if so how? What constructs the links in the chain of their connection?

We have to think ourselves back into his times to sense the radical nature of these questions and the assumptions which blocked them in the thinking of many others. His early answers to these questions[108] were tentative and speculative as well as bold and courageous. *'Why is life short? ...We see the young of living beings become permanently changed or subject to variety, according to circumstances..hence we see generation here seems a means to vary or adaptation.. Therefore generation (**is designed**) to adapt and alter the race to changing world.'* We note the significance of the words in brackets here. *On the other hand generation destroys the effect of accidental injuries which if animals lived for ever would be endless. Therefore final causes of life. Why does individual die? To perpetuate certain peculiarities (**therefore adaption**) and to obliterate accidental varieties and to accommodate itself to change (for of course change even in varieties is **accommodation**)..If species generate other species, their race is not utterly cut off....with this **tendency to vary of generations** why are species **constant** over whole country..Beautiful law of intermarriages partaking of characters of both parents and these infinite in number..According to this view, animals on separate islands right to become **different** if kept long enough apart...Galapagos tortoises. As species, as soon as once formed by separation or change in part of country repugnance to intermarriage settles it. Progopation explains why modern animals same type as extinct which is law almost proved. This view supposes that in course of ages and therefore changes every animal has **tendency** to change. **This difficult to prove.*** He does not use the word evolution, at this stage, but his language is of *tendency* and *accommodation* and *change.*[109] In brackets, we see the word *adaption* appearing, if not for the first time, then still significant in this context. He is finding his way in a transition stage of questions which imply both connectivity and mutability. Then he comes to a special moment of articulation in his questioning. *'If we choose to let conjecture run wild then animals, our fellow brethren in pain, disease, death, suffering and famine – our slaves in the most laborious works, our companions in our amusements – they may partake our **origin in one common ancestor** – we may be all melted together.'* At the end of the *Origin* we shall see him coming full

---

[107] *Life and Letters* 1 p 276

[108] in this notebook of July to February 1837. My emphasis in bold throughout these quotations

[109] We remember that his grandfather Erasmus had pioneered the idea of *development*. See later comments on his work.

circle. The phrase - *pain, disease, death, suffering and famine* will become *'the war of nature, from famine and death'* out of which emerges something *exalted.* Here, then, in his early 1837 notebook he *'lets his conjecture run wild'* to speculate that animals and humans may have their *origin in one common ancestor.* No wonder the next sentences read as follows *'The different intellects of man and animals not so great as between living things without thought (plants) and living things with thought (animals). Organised beings represent tree*

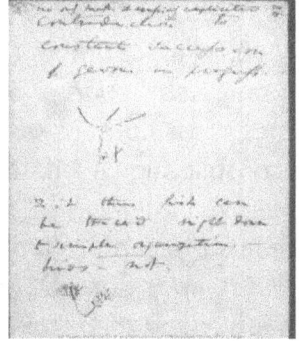 *irregularly branched. The tree of life should perhaps be called the coral of life, base of branches dead; so that passages cannot be seen.*[110] Then he draws a simple candlestick like sketch preceded by the words *'so that passages cannot be seen-this again offers (p.26 top) contradiction to constant succession of germs in progress no only makes it excessively complicated.* The sketch is followed by the words *Is it thus fish can be traced right down to simple organization –birds- not.*

Notebook B On Transmutation of Species 1837-8. p 26. Unpublished. Source

The lines are dotted at the base. He wrote in Notebook C of 1838 *'The bottom of the tree of life is utterly rotten and obliterated in the course of ages. It leads you to believe the world older than geologists think. Cuvier*[111] *objects to propogation of species by saying why have not some intermediate forms been discovered between Palaeotherium, Megalonyx, Mastodon and the species now living? Now according to my view (in S. America) parent of all Armadilloes might be brother to Megatherium – uncle now dead. Species according to Lamarck disappear as collections made perfect. We have not the slightest right to say there never was common progenitor between Mammalia and fish where there now exist such strange forms as...if all men were dead, then monkeys make men, men make angels.'* Then he goes on to anticipate *The Origin* by proposing there must be some kind of law influencing change, without using, as yet, the phrase *'natural selection.'* He references physics and cosmology, as he moves from the idea of separate creation to a law of generation - which is interesting for the ideas in this Trilogy. He uses the example of the *law* of gravity to explain cosmological movement and change everywhere, rather than God creating the *separate* movement of each planet. He is helped in the making of such a transition by viewing the *law* to be at least as *sublime a power* as separate intervention. *'Before the attraction of gravity discovered it might have been said it was as great a difficulty to account for the movement of all by one law, as to account for each separate one; so to say that all mammalian were born from one stock and since distributed by such means as we can recognise may be thought to explain nothing.*

---

[110] Page 25 leading to sketch on page 26

[111] We shall see more of Cuvier and Lamarck later.

*How does it come wandering birds such as sandpipers not new (originated) at Galapagos. Did the creative force know that this species could arrive? Did it only create those kinds not so likely to wander.. ..Astronomers might formerly have said that **God fore-ordained each planet to move in its particular destiny**. In the same manner God orders each animal created with certain forms in certain countries; but how much more **simple and sublime power** – let attraction act according to certain law, such are inevitable consequences. Let animals be created, then by the fixed laws of generation, such will be their successors. Let the powers of transportal be such and so will be the forms of one country to another. Let geological changes go at such a rate so will be the number and distribution of the species!* Only a few weeks later, as he starts the next volumes of his notebooks,[112] we see the speculative, tentative tone of (wild) conjecture consolidating itself. *'Let man visit orang-outange in domestication, hear expressive whine, see its intelligence when spoken as if it understood every word,..see its affection to those it knows, see its passion and rage, sulkiness and..despair; let him look at savage, roasting his parent, naked, artless, not improving yet improvable; and then let him dare to boast of his proud pre-eminence. Man in his arrogance thinks himself a great work, worthy the interposition of a deity. More humble and I believe true to consider him created from animals.'* We can almost hear the tone of the Victorian gentleman observing the behaviour of *'savages,'*[113] and insisting on their *'improvability!'* He is possibly referencing Shakespeare[114] in his use of *'man in his arrogance thinks himself a great work.'* Hamlet says to Rosencrantz and Guildenstern *'What a piece of work is a man! How noble in reason, how infinite in faculty! In form and moving how express and admirable! In action how like an Angel! in apprehension how like a god! The beauty of the world! The paragon of animals! And yet to me, what is this quintessence of dust? '* Darwin may also have been thinking of the Psalms - *'What is man that You take thought of him, And the son of man that You care for him? Yet You have made him a little lower than God, And You crown him with glory and majesty! You make him to rule over the works of Your hands; You have put all things under his feet.*[115] We also note his use of the more distancing *'a deity'* in the phrase *'interposition of a deity.'*

Darwin's notes show that he is aware of the size of the task facing him in working out this law of the Origins and mutation of species. He continues in the same notebook. *'This multiplication of little means and bringing the mind to grapple with great effects produced is a most laborious and painful effort..will never be conquered by anyone..who just takes up and lays down the subject without long*

---

[112] Two other volumes up to 1830

[113] This was the age of such observations in the expeditions of various explorers.

[114] Hamlet Act Two Scene 11

[115] Psalm 8.4-6

dedication.' Then he makes the following revealing note to himself, showing again a connection with the concerns of LE1 of this Trilogy. *'Mention persecution of early astronomers. Then add chief good of individual scientific men is to push their science a few years in advance of their age. ..Must remember that if they believe and do not openly avow their belief, they do as much to retard as those whose opinions they..'* In the introduction to *The Origin,* he was later to make a connection with these early thoughts and wrote. *'On my return home, it occurred to me in 1837 that something might perhaps be made out on this question (of the origin of species) by patiently accumulating and reflecting on all sorts of facts[116] which could possibly have any bearing on it. After five year's work I allowed myself to speculate on the subject and drew up some short notes.'* Again we see the note of humility and his tentative approach as he focused on the facts in order to turn the theory into a law. This could be summed up by the simple and enigmatic phrase *'I think'* attached to his later tree of life which became an icon of his work. Such a tree could be constructed for any group of creatures, but Darwin didn't try to do this. Possibly he was aware of the fossil deficiencies in the Pre Cambrian record and the likely criticism of **Richard Owen**. Darwin believed that the Pre-Cambrian – thirty million years older than the Cambrian – must have swarmed with creatures, but no animal remains were present. As we shall see, many creationists then, as now, were scandalized by Darwin's implication that homo sapiens developed out of earlier primates, but we know this is not the most profound challenge of evolution. Of course, we can trace our ancestry back much earlier than the primates in the family tree sketched by Darwin.

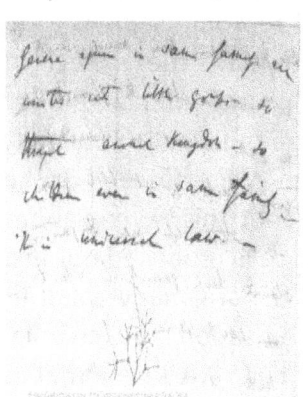

1837 sketch, from his First Notebook B p 36 on Transmutation of Species on view at the Museum of Natural History in Manhattan. P-D Copyright expired

At the side and bottom it says *'I think case must be that one generation should have as many living as now. To do this and to have as many species in same genus requires extinction. Thus between A + B the immense gap of relation. C + B the finest gradation. B+D rather greater distinction. Thus genera would be formed. Bearing relation (next page begins) to ancient types with several extinct forms.'* Then, we find a page (127) torn from his notebook dated 1848, and collated into *'Principle of divergence, transitional organs/instincts'* (1839-1872). The words above the tiny sketch are significant for their reference to a universal law – *'Genera species in same*

---

[116] In his autobiography he says *'by following the example of Lyell in Geology and by collecting all facts which bore in any way onthe variation of animals and plantes..some light might be thrown on the whole subject. My first note book was opened in July 1837. I worked on true Baconian principles and without any theory collected facts.'* Life and Letters 1.p 83

family are united into little groups - so through animal Kingdom - so children even in same family - this is universal law.'

There are several other sketches in his notebooks. In the following, we see the

words *Parents (over crossed out word) of Placentals and Marsupials*. It is probably dated from the early 1850s.[117] There have been many 'trees of life' produced before Darwin and many from ancient mythology and religion. In his time **Wallace** and **Haeckel** produced their own diagrammes. Ever since, as our knowledge changes, many more have appeared. With the discovery of genes, and DNA in particular, the growth of family trees has increased and many can be seen on tee shirts or body tattoos. **Dr. Barth F. Smets,** Professor of environmental microbiology at the Technical University of Denmark, has produced the following diagramme to show horizontal as well as vertical genetic inheritance lines emanating from common

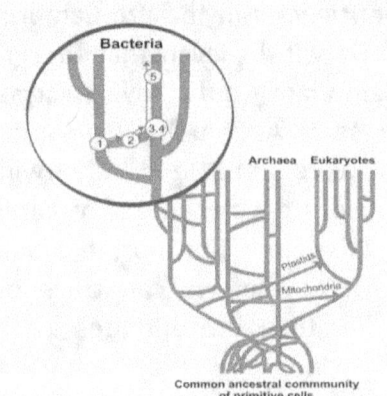

Common ancestral community of primitive cells

With permission from Dr Smets

ancestor primitive cells. It is now accepted that prokaryotes, bacteria and the archaea have the capacity to transfer genetic information causing variation between unrelated organisms through a process of horizontal gene transfer which includes recombination, gene loss, duplication, and gene creation. The latest gene maps list all the genomes sequenced with links to a common ancestor, but they are far too complicated and detailed to be included here.

As we learn to paint the picture of the growth of Darwin's original tree, there were many branches we didn't share with our earliest ancestors and many animals have vanished or become extinct. A did not lead to homo sapiens B, in any simple or direct way, but we know from our DNA that we share much in common with quite unlikely forms, which looked nothing like the primates! What about the sponge? What about early single celled, let alone multi celled animals, joined by the glue of collagen? What would **Bishop Samuel Wilberforce**[118] or the detractors of Darwin in Britain and America have said about this? What about 600 – 560 million years ago, when the first organisation of cells produced not just clumps of cells, but the delicate first animal forms? There are many theories about how this happened including the symbiotic, cellurisation and colonial theories. In all of these, and however early life developed, we know that it

---

[117] Ibid p 184
[118] See later description of the famous Oxford debate with Huxley.

couldn't have happened without different forms of energy being used and transmitted. At Mistaken point in Canada, pre Cambrian[119] fossils such as Bradgatia which might be nearly 600 million years old have been found preserved in underwater volcanic ash. They were frond like plant animals that came from deep water where there was not enough light for photosynthesis. They absorbed nutrients through their bodies. They had no muscles and were immobile. Along with the fossilised, pre-Cambrian Charnia found in Charnwood Forest in Leicestershire, England[120] and others in South Australia, they are the fossil remains of possibly the oldest proto animals on earth. They built up the shape of their bodies in modules of repetitive, fractal branching[121] which created a larger surface area to absorb their food (without mouths). They had the absolute minimum in their genetic coding. These animals seemed incapable of evolving and so did not remain on our evolutionary tree. But now we know to go back even earlier.

Carbon

There was a time when no carbon existed anywhere in the universe and without carbon it is inconceivable that carbon based organic life could appear anywhere. We know that the birth of the first stars gave light to the universe –as Genesis 1.3 itself records, but wrongly located the appearance of light right at the beginning of its story of creation. *'And God said, let there be light and there was light.'[122]* This began the process which would eventually lead to the creation of carbon. In the heat of the stars, hydrogen produces helium and three helium nuclei could produce carbon in the dying stars –their dust carrying the carbon through the cosmos, as we saw earlier in this chapter. Carbon enters the processes of organic life on earth - and perhaps on other planets - through the carbon cycle in the air, in plants and in rocks. Very little is known about the content of the air around the planet earth, in the earliest period of its formation. The development of an atmosphere is related to the story of this early formation[123] and crucial to its later development. In the photosynthesis of carbon dioxide, light on leaves produces carbon and water and sugars. Carbon found in

---

[119] This aeon of time – preceding the Cambrian period, so named because of rocks discovered from this period in Wales - makes up about 85% of earth's history. The pre-Cambrian era includes the Archean and Proterozoic eons and dates from the formation of the earth c 4570 million years ago to the beginning of the Cambrian period about 540 million years ago. The study of geology (Greek ge = the earth; logos – study) doesn't become a professional science until the 18th Century and the formation of the Geological Society in 1807 but has a fascinating history prior to that with contributions from, among many, Aristotle, Pliny, the Chinese Shen Kuo (1031-1095), Abu al-Rayhani-Biruni (983-1048), Ibn Sina (981-1037), Nicolas Steno (1638-1686). Richard Hook and many others. The word *geology* is first used by Ulisse Aldrovandi in 1603 and by Jean-Andre Deluc in 1778, coming into permanent usage after Horace-Benedict de Saussure in 1779.

[120] Found by the school boy Roger Mason in 1957.

[121] See LE1 section on Mandelbrot's fractals.

[122] See also 2 Corinthians 4.6

[123] Perhaps a very large lump of matter hit the matter of the earth as it formed round the sun, causing the material for the moon to form as a result of the explosion. A 'stable' crust may have formed by about 4500 million years ago.

the wood of the trees - in the form of lignin and cellulose – was perhaps eaten by early termites living 30 million years ago, as some do today, taking the wood pulp into their air conditioned mounds and mixing it with a specialized fungus in ways that released carbon into the food chain with animals grazing on the grass, turning cellulose into meat; the ruminate animals breaking down the cellulose in their stomachs and contributing to the extremely complicated carbon cycle so that every living thing becomes a carrier of carbon in one way or another. The backbone of carbon in living things is of course the DNA molecule. Our Last Universal Common Ancestor (LUCA) appeared about 3.5 billion years ago when the living chemistry in rocks produced the biochemistry needed to form the LUCA. Even since, all life has used the DNA codes of ATCG in millions of combinations to produce amino acids and proteins with protons from thousands of light years away in space (and background radiation from rocks) hitting the DNA to cause *random* mutations. These mutations are an inevitable part of life on earth and they produce *innovations*. But to avoid *random* code production in the 'selfish' DNA, natural selection comes into play to influence the survival of selected characteristics coming from the code. The story of how life on this planet developed is part of the story of the whole cosmos. Because the laws of physics and chemistry are the same throughout the cosmos, it is probable that life developed on other planets with similar conditions because of those laws. The story is complicated and extraordinary. Placed alongside the ancient texts of creation theology, in all religions, it is salutary to make comparisons between them and glimpse something of their worldviews and knowledge. In the centuries following the appearance of these texts and their earlier oral forms, their interpretation and relevance have been widely discussed with the changing story of science as a reference point for validity. We are fortunate to know as much as we do, conscious of how much more there is to discover. There is however no going back to the dogmas of creationism – there can be nothing in the science of the future that justifies returning to those pre-modern assumptions and beliefs. That would be to ignore the struggles of significant individuals and teams of people working on different parts of the science, in different parts of the globe.

The conditions for the origins of life

Human beings are evidentially the most complicated species on this planet, but, as we have seen, part of larger cosmological and biological events. Brian Cox has described our species as nuclear stardust, linking our DNA to the collapse of the stars.[124] As we wonder how the first, amazing forms of life appeared on our planet – about 3.9-3.5 billion years ago – the possibility of primitive cells being transported here in the ice of a meteor is still being discussed. The origin of life

---

[124] In his '*Wonders of the Universe*' 2011 and in his '*Wonders of Life,*' 2013, both Collins, based on the BBC series.

on this and other planets links us to the elemental forces in the natural world. Our origins are linked to inorganic matter in the process of biopoiesis or abiogenesis, at a time when the earth was mostly molten. The evolutionary movement is from geo-physics to chemistry, to biology. The physics of each stage for the conditions of life really matters, although the idea of 'stages' is conceptual only to help us picture and measure what is in fact a dynamic and changing process. The planet Earth is approximately 4.6 billion years old. During the initial 0.7 billion years following its formation, the early Earth was heavily bombarded by solar system materials, such as comets and asteroid-sized objects. The energy released by the largest impacts was sufficient to evaporate the oceans and destroy any existing life on the Earth's surface. The first signs of life evidenced by the fossil record came into being approximately 3.5 billion years ago. Life emerged through a complex chain of evolutionary events, dictated by the physical-and chemical environment on the early Earth. The reducing atmosphere (one with only traces of oxygen and therefore a lack of reactive oxygen species which would attack chemical bonds), provided favourable energetic surroundings for the formation of relatively complex polymers from organic monomers which were already present on the primitive Earth. The monomers seem to have come from two probably sources - either formed from terrestrial synthetic pathways or derived extra terrestrially from solar system materials. Over time, simple molecules developed into larger, more complex, biological molecules and eventually to cells. Following further diversification, some cells developed in ways that became metabolically capable of photosynthesis. This caused a cascade of irreversible events, interconnected by bio-geo-chemical cycles. The atmosphere of the Earth changed to that of an oxidizing one and subsequently developed an ozone layer. The introduction of oxygen no longer supported the development of new life forms from the primordial building blocks, but instead enabled the biological development of the early micro-organisms. The ozone layer served as a means of protection, filtering the harmful UV radiation.

These dramatic changes transformed the early Earth into our present day biosphere. The assembly of the first cellular life on the prebiotic Earth required the presence of three essential substances: water, a source of free energy and a source of organic compounds. It is unlikely that life anywhere in the cosmos can exist without liquid water. This water would have evaporated off on planets too near their sun or frozen on planets too far away. The reason for its origin on earth is still not clear – perhaps precipitated from the cooling earth's atmosphere. Was it present in rocky material formed in the Earth's region of the solar nubula, or was it even first 'delivered' by comets? Clearly it was present before the appearance of the first micro-fossils. To polymerize small molecules to become more complex forms, energy sources are needed and free energy was available from lightning and volcanoes as well as the intense UV radiation of the young sun.

Spontaneous generation?

The theories about how this happened have changed over the centuries and raised many theological and philosophical questions. Some argued that any talk of spontaneous generation removed the need for an external creator and others that this was the chosen methodology of such a creator. In our thesis that Love's Energy is the source of the conditions for life, I am proposing that this energy is embedded in *life's longing for life* in the processes of natural generation. Abiogenesis or biopoiesis are the words used to describe the processes by which organic life appeared from inorganic life. The ancient Greeks believed in some kind of spontaneous generation of organic life from non living matter. This belief lasted well into the 19th century. Aristotle observed aphids coming from the dew on plants, flies from decaying matter, crocodiles from the mud at the bottom of rivers. The Egyptian creation myths believed that humans were created or sculpted directly out of the mud silt of the fertile Nile and the Hebrews shared (but also changed) ancient Semitic beliefs that humans were made from the dust of the ground (Adamah in Hebrew). For the time this was an understandable inference. In 1665, **Robert Hook** produced the first drawings of a micro-organism and in 1676, **Anton Van Leeuwenhoek** drew the first drawings of bacteria. This might have initially supported the idea of spontaneous generation as cell division had not yet been observed. Van Leeuwenhoek's later experiments on the incubation of organisms in meat and insect reproduction convinced him, by the 1680s, that spontaneous generation was wrong. In 1668 **Francesco Redi** (1626-97), trained by the Jesuits, and court physician to Ferdinando **Medici** and Cosimo

Medici, showed that no maggots appeared from meat, unless flies had first laid their eggs and slowly the idea of biogenesis took hold. *Francesco Redi. Source and date unknown. Copyright expired*

The great insight in his work was that all life must have come from a previous form of life (*omne vivum ex ovo*). A century later in 1768, another Jesuit and biologist **Fr. Lazzaro Spallanzani** (1729-1799) made his contribution.

Lazzaro Spallanzani photo of original artwork. Author and date unknown. Public Domain Copyright expired

He had studied law at the University of Bologna, where a relative, **Laura Bassi**, was professor of physics. She encouraged his interest in science and helped him with natural philosophy and mathematics. At the age of only twenty five, he became Professor of logic, metaphysics and Greek in the University of

Reggio.[125] He showed that microbes were present in the air. By a century later, in 1861, **Louis Pasteur** (1822-1895) showed that bacteria do not spontaneously appear from a sterile, nutrient rich medium, but come from outside it. Experiments continued and in 1864 Pasteur declared that *'never will the doctrine of spontaneous generation recover from the mortal blow struck by this simple experiment.'* In 1871 Darwin wrote to his great friend, **Joseph Hooker**, suggesting that life may have begun in a *'warm little pond, with all sorts of ammonia and phosphoric salts, lights, heat, electricity, etc. present, so that a protein compound was chemically formed ready to undergo still more complex changes.'*[126] This pond image brings to mind human population growth being compared to the algae spreading across a pond until no more light can get through to grow nutrients.[127] In 1924 **Alexander Oparin** (1894-1980) proposed that spontaneous generation might have occurred once on earth but, after that was impossible, because other life forms would eat the spontaneously produced life form.

Alexander Oparin. Commonly used image in scholarly articles. Source and date unknown. Assumed to be copyright free.

Perhaps, he argued, a primeval soup of life forms could have been created by sunlight before the first oxygen appeared. [128] He was particularly famous for his pioneering early work on the chemistry of the atmosphere and its effect on the 'big birth.' In an early and formative insight he claimed that *'there is no fundamental difference between a living organism and lifeless matter. The complex combination of manifestations and properties so characteristic of life must have arisen in the process of the evolution of matter.'*[129]

J.B.S. Haldane taken in Oxford 1914 PD-US.

Perhaps, argued the British born evolutionary biologist and Marxist **John B.S. Haldane** (1892-1964),[130] the earth's first pre-biotic oceans could have acted like a primeval soup in which the first organic life forms would have appeared in a process of biopoiesis – evolution from self replicating but non living molecules. Then came the crucial **Miller-Urey** experiment in 1952. A mixture of water, hydrogen, methane, and ammonia was circulated through an apparatus and electrical sparks were passed into the mixture. After several days, the carbon in the mixture had formed organic compounds including amino acids as the basis for the production of proteins. As late as 2007, twenty more amino acids were found in the original phials used for the experiment. The

---

[125] In 1768 Spallanzani was elected a Fellow of the Royal Society.

[126] *The Life and letters of Charles Darwin*, 1887. **3**. London: John Murray. p. 18.

[127] Beneath all the environmental challenges, raised by so many groups, overpopulation is a common factor which they mostly avoid discussing. The maths of exponential growth and the rule of 70 seem to be uncontrovertable.

[128] *The Origins of Life* 1924

[129] Source unknown. See also Oparin, A., and V. Fesenkov. Life in the Universe. New York: Twayne Publishers (1961).

[130] Who became an Indian citizen after the Suez crisis.

basic instincts of Oparin and Haldane – that the conditions for life were to be found in primeval chemical reactions in the early earth - had been proved correct and the discovery inspired more attempts to find evidence for other biologically crucial compounds in the primeval earth. Oparin and Haldane both took seriously the hypothesis that, in the early conditions of a newly formed planet, the chemical conditions were right for the synthesis of organic compounds out of inorganic matter. The origin of life, in the Big Birth is still a matter of debate, but it locates early organic development within the geology of cosmological events and illustrates the need for a radical understanding of autonomous physical processes which are connected, as well as being free in their development. So what picture can we build of the appearance and history of life on this planet? The scientific version goes back much further and shows a more complicated but just as wondrous story as that used by the creationists, in all religions. The basic timeline of a 4.6 billion year old Earth, with very approximate dates, might be set out as follows:

3.6 billion years of simple cells (prokaryotes),
3.4 billion years of stromatolites demonstrating photosynthesis,
2 billion years of complex cells (eukaryotes),
1 billion years of multicellular life,
600 million years of simple animals,
570 million years of arthropods (ancestors of insects, arachnids and crustaceans),
550 million years of complex animals,
500 million years of fish and proto-amphibians,
475 million years of land plants,
400 million years of insects and seeds,
360 million years of amphibians,
300 million years of reptiles,
200 million years of mammals,
150 million years of birds,
130 million years of flowers,
65 million years since the dinosaurs died out,
2.5 million years since the appearance of the genus Homo,
200,000 years of anatomically modern humans,
25,000 years since the disappearance of Neanderthal traits from the fossil record.
13,000 years since the disappearance of Homo floresiensis from the fossil record.

Professor Ian Stewart in his series 'How to grow a planet; Life from Light' [131] describes it in the following simple and vivid way. 3-2.5 billion years ago nothing

---

[131] 7 February BBC2 'How to grow a planet'

grew because the atmosphere around Earth was too thin to protect the surface from the UV Light and direct heat coming from the sun. The oxidisation event or the biologically induced production of free oxygen took place at about 2.4 billion years ago. This can be traced by looking at the oxygen trapped in the iron sulphide banded formation of rocks from this period. The way oxygen was produced is still debated. The first burst of life came from in the sea with bacteria which were purple in colour and then, after living off the different colours of light they became green. Cyanobacteria, which were blue green in colour, produced oxygen. Photosynthesis came from these first cyanobacteria, from which chloroplasts evolved. Photons of light from the sun hit the cells of chloroplasts and two photons of energy spit water molecules into Hydrogen and Oxygen. Very early plants used hydrogen and carbon dioxide to grow and produce sugars producing oxygen as a waste product. The oxygen built up over thousands of years and created an ozone layer to reduce UV rays. Plants colonised the land for the first time 400 million years ago. Hot springs and geysers pushed out the bacteria. Primitive plant forms were still tied to the water's edge until they developed roots which die, combine with broken rock and sand to form soil. Without plants the appearance of the first animals from the sea onto land would not have been possible. Animals appeared and evolved the ability to extract oxygen from both the sea and air, for example the horseshoe crabs. About 400 million years ago, the first insects and amphibians appear on land. Plants became victims of their own success, using up the Carbon dioxide too quickly so that larger plant structure evolved including leaves to breathe more carbon dioxide through stomata on the underside of leaves. Leaves provided shade from the sun to other creatures and plants evolved to grow taller in competition for the light of the sun. Plants developed the capacity to grow in height and produce wood to support that growth. Trees appeared. Photosynthesis and the process of energy and light from the sun produced the evolution of plant life in ways that increased the supply of oxygen, gave life to other creatures and helped in their evolution, not least the size of plant eating dinosaurs who evolved longer bodies and taller necks in order to reach up to the height of trees - in turn competing for access to the light and energy of the sun in the densely packed plant life of forests which now covered much of the planet. Again and again in this story, we discover the presence of and need for different forms of energy as the universal constant in the connectedness between every A and B in the processes of the development of life in its increasing complexity and diversity. If organic life appeared on earth somewhere between 4-3 billion years ago, it happened at the interface between the energy constituting water and the atmosphere, or near the energy of volcanic vents in the deep sea. The earliest known fossils are marine photosynthetic eubacteria (Greek for canes or rods). These remain as the most abundant and diverse organisms on the planet which lack a membrane bound nucleus. They proliferate by division and some by sexual reproduction. Some photosynthesise as their way of responding to and creating energy. Some obtain their energy from minerals. They live everywhere including in acidic or toxic environments and inside other organisms sometimes causing disease. Chromalveolates are a very diverse group of unicellular or multi-cellular eukaryotes, including, for example, the forests of brown algae found in coastal waters. DNA analysis shows they are descended from ancient common ancestors, which engulfed photosynthethic red algae to form permanent relationships and connections capable of absorbing and fixing the energy of sunlight. The study of these and others of the earliest life forms – arcaea, eukaryotes, deuterostomes, lophotrochozoans, protostomes - includes DNA tracing and the recognition of how energy is is exchanged as part of the process of relating, adapting and changing.

Energy exchange
As we have seen all living things require energy to maintain internal organisation. Internal organisation is composed of structural relationships at the molecular, cellular, bio- chemical and sub particle levels. Structure is an exchange of physical connectedness which involves the passing on of energy embedded in relationships and relationship messages. It is the way through which parts of the Other relate to other *others*, both within and around the structure of the organism. At some point, a crucial exchange known as the 'Fateful Encounter' took place between Prokaryotes and a free-living bacteria. Mitochondria were once Prokaryotes themselves.[132] While eukaryotic cells evolved, prokaryotes remained simple and only once made the transition to more complex eukaryotes. Apparently a bacterium was engulged by a larger cell and amazingly the bacterium not only survived, but was absorbed or domesticated. The event forged a new structural partnership and energy exchange, from which came all later eukaryotes with mitochondria as their energy centres causing chemical reactions between proteins (found inside the mitochondria's inner membrane) and oxgen that produce ATP – the molecule which acts as the energy vehicle of a cell. Energy is embedded in the process of structural exchange which is a dynamic process in all living things. To live is to move, relate and adapt, using and sharing energy. To live is to exchange, but amazingly this happened naturally within the physics of living things, even before there was ever the possibility of consciousness to control such exchanges. Exchange means transformation in evolutionary terms. It also involves the subtlest and most basic transformation of energy into matter and different forms of energy -matter. Life feeds off negative entropy. As every A and B relate, there is always an exchange of energy of some kind. Life is a system of exchange through which entities are able to decrease

---

[132] See the work of Dr. Nick Lane, already mentioned

their internal entropy and transform energy into different forms. As we have seen in LE1, this occurs cosmologically as well as biologically. *Metabolism* is a chemical structure of process reactions which maintain, use and transform the energy of life. *Catabolism* breaks down organic matter to harvest energy in cellular respiration. *Anabolism* uses energy to construct components of cellular life through nucleic acids. Metabolic pathways transform one chemical into another through a process sequence of enzymes. Enzymes are crucial to metabolic exchange allowing organisms to drive desirable but thermodynamically unfavourable reactions by coupling them to favourable ones - acting as catalysts to allow these relationship reactions to happen efficiently. They also facilitate the regulation of metabolic pathways in process response to changes in the cell's internal structure, or environment, or messages from other cells. Surprisingly metabolic pathways or relationships seem similar across very different species. Carboxylic acids, in the citric acid cycle, or Krebs cycle, are present in all organisms, even those as diverse as the unicellular bacteria Escherichia coli and multi-cellular organisms as large as elephants. All share a common evolutionary history and a similarity in relationship eschange and connectivity.

Love's energy is embedded in relationship exchanges of all kinds. It is present in the most basic physical processes and structures of all living things. The particularity of living things is the way by which energy is transformed, reaches out and embeds itself in ways that make internal and external relationships and new forms of connectedness possible. Without this energy/ exchange based relationship, in its many, different forms, there is no life. The energy in sunlight is captured and transformed in many ways and passed on in embedded relationships which produce harmful as well as beneficial results and side affects – but all within a connected bio-eco-system. Light/heat energy depends on its own internal processes of exchange and transformation, its own particular environment in response to which reactions take place, absorbing and transforming the internal and external relations and structures of matter. Without the basic relationship between energy and matter, there could be no living things. Just as we can trace evidence of the big bang from what we can see now, in, for example, the Cosmic Background Radiation map, or Hubble's confirmation of red shift, so we can work backwards down the evolutionary story until we get as close as possible to LUCA or whatever came before it. As we saw in LE1, physicists understand the process back to the smallest fraction of a second after Point 0 of the Big Bang, but not Point 0 itself. To understand the physics of this 'moment,' we might need to engage with deeper questions than physics can answer. The question about the nature of Point 0 quickly becomes a question about the conditions for Point 0 to appear. Some cosmologists see the answer to that question by positing multiverses and other big bangs as the cause of Point 0 in this universe. This may only postpone the question, which itself may need metaphysics and physics to work together. Similarly, the Point 0 of the Big Birth is problematic, but for other reasons. Now the problem isn't what preceded it, but at which moment in time 'it' occurred and how. The phrase 'Big Birth' is misleading because it implies a singularity, when in practice it is obviously more complicated than this. There may not have been one common ancestor, but there was certainly a common process. Understanding the relationship exchange within this common ancestor process is crucial to understanding its structure and formation. It had to have RNA features which could be sustained and replicated. Whatever life form first appeared from the matter and energy around it, this form had to possess genetic information that could be passed on, adapted, shared, exchanged, through the ongoing internal and external environment relationship dynamic of energy and matter that made life possible. Perhaps the relationship between organic and inorganic life was much closer than we have imagined. Theologically, or at least in the thesis of Love's Energy, this relationship was formed within the energy/matter connectivity which creates the possibility of the other as an *entity potential* and an *entity state.* Within and from this *potentia* and *propensity* the diversity of other *others* can appear and relate. This potential of early genetic information made possible heritable patterns and processes or 'instructions' for functioning and reproducing - the means to replicate and carry out those instructions, otherwise no descendents or adaptions would have been possible, however many strange life forms appeared over whatever period of time. These systems for replicating the genetic material of its own *potentia* had to allow for some 'random' variation in the heritable characteristics so that new traits could be selected and species diversity appear.

From observable variations in the processes of natural selection and our best knowledge of genetics, we project back to the idea of *commonality* and *connectedness* within a dynamic energy *potentia* – perhaps as a biological equivalent of the point of infinite particularity and density in cosmology. We assume the big birth, like the big bang, was a point of connected, structured, energy exchange particularity from which other life could develop and replicate its *connectivity* and *diversity*. If so then *connectivity* and *diversity* are themselves in a dynamic, chemical and physical relationship within their own energetic *pattern paradox.*

It is assumed that common universal traits could not have evolved separately. **Leslie Orgel** of the Salk Institute for Biological studies compares this to the discovery of two very similar screen plays differing only in a few words. It would be unreasonable, he argues, to assume that each was created independently by two different authors. Either one script was a changed variant of the other, or both versions were slightly altered copies of a third (original). Whether this analogy works is another matter. To be consistent, natural selection must have always been part of the process from the time the first organisms appear. There must be a sense in which natural selection is present in the energy-matter relationship *potentia* from the beginning of the big bang, as well as the big birth, although operating in different contexts. As the first proto organisms appeared, there was *plurality* as well as *similarity* across their connectedness. It is unlikely that one single organism appeared without others of the same or similar form appearing within the same conditions for life. The chemical energy exchanges and relationships between the elements in the atmosphere, the sea and land naturally produced new organic life forms as expressions of their own sub-particle existence. It happened as a natural consequence of energy stretching and structuring itself in all directions and being transformed into the matter of an adapting, internal and external environment. Understanding the way organic life subsequently triggered its own replication and adaption is the exciting science of the 21st century, building on the thinking and observations, theories and counter theories of previous centuries. Some faith communities have adjusted their creation theologies with time; others did not. Any natural theology for our times needs to be rooted in our best understanding of how organic life appeared and how the embedded processes of energy matter transformation function. This is an exciting time to be connecting the theology and science of these processes. Any religion wishing to articulate belief in God as transcendent and imminent has to take the challenge of connectivity seriously. In a pre-Copernican and Newtonian world, the main narrative of life was one of fairly fixed and separate units, either at a cosmological or atomic level. It is has been argued that this influenced our understanding of the individual from the Renaissance through the Enlightenment to at least the end of the 18th century. If the universe was based on certain laws which explained the mechanical relationship between things as separate forces, laws, objects or events, then individuals fitted in to this narrative and had their own ordered positions, hierarchies and relationships. It is hard to think ourselves back into the implications of such a world, although some of its assumptions and practices are still with us. Of course, families and groups fought hard to protect their own sense of tribal and religious connection. Cultural myths take decades, if not centuries to change and often lag behind new scientific thinking. In **Judy Cannato's** book *'Radical Amazement'* [133] we are given an eloquent and inspiring description of the way the story we use about the universe has moved on since the days of Newtonian physics. Love's energy cannot simply be the longing of Love for the other to exist in a way that leaves no relationship between Love and the other. These two apparently separate narratives can and must be connected. The first, as we have seen, holds on to the radical freedom of the Other as the outcome of Love's longing for 'it,' the

---

[133] Sorin Books, Notre Dame 2006

universe and its processes of energy transfer to exist. We dare not reduce the nature of Love in describing this energy or longing, by reducing the autonomy of the Other. Any reduction of this kind would turn love into something more controlling and interventionist, leaving not just matter, but organic life and the human species in particular, living in a universe where events were determined outside of their own autonomy for free energy exchange, structural development and random incluences. Many religious thinkers of the past promoted the idea of pre-destination and its assumption that A controlled B even before B could know anything about it. Ironically it was not beyond the same group to blame B for its 'sins' even though its 'freedom' to make mistakes was apparently pre-determined.[134]

Variables, catastrophes and adaptions

When we consider the amazing series of events which led to the formation of the stars and planets and the appearance of organic life on at least one of these planets, we come up against the real possibility that none of this might have happened if the specific *conditions* for life had been any different. It seems very unlikely that such events and conditions were inevitable, because many of them were catastrophes. Perhaps that is too anthropomorphic a way of describing these cosmological and natural events. Let's take just one example. It seems that 640 million years ago, this planet was a snowball. Ice and snow reflected most of the heat from the sun away from the earth's surface. The more heat reflected away, the more the process of cooling developed. At this date, the only forms of organic life on the planet were probably bacteria on the seabed. The heating or cooling of the atmosphere around the planet very much depends upon the balance between the nature of global warming or cooling at the time, and the part that carbon dioxide and other emissions play in that. Too many emissions and the planet warms up. Too few and the planet cools down. It seems that over about 20-25 million years, the amount of carbon dioxide released was increasingly reduced and as a result, ice and snow spread over most of the planet. The edges of the ice reached the equator about 640 million years ago. The extraordinary thing is that bacteria managed to adapt and survive at these temperatures. The same bacteria have been found today, deep in ice caves in northern areas of the globe. They have survived even where no light or energy has penetrated through the ice. If these bacteria had not survived and adapted to their conditions, all organic life would have been destroyed. It was the eruption of volcanoes from deep within the ice – sometimes many kilometres deep – that unfroze the snowball. If planet snowball was a potential catastrophe, so then was planet volcano. The eruption of volcanoes put carbon emissions back into the atmosphere, so helping the planet to warm up again and the ice to recede. This is

---

[134] See the discussion of this in LE2

the time when the one super continent grouped around the equator area of the planet, moved away and broke up into other continents. This is the time when single cell and multi cell creatures first appeared in the sea, presumably kick-started by the new heat and energy that broke through this most severe of all ice ages.

Cosmologically, the conditions for life on any planet depend on huge variables such as distance from the star, size and mass, rate of spin, forces of gravity, composition of the atmosphere, temperature and so on. Organically, the conditions for life to appear and develop are just as inter-related, fragile and complicated. We take too much for granted, as an expediential increase in carbon emissions makes possible an average global warming of anything between two and perhaps six degrees over the next century. This, we believe would produce sea level rises which would flood low lying lands, including some of the financial capitals of the world – London, New York, Shanghai, Hong Kong. Although we plan for large scale emergencies, no-one can factor in the total human, social and financial cost, let alone the political chaos that would result. This is enough to remind us that we live on a fragile, but extraordinary planet and that controlling our mitigating our human impact on the delicate balances which sustain life is one of our greatest challenges. Perhaps the main threat is not to the planet which has survived even greater catastrophes, but to human life and stable societies. The social and economic consequences of migration are huge, leading to the probability of increased racism and exclusion, stretched social and educational services, as well as the positive benefits that it has brought through the centuries. Environmental migration, however, would be on a scale that increases by a factor of perhaps 10, with the movement probably being generally from the south to the north, from less congested to more congested societies. Economic and social stability in any society is fragile and could not easily bear such a huge increase in migrations. As we face the possibility of this environmental tsunami, we can look back on the different periods of 'catastrophe' in the history of our planet, and understand what an extraordinary story of sustainability, fragility and adaption this has been. The way Homo sapiens has manipulated its own environment through technology, rather than significant, biological, evolutionary change is, in itself, an extraordinary part of this story. Our DNA connectivity with all living creatures reminds us of both what we have in common and the importance of small differentiators in our physical make-up. The growth of the human brain which preceded and was kick-started by the appearance of language is itself an extraordinary story. It seems that early Homo sapiens and some of its earlier primate and other predecessors discovered the evolutionary value of cooperation, as well as competition in situations of finite resources.[135] The more cooperation, the larger the tribe became. The larger the tribe, the more need for social organisation emerged, not least to gather and manage those finite resources. The more social organisation, the more need to move beyond basic sounds to more sophisticated communication over greater numbers and distances. Once language appeared, this complexity could be better organised. With language came the ability to learn, remember and pass on information, reflect and communicate across and between groups, and so develop our sense of connectedness across differences in time and space. In describing a scientific story of life on earth and its relationship to the formation of this planet and its solar system, we use a language that is very different from any ancient, origin narrative.[136] It seems pointless to accuse them of ignorance, or blame them for painting a different kind of picture, particularly to discredit religion out of hand – as happens in the work of Richard Dawkins. Nor, as we have seen, is there any point in defending the scientifically untrue in any religious text because it is the 'Word' of Divine Revelation, or because we believe what it says will be proved true in the longer run against our current state of scientific knowledge. Far better that we should engage with the narrative's meaning in its internal and external context and therefore as something different from, and often deeper than, a 'scientific' description of the processes involved. The story of life will always remain one of humanity's most treasured narratives. There will be many new ways of telling it and many new reasons to do so. At the heart of this story is the precursor of the cell which appeared in our early oceans some 3.5-4.0 billion years ago. This book is about the process of evolution which links the human species on this planet with those early cellular precursors and beyond. As humans walk around their planet thinking about this, 70 billion human cells are born and die in our bodies each day. We have learnt how cells are the basic building blocks of life and, like atoms, are divisible into the physics and chemistry of sub particles. We have learnt that they come in all shapes and sizes (e.g. nerve cells are very long), but there are, as we have seen, basically only two types of cells - the eukaryotic, which is comparatively small and lacking a nucleus and the prokaryotic. The story of how we learnt all this is fascinating and compelling and part of the later chapter on Cells, DNA and Genes. We have seen how, in 1665, **Robert Hook**, the proud polymath and inventor of the earliest microscopes, looked at a piece of cork and used the word *cells* from *cella* or small room, because what he saw reminded him of monastic cells. We have seen how, at about the same time, **Antonie Van Leeuwenhoek** (1632–1723) taught himself to make simple lenses and was probably the first to see living cells of bacteria

---

[135] See later chapter on Altruism

[136] Remembering that in the case of Islam it is claimed that the words were dictated directly by an angel in the language of Arabic. Sso if there is anything scientifically untrue, then it must be the Angel's fault not the scribe's! There are still many Christians who claim that the bible is literally the Word of God, however it was transmitted.

from his own mouth, protozoa and vorticella from rain water and sperm cells. In 1839, **Theodor Schwann** and **Mathias Jacob Schleiden** asserted that all plants and animals are made of cells which must be a universal unit of development. In 1855 **Rudolf Virchow** took the next step and said that all cells emerge from the division of other cells.

The future of cell and gene research

In the century and a half since Virshow, laboratories all over the developed world have been working on cells, genes and their DNA coding. There has undoubtedly been unethical work on developing viruses or cell mutations for military use. Some of the countries that are UN signatories against the development or possession of biological WMD, continue investing in research, even as they criticize other countries for allegedly possessing such weapons. Meanwhile, there is no doubt that stem cell research is one of the most promising branches of therapeutic medicine and that this century will see radical new procedures and knowledge. As we learnt to use natural remedies for healing from plants and then developed a pharmaceutical industry on the back of that knowledge, so, in the future, we may develop cell manipulation not only for prevention and cure, but even the mutated development of the human condition itself. In Dan Brown's fictitious novel *Inferno,*[137] the head of the WHO is given the following words to say. *'Normally the evolutionary process – whether it be a fish developing feet or an ape developing opposable thumbs - takes millennia to occur. Now we can made radical genetic adaptations in a single generation. Proponents of the technology consider it the ultimate expression of Darwinian 'survival of the fittest' – humans becoming a species that learns to improve its own evolutionary process.'* If this happens, it will go well beyond any process of evolutionary development based on natural selection. It will no doubt be accompanied by the manipulative self-interest of corporate profit and political competition, backed up or initiated by military and intelligence service arguments for national security. It will probably be used by terrorists and those we call our enemies and who see us as their enemies. Along the way, science will understandably protest its independence and moral neutrality, even as it accepts the public and private research grants that indirectly come from who knows where in an interconnected, global economy. Certainly, scientists of the future will understand the cell in living organisms in ways that make our present levels of knowledge seem antiquated. Computers that look nothing like our present models will be run by molecular power and our bodies, health and communications will be supported by implants and biologically based technical enhancements beyond our imagining. In the entangled bank of our creativity, homo sapiens will have evolved through its own interventions so far beyond the

---

[137] Bantnam Press 2013. P 293

first appearance of cellular based life as to be unrecognisably related to it, but the connectedness will remain for those who are willing to perceive and value it.

Different aspects of A and B connectivity.

That story of connectedness between A and B includes the processes of separation and specialisation, of order and disorder, of living and dying. It includes these and other processes because this connectedness is constructed out of the freedom of Love's energy, rather than external compulsion or control. It comes from within, not without the movement from A to B which can, as we have seen, involve many processes and dimensions in physics, quantum theory, the natural and social sciences and theology. We have seen that in the movement from A to B, in cosmology and evolution, there is a connectedness between the changing nature of both A and B. B evolves out of A through processes that are present in A, but are not determined to result in B, or the intermediate stages in between A and B. The 'laws' of physics or biology which explain the existence of A and B apply equally to both, but not in any static, mechanistic way. There is indeterminacy in the processes between them and they are both subject to random events in their environment. In the movement from A to B there is nothing determined about the nature of B, though, without A, it would not exist in space and time. Quantum taught us that both A and B are made up of their own worlds of connected complexity, where the usual descriptions of the nature of each as separate entities or states may not apply. To understand the physical reality of each we have to accept behaviours within each which are unpredictable in the classical laws of physics. The different As and Bs which exist within each A, and within each B, may not be known equally at the same time, but all are connected, whatever they are and wherever they appear. Theologically, we have seen that Love's energy A moves in the direction of the Otherness of creation B, because of the nature of A as Love. It does so in a way that treats B as the autonomous Other, free to develop itself in ways that are implicit within its own ontological and physical autonomy. This is consistent with what we have learnt from quantum. We have seen that in Christian belief, we can talk of the presence of A as Love's Energy within the processes of B as the Otherness of the universe, without compromising its autonomy. We have seen that A acts in the same way towards B as it does from within B. We have also claimed that it is the same Love's energy working in creation as we see in the Incarnation. The Bible talks about God in Christ as being the Alpha and the Omega. These were the first and last letters of the Greek Alphabet. So, the writer of the Book of Revelation [138] was saying to his contemporary audience that the meaning of Christ was like these two letters. It encompassed and embraced everything in every letter in between. All were connected as part of the same

---

[138] Revelations 1:8, 21:6, and 22:13).

process.These two simple letters became symbols of the very meaning and presence of Christ. They appeared within the symbols for the cross and on halos round the heads of biblical figures and the saints in iconography. The early church writers 'played' with these symbols to capture the all inclusive significance of what was being said. In so doing, they managed to talk about the significance all processes of creation in relation to God in Christ and the telos - end and purpose of all things. For example, the well travelled, **Titus Flavius Clemens**, known as **Clement of Alexandria** (c.150-215) was an educated Greek philosopher who was converted to Christianity and then became the teacher of the same Origen referred to in LE2. He said that the Word (the Logos of God in Christ) was *'the Alpha and the Omega of Whom alone the end becomes beginning, and ends again at the original beginning without any break.'*[139] God, or 'God in Christ' was seen as the the meaning of the beginning and the end, the very A-Z of life, as we might say, including everything in between. All that I have tried to say about the movement between and connectedness of A to B in this book could be set against this theological backcloth of A and Ω. However much we change ourselves and our environment, our connectedness with all life will remain a constant and this is one of the wonders of our universe. When placed alongside the story of the coming into being and development of cosmology and biological life, the phrases Big Bang and Big Birth seem rather crude to say the least. The word 'Big' captures the significance of the events and the processes involved including the energy behind and within them. We know that the cosmos began not so much with a *Big* Bang but with a point of very small, but very intense, density of energy. This created the energy matter from which everything, exploded, expanded and developed into billions of stars and their systems. We know that living organisms developed not so much from a *Big* Birth, but from the smallest of proto cell conditions from which all life developed over millions of years into countless forms of divergent, living energy-matter. As Darwin put it at the end of his great book, *'There is grandeur in this view of life, with its several powers, having been originally breathed into a few forms or into one; and that, whilst this planet has gone cycling on according the fixed law of gravity, from so simple a beginning endless forms most beautiful and most wonderful have been, and are being, evolved.'* [140] So, now It is time to turn to Darwin, in some detail, because of his huge significance for our understanding of natural selection and its implications for those who would talk about God as Creator of the *grandeur* in cosmological and biological *'endless forms most beautiful.'*

---

[139] Stromata IV, 25
[140] Last page of *'The Origin of Species'* Charles Darwin.

CHAPTER 5
Charles Darwin

Sub Headings

Naturally selected shock and horror; From Meme to the Web; Evolutionary Geology; Darwin's Dillemma; From Geology to the Lunar Society; The Power of Propensity and the Living Filament; The Life Force of Transformation and Spontaneous Generation; Another and the same; From Malthus to Vestiges; From Vestiges to the self evolving power of nature; From the self modifying power of nature to Adam Smith and Karl Marx; The Doctrinal Dispute between Darwin and Owen; Wallace; Evolutionary Teleology.

In the 1840s, **James Bateman** (1811-1897) bought the old Rectory, Biddulph Grange and, with the help of the painter and naturalist **Edward Cooke**, transformed the estate's gardens to create separated areas depicting scenes from Scotland, China, Egypt and Italy. These were myths and illusions, but very popular at the time. In January of 1862, Bateman, an expert on orchids sent a box of specimens to Charles Darwin. Darwin wrote back, but that letter has been lost. Then in February of the same year, Bateman's letter back to Darwin included the following sentences. *'I was very glad indeed to hear that the Orchid flowers were so acceptable.. I am sorry to say that my knowledge of Orchids almost ends where yours-according to your own modest representation-would seem to begin i.e. I know them much better systematically than structurally..I much wish you would take up the subject of the marvellous changes-I might almost call them metempsychosis-to which Orchids are prone. Though by no means a convert to your theory as to the `Origin of Species' I wish the matter to be thoroughly ventilated and cannot but think that facts of great significance may be gathered in the direction I have indicated. If you put me in the witness-box I shall be happy to tell all I know.'* There would of course be many commentators, interpreters, supporters and opponents who would, in their different ways, climb into their own *witness boxes* then and ever since. I wish Darwin had taken Bateman up on his offer – perhaps his views on *the subject of the marvellous changes* would have made its *own modest representation.*

Meanwhile, in that same decade of the 1860s, the illustrator and watercolourist **Myles Burket Foster** (1825-1899) was looking not just at gardens, but the English countryside in his own particular way. He produced very popular paintings of an idealised rural arcadia[141] which covered up, or sentamentalised the reality of rural poverty as people left to work in the towns and cities. At the same time, the

---

[141] Cadbury's used these for their chocolate box tops in the 1860s

cultural critic, watercolourist[142] and philanthropist, **John Ruskin** (1819-1900) became the first Slade Professor of Fine Art at Oxford, as well as writing on geology,[143] botany, literature and the political economy. In the same year as Darwin published the *Descent*, Ruskin began his letters *'to the workmen and labourers of Great Britain'*[144] and wrote a poem decrying the incursion of rail roads and the Headstone Viaduct in particular, which included these lines. *The valley is gone, and the Gods with it; and now, every fool in Buxton can be in Bakewell in half an hour, and every fool in Bakewell at Buxton; which you think a lucrative process of exchange – you Fools everywhere.'* The 'countryside' was so often identified with tradition and religious values and seen as *'natural'* by contrast with urban areas. In reality the former was its own construct and had been changed over time by natural and human processes. It seems that at times of change and uncertainty people create their own sense of stability and meaning by turning to myths about the past so that it can become a form of escape from the present. Nearly a century earlier, while **Priestley** was 'discovering' Oxygen in the library of Bowood House, **Edmund Burke** wrote,[145] *'People will not look forward to prosperity who never look backward to their ancestors.'* One of Darwin's great achievements was to challenge our assumptions and change our view of *natural* processes in the past and connect our present place and development within its inheritance as we look *backwards to our ancestors.*

In LE1, based on the theme of Love's Energy, we asked the question in cosmology and physics – how does A become B, or how does B become different from A, what is the nature of their connectivity; what constitutes that nature and how does it express itself externally; what triggers the processes involved; what are the carriers of those processes (or as Darwin sometimes said the 'transport'); what are the causes behind those carriers; how does cause relate to effect and in turn influence the next cause? We can only imagine the huge leap it took to move from an answer that included some kind of Divine will or intervention to one that focused just on internal processes alone - albeit one not fully understood and still based on the overall assumptions of natural theology. These are the questions that haunted so many great thinkers of the seventeenth, eighteenth and early nineteenth centuries. We know they are significant questions. We look back on this time with the hindsight of more recent progress

---

[142] In 1843, he produced the first volume of his *Modern Painters* defending J.M.W.Turner and arguing that the first responsibility of the artist is to be true to nature. See my comments on Turner in LE1. In the 1860s Ruskin became the first Slade professor of Fine Art at Oxford Univeristy,

[143] Which he saw as a hammer blow to his own faith and any literal truth in the Bible. He was in later life critical of Darwin's theory.

[144] *Fors Clavigera* (1871–1884)

[145] in his 1790 *Reflections on the Revolution in France*

in DNA and genetics. We continue to ask how our knowledge helps and changes our understanding. We look back with sympathy, because there are many in our own times who still wrestle with the implications of those discoveries in their questions and doubts about belief in God. Scientifically, we are still searching for certainty about how the Big Birth happened and how life in all its diversity appears and changes from within. We should not take these questions for granted or gloss quickly over their implications. This chapter on Charles Darwin is central to this part of the Love's Energy trilogy. Much has been written about Darwin and his times. I will attempt to paint a picture of other writers and thinkers before and after him, whose work is connected with his achievements and the questions it raised - not least the work of his grandfather **Erasmus Darwin**. I believe those questions resonate still, with something quite formative in the history of our changing understanding about the human condition. Darwin shifted the ground in those questions and shone a light on the internal processes of change within the natural world. There is no going back to the darkness before that light shone,[146] though much remains in its shadows about the meaning of the connected *pattern paradoxes* of creation. In this telling of a well known story, I will occasionally pause to reflect on those *pattern paradoxes* about the meaning of what many, before and during Darwin's time, called the *life force* which *animates* us. Such a story also moved through the internal processes of human thought, with all its connections and diversity, as different thinkers[147] adapted to each other's ideas, in the changing environment of their different assumptions and perceptions, purposes and contexts. As we shall see, the *pattern paradoxes* continue, not only in the natural sciences, but even in the world of Logic and mathematics, where questions of *uncertainty* and *certainty, completion* and *consistency* still compete and struggle to survive! *Love longs* for completion. To that end, Love's energy allows for the natural, internal processes of nature to move and change in ways that aren't always *'consistent'* with that longing, but *completed* in it and by it.

Naturally selected shock and horror

'It was with something like horror, therefore, that great numbers of honest and religious spirited men followed the work of the great English naturalist Charles Darwin... The Darwinian movement took formal Christianity unawares, suddenly. Formal Christianity was confronted with a clearly demonstrable error in her theological statements. The Christian theologians were neither wise enough nor mentally nimble enough to accept the new truth, modify their formulae and insist

---

[146] As depicted in many of the paintings of **Joseph Wright** painted before the work of Darwin

[147] As in LE1 and LE2, I am telling this story at second or third hand, at least, with a debt of gratitude to those who have done the detailed research on the lives and work of these individuals. All I have done is to reflect on that knowledge from within the thesis of Love's energy and to attempt various connections on the way.

upon the living and undiminished vitality of the religious reality those formulae had hitherto sufficed to express. For the discovery of man's descent from sub human forms does not even remotely touch the teaching of the Kingdom of Heaven.'[148] This was **H. G. Well's** (1866-1946) reaction to the inflexibility and stubbornness of what he describes as *formal Christianity,* although of course many theologians then and since have been, in his words, *mentally nimble enough to accept the new truth* he outlined.

Wells. P-D. Author unknown. Copyright expired

He continued his analysis by saying, [149] *'It was the orthodox theology that the new scientific advances had compromised but the angry theologians declared that it was religion. In the end men may discover that religion shines all the brighter for the loss of its doctrinal wrappings, but to the young it seemed as if indeed there had been a conflict of science and religion and that in that conflict science had won.. a real demoralization followed.There was a real loss of faith after 1859. The true gold of religion was in many cases thrown away with the worn out purse that had contained it for so long and it was not recovered.. the old faith of the kings, owners and rulers of the opening twentieth century had faded under the actinic light of scientific criticism. so the Darwinian crisis continued that destruction of Christian prestige which the narrowness of priest-craft and the consequent division of Christendom among the monarchist and national Protestant churches of the Reformation had begun and at a time when man's need for pacifying and unifying ideas was greater than it had ever been.'* He was writing this work towards the end of WWI and from within the context of its social consequences. He therefore viewed the impact of Darwinism from that perspective and his own agendas. We may not agree with him that *priest-craft* and *the division of Christendom* could be blamed for quite so much, but this is how strongly he and no doubt many of his readers felt about the influence of Darwin on broader issues. What Darwin and his followers taught is that natural selection undermines the basic idea of God's intervention in the birth and growth of natural things. Those religious people who continue to talk about God's intervention have, therefore, still to face this serious challenge. In every age they must find their way of articulating belief in ways that do not ignore the findings of science. Darwin used the phrase *'through natural selection'* to describe the way species grow, change and die. This was well known but a cultural hammer blow to those whose paradigms depended upon fixed immutability. *The Origin of Species* provided a clear, straight forward and logically constructed argument. It took care to take the reader slowly through the evidence. It de-mystified science

---

[148] *The Outline of History,* H.G. Wells, The Waverley Book Company 1920. Volume 11. Revised and Corrected Edition. Pp 520- 521. This had enormous influence at the time and sold 2 million copies
[149] Ibid pp 523 and 524

and made it available to the ordinary person. It focused on simple, apparently common place observations – the breeding of pigeons to produce a longer tail for example. It reminded us of the focus on detail found in Aristotle rather than the more abstract ideas of Plato. In its approach to the particular, it might appear to the sensitive theologian as incarnational in style. It put science at the heart of everyday life and natural processes – which religion had often failed to do. It acknowledged other points of view and our extensive ignorance, opening the door to new science and new discoveries, which were of course to follow, particularly with DNA and genetics. Its tone is humble, patient and cautious. Its breadth is impressive and its way of developing an argument within the evidence is compelling. At the time, comparatively little was known about what Darwin later called 'The Descent of Man'. The significant fossils had not been found. It took Darwin another 12 years to have the courage to write on this subject. In *The Origin*, he had hinted that his central argument could be applied to the origin of Homo sapiens as well, and that was certainly the implication drawn in the public debate which followed. Was he being cautious and tactical about the timing? Had he not done sufficient research on the higher primates to include this in *The Origin*? Would this have focused too much attention on the *ape to man* implications and detracted from his main argument. Were his friends arguing for or against it? This is a subject of much debate, but it is likely that there were many factors involved. Certainly, he knew the stakes were very high, as was the pressure to publish what he already had. Certainly, it was true that most of his research had been on plants and the 'lower' animals from birds such as finches to his years studying barnacles. Certainly, he spent much time revising *The Origin* and issuing new editions to answer his critics.

As we saw, the broader Victorian debate came to accept that natural selection was correct, but within some kind of divinely ordained process. The later editions of the book referred to God as creator several times, and could be understood as providing a version of how God acted within natural theology. From this point of view, Darwin was offering a compelling argument for a new law in the processes of natural theology. Others saw it as threatening divine revelation in the Bible and undermining any real basis for natural theology in the name of science. Others feared it was challenging the very moral architecture of a Christian society. This is controversial territory. It may be that Darwin referred to a Creator in later editions to placate certain people including his wife[150] and some of his critics and friends. As we shall see, there is ample evidence to show that he lost his earlier faith, not least because of personal loss in his family as well as through intellectual struggle. Darwin realised, unlike Wallace, that the argument from divine will or design was dead, even if he didn't see the purpose of his book as an

---

[150] See Ernst Mayr's 'One Long Argument: Charles Darwin and the Genesis of Evolutionary Thought' Harvard, 1991

atheistic attack on religion. There are many public and personal layers to unveil as we try to understand Darwin's own religious position and its influence in The Origin. He wrote to **Hooker** in March 1863 *'It will be some time before we see slime, protoplasm, etc., generating a new animal. But I have long regretted that I truckled to public opinion, and used the Pentateuchal term of creation, by which I really meant 'appeared' by some wholly unknown process. It is mere rubbish, thinking at present of the origin of life; one might as well think of the origin of matter.'*[151] Darwin was no Richard Dawkins, although in the late 20th century, his work was being used by some atheists as a rallying call to liberate British society against the state church, the monarchy and God. In the Established Church of England, there were many who valued the work and made no fuss about it, but most Anglican priests disliked the book, including, as we shall see, many academic clergy. From a longer perspective, this was a turning point in the debate between science and religion, but this was not Darwin's main motive in writing. Many questions arise about intention and purpose. Although we have the record of so many letters and contemporary accounts, the debate continues. Was his caution affected by what different sides would do with his theory or more from fear of personal attack from those he respected? Did he fear an extreme creationist reaction? How would he respond now, if he visited the Creation Museum in Cincinnati where dinosaurs and humans are portrayed as living at the same? How would he react if he discovered that so many creationists still reject his work and see him as the arch enemy of all religion? What would he say to the ultra evolutionists who use science to 'prove' their atheism? Maybe he would have agreed with Professor **Daniel Dennett**[152] and others that this was a category mistake.

From Meme to the Web

How would he have debated with supporters of Meme theory[153] who take natural selection to the next stage – that all life is but a host for packets of information, striving for survival and communication, and doing this in the human species through culture, politics, science and religion. Following Dawkins, who coined the Greek phrase in *The Selfish Gene,*[154] they see *memes* as the equivalents of gene like cultural replicators, but working on the basis of Darwin's natural selection. Memes can therefore mutate into harmful and destructive messages, as well as positive ones. They can penetrate individual, community and transnational consciousness, as viruses do in any living organism. They can

---

[151] Francis Darwin's (his son) *Life and Letters of Charles Darwin*, 1887, 3:17.

[152] The Co-director of the Centre for Cognitive Studies and Professor of Philosophy at Tufts University. Often associated with the new atheism of Dawkins, Sam Harris and the late Christopher Hitchens.

[153] The belief that cultural norms spread and evolve by natural selection through the processes of variation, mutation, competition and inheritance – each influencing reproductive success or failure.

[154] First published in 1976

divide and multiply their nature, as cells do in healthy or cancerous growth. For a man who believed in connections, within and between all life forms and produced so much correspondence, how would he have reacted to a webbed, internet world? We are beginning to understand how his own big idea evolved *through an unknown process* - as he might have said - out of a complicated *web* or even a *swarm* of related and *selected* ideas - either dialectically, or because they were variants and mutations in its evolution! The more we study his work and life the more connections appear with social and political, environmental factors that were part of *'the entangled bank'* of his times and the ideas which led up to and created them. As we contemplate *that entangled back* of which he was so crucial a part, I will try to focus on the *intermediate stages,* as he might have said, of the development of his argument as it appears in other writers. I hope that, as we pause to consider their work, connections will appear, if not a pattern. I hope that in each case I will say just enough to remind readers of their contributions, or to introduce other readers to the significance of their ideas. My selection of relevant individuals is of course partial and many more could and should have been added to the story.

Charles Darwin, photo probably 1854 by Mssrs Mauli and Fox (according to Francis Darwin) although their book destroyed by fire. US-PD and copyright expired

Darwin finished his great work in the late summer of 1859. As we have seen, it came out of years of emotional and intellectual anxiety, as well as detailed study and research. The personal agonizing, as well as the intellectual results, reflect the tensions in the science of the age and all that had led up to it in the previous century of new ideas being tested against intransigent opposition. **Charles Lyell** and **Joseph Hooker** were his closest confidants during these difficult years. In 1858, a year before publication, Darwin wrote that Hooker was *'the one living soul from whom I have constantly received sympathy.'*[155] Darwin consulted them both when Wallace's unexpected and devastating letter arrived.

Joseph Dalton Hooker (1817–1911) was arguably the most important British

botanist of the nineteenth century, a traveller[156] and plant-collector and clearly one of Darwin's closest friends, who became director of the Royal Botanic Gardens, Kew.

Joseph Hooker by Henry Joseph Whitlock. Public Domain. Copyright expired

Their friendship had begun after Hooker had helped Darwin with the notes he made on the Beagle and the classification of his specimens. Darwin showed Hooker his early thoughts on

---

[155] See Letter 2345 in Darwin Correspondence Project Darwin to Hooker
[156] Including Morocco, Palestine, America, Himalayas and India and the Antarctic

the transmutation of species in 1844, and continued to confide in him from that point in the development of his idea, through to its publication. We can only imagine the pressures on Hooker at the time, not least in his conflicts with **Richard Owen**[157] and so he must have empathised with the agonies felt by Darwin for this and other reasons. When Darwin made his great confession to Hooker, saying *'species are not immutable,'* he felt that making such a statement was *'like confessing a murder.'* This shows the extent of his awareness and concern about the controversial implications of his work. This goes to the heart of this story and why it is still so important today.

We have echoes of this in the contemporary world where religious opposition to the teaching of evolution in schools brought such vitriolic and ideological opposition from fundamentalist Christians in the United States. This opposition had its own irony in a country which prides itself on toleration and freedom including religious freedom which was so important to the founders of the United States. No wonder **Richard Dawkins** has reacted so vehemently to such attitudes in his media programmes and *'The Selfish Gene'* and above all *'The God delusion.'* In Darwin's time, the weight of Anglican conservatism was the context for his hesitation and fear. The dominant belief paradigm held by bishops, theologians, church members and as importantly most scientists was that humans were created separately. They were not just a product of nature. Darwin was afraid of *'the black robed beasts'* as he sometimes called the clergy and some of his fellow scientists who still took a divinely, created order of species for granted. Even if he hadn't been such a patient thinker and observer, perhaps this fear alone would have made him cautious and diligent in collecting more evidence, as carefully and thoroughly as possible. He was to write in *'The Origin,'* *'Authors of the highest eminence seem to be fully satisfied with the view that each species has been **independently created**. To my mind it accords better with what we know of the laws impressed on matter by the Creator, that the production and extinction of the past and present inhabitants of the world should have been due to secondary causes, like those determining the birth and death of the individual. When I view all beings not as **special creations,** but as the lineal descendants of some few beings which lived long before the first bed of the Silurian system was deposited, they seem to me to become **ennobled**. Judging from the past, we may safely infer that no one living species will transmit is unaltered likeness to a distant futurity.'*[158]

---

[157] Then in charge of the nature collections of the British Museum. Rivalry with Hooker came to a head and the issue was even taken to Gladstone and Parliament
[158] Chapter 14, *'The Origin'* 2nd Edition J Murray 1860 p 489

We note his phrase *'authors of the highest eminence.'* We assume this included many of the leading academics of his time, many of whom were clergy. We sense the tone of genuine respect, plus the tinge of sarcasm. Darwin was still sufficiently part of the beliefs of his own times to use the language of a 'Creator.' For Darwin the idea of 'laws' *impressed on matter* still worked. He was writing from within a basic, theological paradigm, but he was shifting it away from the assumption of previous centuries – that *'each species has been independently created.'* This is the crucial difference. The Genesis 1 story of creation does not specifically rule out a creating process through the *'laws impressed on nature.'* It does imply, if not specify, the separation of species creation, not least through its chronological structure of different days. Even if each biblical day was a thousand or million of our days, each group of species – the plants, the sea creatures, the animals receives special mention as part of the purposive, creative action of the Creator. In the Genesis account there is no clear justification for assuming a divine, *hands on intervention* in the creation of each, individual species, yet this seems to have been the theological view extracted from it, certainly in the time of Darwin and his opponents. Darwin was asserting something essentially controversial, if not new – that any particular species develops from a previous one in the past and not from a new and direct, divine intervention. The laws *'impressed on matter'* may come from the Creator but, once *impressed*, each species development takes place through a natural, internal process rather than an external, divine intervention. Because of Darwin, we have to find a new way of explaining what we think we mean or claim, as believers, when we say that God created all things. We cannot by pass his formative, scientific, step forward. We have to talk about our first order belief in a created world through a new kind of natural theology based on the science. Without this, the idea of God as Creator will be increasingly eroded. Even for those who no longer talk about God as Creator, but only Christ as saviour, there is a problem. The classical creed of the Christian church and certainly St. John's Gospel implies a relationship between the creation of all things and the agency of the second person of the Trinity in the process. Of course the Trinity has always been a difficult concept and not only for Muslims! It would be ironic if, after all the early centuries of doctrinal disputes, the idea of God as Trinity finally fell apart because Christians themselves contributed to its erosion, as they failed to adjust their view of God as Creator as well as 'Father!'

Evolutionary geology

For Darwin, God is still in the process through a *'law impressed on matter,'* presumably once established for all time, rather than through continuous or separate interventions and top ups. We need to take up this challenge of finding ways of talking about the presence of God through the process which God *'impressed on matter.'* Darwin gave us the biology of matter just as Newton and his successors have given us the physics of matter. There are many other locations for the relevance of this phrase. Darwin also studied geology, but in the last hundred and sixty years, or so, this science has developed well beyond his knowledge. In many ways, Darwin's idea of natural selection also applies to the history of the planet Earth and to all planets. We know that through natural selection everything changes and is connected. If we apply this to geology, we know,[159] that there is a *natural* explanation for the way rocks are created and move; continents shift; new mountain ranges rise and sink and weather patterns change. The formation of the planet itself was a dynamic and destructive process of *'naturally,'* gradated change and selection and, until the conditions were right, there was no organic life present. While many rocks came out of the catastrophic events of cosmological and planetary change, others were created over millions of years by the death of millions of plants and animals such as plankton, laid down in the sedimentary levels of climate and other kinds of change. I was recently walking on Barry Beach in Wales after the storm surges of early 2013, and saw groups of geology students from Cardiff University making maps. The tides had shifted the level of the sand on the beach revealing the remains of a Victorian jetty with wooden barrels filled with concrete and wooden posts. The waves of water had also brought in new stones which now littered the ends of the beach and smashed rocks of stone out of the cliff face - which was from the Triassic period when layers of sedimentary rock were laid down from the time that area was desert. On the promontory called Friar's point, just a few hundred yards away, the limestone was formed by minerals and the shells of tiny animals about 150 million years before the Triassic. That which we now perceive as solid rock was once living animals! The life cycles of natural selection over millennia can be related to the geology. I asked three different groups of students about the age and composition of these different formations and they all gave different answers! I asked them about the Cambrian boundary and they looked up those dates on their phones – a good example of how knowledge is now stored and accessed!

Some, like the American planetary scientist, **William Kenneth Hartmann**, have argued that without an apparently *random* 'catastrophe event' creating the moon, our planet would never have developed its tilt or the conditions for life to appear There were many catastrophic events in the history of the formation of our planet and they include the extinction event of a meteor strike and its affect on global climate. As we shall see later in this chapter, the argument between the catastrophists and the uniformitarianists continues to this day and Darwin was mostly influenced by the latter in the work of **Hutton** and **Lyell**. The continents are continuously moving and influence the climate as does the Gulf Stream. Ice sheets ebb and flow over millennia of years and the processes of the release and

---

[159] And our knowledge was much helped by the work of Hutton as we shall see later in this chapter

absorption of carbon continues. Volcanoes, earthquakes and plants have been and remain crucial parts of the connected, but differentiated process of change. Our distance from the sun and its variating behaviours, the fluctuations in the earth's axis, gravity and climate patterns are all related. It's only over the past few millennia that there has been any comparative stability in the climate – enough for human civilization – to develop. With human life comes a faster pace in climate change. As with mutability in different species, as observed by Darwin, so geologists study planetary mutability, its causes and effects. Again change of this kind only happens over millions of years, and it is still happening. Any interpretation of creation as created the same then, as now, fixed and immutable, is, therefore, as inappropriate in geology as in biology.

If Love's Energy is embedded in the processes of energy-matter, then it encourages an open minded exploration of biology, geology, chemistry and physics. As we saw, Darwin used the idea of secondary laws as a compromise with the natural theologians. A law is true for different and all situations until it is disproved. It provides a framework for consistent assumptions that can be revised and adapted. It is not so much a bulwark against new discovery, as a platform from which to make new discoveries. As in Darwin's world, so in the latest genetics and quantum worlds, religious people have to articulate their belief in God's presence through the scientific processes we are coming to understand - whether as proven 'laws' or open questions. The theology or spirituality of this presence has to be articulated in ways that are consistent with first order beliefs and the latest science. Matter and energy are the focus of all science. Any spiritual constructs of creation which ignore matter and its processes will not be good theology, although much of this type has prospered in the churches, even now. Ideally, theology with its own commitment to the truth should be helping to drive science forward, but the record speaks of the opposite tendency – with religion fighting rear guard actions in defence of the indefensible, scientifically speaking. It has often used its ecclesial heavy weights (Darwin's *'authors of the highest eminence'*) to guard tradition and dismiss new science, blinding itself in the process to the openness within what is most central, inclusive and creative in its own belief systems. This is, of course, a scandal for many reasons, not least the harm or threat of harm done to different scientists through history and the reduced credibility of its own position. Darwin was fully aware of the cultural power of the guardians of tradition as he used the idea of 'secondary' laws as a way of understanding how the processes of creation worked. The secondary for him *'accords better with what we know of the laws impressed on matter by the Creator.'* With our 21st century hindsight and from within our debt to Darwin, it is hard to see why this was so controversial at the time. He wasn't going much further than his grandfather's theological statements about development[160]. A more controversial step would have been to directly deny any idea of a Creator's involvement in natural processes, but Darwin chose the more subtle path of admitting involvement through a secondary law, but then continuing as if nature did all the work anyway! The idea of Law as a vehicle for a Creator's involvement raises many questions;

> Does this law imply a willed design of natural selection by the Creator, including all that is violent in species behaviour?
> Is this 'species violence,' as **Carolyn King**, a biologist and theologian, has argued in '*Habitat of Grace*' an expression of the doctrine of original sin that Western parts of Christianity have so often used; or is this too much theological/moral projection on the non-human world?
> Does the idea of law prevent us from seeing the more dynamic and unpredictable, chance events in species gradation, extinction or development?
> Does the idea of law remove all notions of freedom, innovation and novelty in those processes?

Darwin slowly found the confidence to contemplate natural selection, free from the dominance of 'evolved' theological assumptions, although he inherited many of these, not least from his grandfather. He could say with conviction that when viewed '*not as special creations, but as the lineal descendants of some few beings which lived long before the first bed of the Silurian system was deposited, they seem to me to become* **ennobled**.'[161] In moving us forward and in facing the opposition of most of his Cambridge and other colleagues, Darwin saw no reason why the rejection of '*special (independent) creations*' should undermine the theological significance of species development. Again, perhaps he was intuitively following the approach laid down by his grandfather, who, as we shall see, wrote '*What a magnificent idea of the infinite power of THE GREAT ARCHITECT! THE CAUSE OF CAUSES! PARENT OF PARENTS! ENS ENTIUM!*' For Charles, to see adaption as a process of lineal descent meant that nature '*becomes ennobled.*' Later, in this same passage and just before the most famous paragraph of all about the 'entangled bank,' we find an expression of his confidence about *progress to perfection* in *futurity* that we may not share.[162] Perhaps he was expressing a Victorian belief in the inevitability of progress, as well as his understanding of the process of natural selection. It was an

---

[160] Cardinal John Henry Newman (1801-90) of Oxford Movement fame was also writing his theological theories of *development.*
[161] The Origin Second Edition J Murray 1860 p 489 and all later editions. My emphasis in bold
[162] And also perhaps some ignorance of the effects of the *snow ball* state of the earth

understandable assumption, given the idea of *lineal descent* that many of his predecessors had developed. Building on the work of his grandfather, he wrote '*As all the living forms of life are the lineal descendants of those which lived long before the Silurian epoch, we may feel certain that the ordinary succession by generation has never once been broken, and that no cataclysm has desolated the whole world. Hence we may look with some **confidence to a secure future** of equally inappreciable length. And as natural selection works solely by and for the **good of each being,** all corporeal and mental endowments will tend to **progress towards perfection.**'[163] In the theological task of affirming value in creation, there is a need to be realistic about the nature of species behaviour. Darwin was no romantic about natural processes, but still used *progress towards perfection* to capture what he also called the '*wonder*' of species diversity and *development*. Many detractors saw in the '*Origin*' the inherent risk of removing all that was 'ennobled,' particularly in human beings, because of their evolutionary link with lower creatures. The question is why? We know something of the paradigms which closed their minds to making these kinds of connections, finding them such a powerful threat to their idea of God, but what was really at stake? Why did they interpret the Genesis texts and their belief in a Creator in ways that blocked the insights of Darwin, rather than welcomed them? It wasn't Darwin's fault that such attitudes existed, however sensitive he was to them.

Darwin's Dilemma

He certainly and humbly, or at least graciously, acknowledged '*that many and grave objections may be advanced against the theory of descent with modification through natural selection... I have endeavoured to give to them their full force.*'[164] He could well imagine the effect his idea would have when so clearly stated and convincingly researched, and he continued to demonstrate his sensitivity to its implications '*Nothing at first can appear more difficult to believe than that the more complex organs and instincts should have been perfected not be means superior to though analogous with human reason but by the accumulation of innumerable **slight variations**, each good for the individual possessor.*' He readily admitted how difficult it was '*even to conjecture by what gradations many structures have been perfected more especially amongst broken and failing groups of organic beings.*' He asked, '*why do we not see these linking forms all around us? Why are not all organic beings blended together in an inextricable chaos. With respect to existing forms we should remember that we have no right to expect to discover directly connecting links between them but only between each and some extinct and supplanted form.*' We know there was a problem initially in finding any fossil confirmation of mid way gradations in the

---

[163] Ibid pp 489 and 490. My emphasis in bold

[164] Excerts from Chapter 14 of *The Origin*

geological record. Again, he generously admitted this. *'Why does not every collection of fossil remains afford plain evidence of the gradation and mutation of the forms of life? We meet with no such evidence and this is the most forcible of the many objections which may be urged against my theory. Why, again, do whole groups of allied species appear, though certainly they often falsely appear, to have come in suddenly on the several geological stages? Although we now know that organic beings appeared on this globe, at a period incalculably remote, long before the lowest bed of the Silurian system was deposited, why do we not find beneath this system great piles of strata stored with the remains of the progenitors of the Silurian fossils? For on my theory such strata must somewhere have been deposited at these ancient and utterly unknown epochs in the world's history.'*[165] In his final chapter, designed to meet with various objections to his thesis, Darwin admitted that the greatest problem was the lack of fossil evidence for the intermediary stages of graduated change that would otherwise have proved natural selection. He devotes more space to this than any other objection and was probably thinking of what **Sedgwick** and **Owen** (see later in this chapter) might say. As ever, he reacts with humility and honesty to the *variations* in the *connected or linked pattern* of thinking all around him, which were *adapting* and *mutating* in their own *competitive struggle for survival*! Perhaps he saw his own contribution as an *intermediary link* between what has gone before and what will follow in the intellectual, sedimentary record; one that is not so much a *doubtful form* of *variety,* but its own species! When he wrote– *'only a small portion of the world has been geologically explored,'* he must have been thinking of Sedgwick and Owen and the vulnerability of his own theory with its *'enormous blank intervals.'* This is part of his response. *'We should not be able to recognise a species as the parent of any one or mores species if we were to examine them ever so closely unless we likewise possessed many of the* **intermediate links** *between their past or parent and present states; and these many links we could hardly ever expect to discover owing to the* **imperfection of the geological record***. Numerous existing doubtful forms could be named which are probably varieties; but who will pretend that in future ages so many fossil links will be discovered that naturalists will be able to decide on the common view whether or not these* **doubtful forms** *are varieties? As long as most of the links between any two species are* **unknown,** *if any one link or* **intermediate variety** *be discovered it will simple by classed as another and distinct species. Only a small portion of the world has been geologically explored. Only organic beings of certain classes can be preserved in a fossil condition at least in any great number...Successive formations are separated from each other by enormous blank intervals of time for fossilferous formations thick enough to resist future degradation can be accumulated only where much sediment is deposited on the subsiding bed of the*

---

[165] Ibid p 548. My emphasis in bold

*sea. During the alternate periods of elevation and of stationary level the **record will be blank.** During these latter periods there will probably be more variability in the forms of life; during periods of subsidence, more extinction.*[166]

It must have been agonizing for him to continue with a belief in mutability without pre-Cambrian fossil evidence. 542 million years ago, fossils suddenly appeared, but they were all shell fossils. It is only in the 20th century that soft bodied fossils have been identified before the Cambrian period, although **John William Salter** (1820-1869)[167] had identified some pre-Cambrian forms in 1856/7. **Callow** and **Brasier** came up with an answer in 2009. *'The study, carried out by **Richard H. T. Ballow** and Martin D. Brasier at the Department of Earth Sciences at the University of Oxford, focused on a rock formation from Shropshire, England, known as the Longmyndian Supergroup. These rocks had been examined in Darwin's time by the geologist J. W. Salter, who suspected them of containing records of Precambrian life, but he was unable to identify anything beyond 'trace fossils': unusual markings which may have been left behind by organisms. The study used Salter's collection as well as fresh samples from the Longmyndian Supergroup, and identified microscopic fossils of exceptional preservation. The fossils represent a wide array of microbial life from the Ediacaran period, the period immediately preceding the Cambrian (630 – 542 Mya). They were preserved in a number of ways. Some had been compressed under layers of sediment until they formed a thin film of carbon residue on the surface of the rock. Others were preserved in three dimensions and are thought to have undergone permineralisation, a process where water containing minerals seeps into the spaces within an organism and evaporates, leaving behind mineral deposits which build up into a hard fossil. Some had also been preserved as impressions and moulds within layers of sediment, appearing as sharp ridges on bedding planes, or as their equivalent negative impressions.'*[168]

The Australian geologist **Reginald Claude Sprigg**, (1919 –1994) had pursued this pre-Cambrian question in the face of much cynicism, and in 1946 discovered the Ediacar biota, a collection of some of the oldest animal fossils ever found. They were preserved in coarse grained, sand stone. There were also findings by school children in 1957 in Charnwood forest England of animals dating up to 90 million years before the Cambrian.[169] The pre-Cambrian world was radically different,

---

[166] *The Origin* Random House 1979 edition p 439-440. My emphasis in bold

[167] Employed by Sedgwick to arrange fossils in the Woodwardian museum in Cambridge and to accompany on his field trip to Wales. I do not know whether he met Darwin on that visit

[168] Richard H. T. Callow and Martin D. Brasier. 'A solution to Darwin's dilemma of 1859: exceptional preservation in Salter's material from the late Ediacaran Longmyndian Supergroup, England.' *Journal of the Geological Society*, Vol. 166, 2009, p 1-4

[169] The Burgess Shale of soft bodied fossils was discovered by Charles Walcott in 1909 in the Canadian Rockies where he worked until 1924 amassing over 65,000 specimens. It comes from the middle Cambrian of about 505 million years ago. In

not least because of the single continent, land mass straddling the equator – with its different currents and climate. The plate tectonics were a major driver of biological evolution. 750 million years ago, the splitting of the single continent triggered two massive periods of glaciations and the appearance of new life forms followed within five million years. The ediacara biota existed for a comparatively small period of time between 635-543 million years ago. The varied animals found didn't have mouths, stomachs or internal organs. They had no mobility and lived on the sea floor without the advantage of light. They were connected with an earlier form of bacteria from 3.8 billion years ago which developed photosynthesis, thus improving oxygen levels.

Biological evolution may itself have contributed to the freezing of the earth. As soon as ediacara biota appeared, 'snow ball' freezing took place on a scale not seen since. Before this, there is fossil evidence of the body mass of more multi cellular algae, from about 1000 million years ago. These animals controlled the carbon cycle in a new way. In the boundary period before the Cambrian, skeletal fossils appeared which seem to have been affected by both external environmental changes and internal biological changes, coming from innovation within the ediacara bioto, which first took place about 560 million years ago and then 550 million years ago. Some animals of the Ediacaran period did evolve into sponges and algae. At the pre-Cambrian boundary, it seems that the ediacara experienced the first massive and global extinction, perhaps caused by lower levels of oxygen in the sea, by internal limits to adaption and the probable appearance of more mobile predators. The fossil record shows that life developed in fits and starts over millennia of years. The struggle between prey and predators may itself explain the huge diversity of life in the sea – well before there were any animals or plants on the land. In the Cambrian explosion of diversity, the development of fins, legs, mouths, guts, and eyes were crucial, as was increased size, mobility and armoured protection (including spines). The way this predator-prey interaction encouraged oscillating forms of evolutionary change was its own *pattern paradox*. Strengthened skeletal structure provided more support for the body in different crustaceans and in this same, Cambrian period the arthropods [170] (invertebrates with an exoskeleton, segmented body and jointed appendages) developed – ancestors to our lobsters, crabs and shrimps. Of the arthropods, the Trilobites (three lobes) were the most well

---

1962 Alberto Simonetta realised how much more there was to uncover. Stephen Gould's *'Wonderful Life'* of 1989, publicised the significance of the diversity of the species and the differences from body forms that survive today although Conway Morris disagreed.

[170] From Greek *arthron* – joint and *pous* –podos – foot or leg,

known and successful group of extinct marine animals.[171] They were the most advanced form of life in the seas with at least 50,000 different species.[172] Their complex and diverse eye structures were part of their exoskeleton and made from a crystallised form of rock! Their sophisticated 3D stereoscopic vision - some with 5000 different lenses - made them excellent hunters and dramatically shows the power of adaption to different environments.

Genetic paleontology is a comparatively recent science. Darwin would have greatly relished its findings! It proves that contemporary life forms are related to some life forms in the pre-Cambrian, even though it was so radically different an environment in the sea from on land.[173] The movement of animals onto the land required huge adaptive ability to breathe air,[174] be protected from the heat, adjust to pressure changes and different kinds of mobility. There are many theories about different fossil finds and how and when this happened. In the Burgess Shale only one animal fossil has been found capable of making that adaptation – the *Aysheaia* c.540 million years ago, although the Tiktaalik roseae[175] was an important intermediate step - a large shallow water fish with a skeleton that evolved not only legs but a stronger neck and head. Fossils for Picaia – tiny back boned creatures have also been found in the Burgess Shale and may be the ancestor of all animals with an internal skeleton including humans. We know Darwin persevered with the problem of geographical distribution and the theory of descent with modification from common parents *'in however distant and isolated parts of the world they are now found, they must in the course of successive generations have passed from some one part to the others. We are often wholly unable even to conjecture how this could have been effected.'*[176] This was courageous thinking, done without the knowledge we now have of a unified land mass prior to the breaking up of continents. But his confidence lay in a simple certainty that *'the truth of these propositions cannot, I think, be disputed.'*[177] But they most certainly were, as if all heaven depended on it! It is hard to imagine what he must have felt – having done so much detailed work and hard thinking – when caught in an era so dominated by religious views prejudiced against even the thought of such a proposition. We remember, for example, that, in his day the University of Cambridge demanded all students sign

---

[171] From the early Cambrian 521 million years ago through the lower Paleozoic, Devonian eras to their mass extinction at the end of the Permian era c 250 million years ago. See the beautifully written book *'Trilobite'* by Richard Fortey, First Vintage Books Edition, 2009.

[172] Fossils found in the Atlas Mountains of Morocco dating 150 million years after the Cambrian.

[173] Many if not most became extinct.

[174] Gills can only absorb oxygen when wet.

[175] Fossil found on Ellesmere Island, in the Nunavut Territory of Canada,375 million years old

[176] The Origin Fourth British edition 1866, page 546

[177] This sentence appears in every edition of his last chapter *Recapitulation and Conclusion*

the 39 articles of religion of the Church of England and about half of the students would become clergy.

From Geology to the Lunar Society

This was an age when clergy dominated the intellectual and cultural scene, many of whom were heads of colleges. They also controlled the 'orthodoxy' of what was taught in ways that parallel what happens in some contemporary madrasas and mosques, particularly in conservative areas like Iran and Saudi Arabia. It is worth reminding ourselves of their extensive and dominant influence in the world Darwin inherited, and also to contrast that world with our own where very few clergy hold comparable, academic posts in the natural sciences, or even show any interest at all. In Darwin's times and before, it was natural for academics to assume belief in God and so the institutions of the day and their cultures reflected that belief. St Mary's Cambridge was a geographic and symbolic centre of religious and political, intellectual power. **Rev Adam Sedgwick** (1785-1873), the geologist, was typical of this Cambridge theological milieu.

Sedgwick before 1890. Original painting dated 1832 (see: J.W. Clark and T.M. Hughes, The Life and Letters of the Reverend Adam Sedgwick, Cambridge University Press, 1890, vol. 1, replica of Frontispiece) Public Domain, Copyright expired

Darwin was one of his students, and in 1831 was taken by him on a formative field trip to Wales. Throughout his life, Sedgwick opposed Darwin, as he did the 'Vestiges'. His reaction to the 'Origin' expressed the emotions involved. '*If I did not think you a good tempered & truth loving man I should not tell you that... I have read your book with more pain than pleasure. Parts of it I admired greatly; parts I laughed at till my sides were almost sore; other parts I read with absolute sorrow; because I think them utterly false & grievously mischievous— You have deserted—after a start in that tram-road of all solid, physical truth—the true method of induction—& started up a machinery as wild I think as Bishop Wilkin's locomotive that was to sail with us to the Moon. Many of your wide conclusions are based upon assumptions which can neither be proved nor disproved. Why then express them in the language & arrangements of philosophical induction?*' [178] Ironically this was a man who had said to the Geological society in London, in 1831 '*No opinion can be heretical, but that which is not true.... Conflicting falsehoods we can comprehend; but truths can never war against each other. I affirm, therefore, that we have nothing to fear from the results of our enquiries, provided they be followed in the laborious but secure road of honest induction. In this way we may rest assured that we shall never arrive at conclusions opposed to any truth, either physical or moral, from*

[178] Letter 2548 Darwin correspondence project Sedgwick to Darwin. November 1859.

whatever source that truth may be derived.' In Cambridge, Darwin started work on beetles, not birds. The thousands of types of beetles raised the question of God in relation to different species and diversity within species. His teachers assumed that God had made each of these separately and directly, although Sedgwick acknowledged the diversity of species. Darwin's first drawings of beetles (he was the first to discover or identify some) are still held by Christ Church library, as are his 150 letters to his second cousin the **Rev William Darwin Fox** (1805-1880) with illustrations. *William Fox. Author unknown. P-D. Copyright expired*

'My dear Fox I am dying by inches for not having anyone to talk about insects.' We have much to thank William for – as Charles noted in his autobiography *'I was introduced to entomology by my second cousin W. Darwin Fox, a clever and most pleasant man, who was then at Christ's College, and with whom I became extremely intimate.'* Fox and Darwin overlapped for some time in Cambridge. Their correspondence continued until Fox died in 1880. It provides unique insights into Darwin's personal and family life, his fears and hopes as well as Fox's own life as a country clergyman who married twice and had seventeen children! Darwin confided in him perhaps more than with any other, despite the breadth and divergence of his scientific and other correspondents. Willliam never really accepted natural selection, despite his continuing interests as a naturalist, but he did introduce his cousin Charles to Rev John Stevens Henslow who held open evenings for students at Cambridge. Darwin learnt a great deal from him on their walks together around Cambridge. As we saw earlier, Henslow held the chair of mineralogy at Cambridge. He was deeply religious and Darwin admired his unselfish character and saw him as a role model of what an Anglican priest could be - knowledgeable, balanced, open, warm and moral. Henslow introduced Darwin to plants and *'ecology,'* although that name had not been invented. They walked on heavily, grazed landscapes with many species of plants and grasses. Henslow suggested there might be patterns in nature if only we knew how to discover them. He saw plants as living organisms, not just dried specimens. He believed plants could change within the nature of their species. He raised questions in Darwin's mind that continue today about the nature of species diversification. Darwin called this *'the mystery of mysteries.'* What is a species? What are the intermediate forms? We remind ourselves that these men did not have the advantage of knowing about genes. The questions they asked of each other were crucial to the classification of fossils and living species. Walks with Henslow sowed the seeds of Darwin's search for an idea. He was developing his evidence based techniques and questions into a pattern thta would stay with him for the rest of his life. Later he was to say *'at last I have a theory to work with.'* He realised that facts alone were useless, without a theory that made sense of

them, and provided a framework for looking at new and different facts. Paley, to some degree like Newton, had seen the universe as a gigantic and divine machine and recognised his duty to understand its inner workings. Darwin inherited this passion. When he left Cambridge, he expected to become a country squire and clergyman with a wide classical and natural education. He could well have become a respected church leader with an academic interest in natural philosophy. He could well have given up many of his questions and free thinking. He would have hunted, talked and read widely. He could well have reproduced the assumptions and work of Paley and Sedgwick in the life of a country parish, but he didn't. He might well have written a good book about Horace, Euclid or beetles. The trip to North Wales added practical, geology field skills to Sedgwick's lectures. He wanted to learn how to observe and record. Again it was John Henslow who made the suggestion that he should accompany Sedgwick to test some rocky outcrops to prove or disprove **George Bellas Greenhough's** (1778-1855) geological maps of England and Wales. This was when Darwin made his own, first discovery. He showed that Greenhough's map was wrong. When he returned to Shrewsbury, he opened the letter from Henslow proposing that he should join **Captain Robert FitzRoy** on the Beagle's journey to South America. Henslow had intervened twice in Darwin's life with profound results. He, like Fox and others, influenced his choices, interests and compass points, shaping the map of the world within which Darwin would discover and explore his own freedoms and ideas. Rev William Paley was, as we

saw, a competent biologist and theologian, describing the observed world on the basis that its wonders must have been designed by God. *William Paley 1831 Engraving. Author unknown. P-D. Copyright expired*

In this sense, he was typical of many of his contemporaries and the age in which he lived. The title to his book on Natural Theology, written while he was Archdeacon of Carlisle in 1802, reveals much about his times and their assumptions, as well as the content and approach of this formative work. It was called '*Evidences of the Existence and Attributes of the Deity, collected from the Appearances of Nature.*' His famous, teleological watch maker argument for God was highly influential. In 1785, his *The Principles of Moral and Political Philosophy* was published and it became a key ethical text book at Cambridge. Darwin began as a Paleyite. By doing natural history, he believed he was doing theology. The two approaches to knowledge seemed inseparable at the time. Paley gave Darwin the ability to reason empirically and deductively. Darwin read Paley's natural theology for pleasure and said he knew the argument by heart. 'Evolution' at the time was never used to refer to species development. Darwin and his large family were part of a pattern of intellectual thought that encouraged the making of connections across an enormous range of intellectual interests. At the centre of these interests, we find a common commitment to the causes of natural theology and changing ideas about its effects in the natural world. Charles Grandfather, **Erasmus Darwin** (1731 – 1802) has a very significant place in our story and his work was  highly influential. As we pause now to consider something of his contribution, we also note his connections with and influence on other natural philosophers of science. *Portrait of Erasmus Darwin by Joseph Wright of Derby P-D. Copyright expired.* He was a member of the Darwin-Wedgwood family and founder member of the famous Lunar society which emerged out of his personal friendship with **Mathew Boulton** (1728-1809) which began in 1757. Boulton was an English manufacturer and business partner of the Scottish engineer James Watt and famous for founding the Soho Mint. Together they produced many Boulton and Watt steam engines. Boulton, Erasmus Darwin, Rev Joseph Priestley and Josiah Wedgwood (grandfather of Emma Darwin) were at the heart of this new Midlands 'think tank,' which met on a monthly basis at the time of the new moon. It straddled the natural and other sciences, the arts and natural theology and was influential in producing many practical ideas which laid the foundations of the Industrial Revolution. Both Erasmus Darwin and Mathew Bolton admired the astronomer, geologist and naturalist **Rev John Michell** (1724-1792), who became a Fellow at Queen's Cambridge, and was elected a Fellow of the Royal Society at the same time as Henry Cavendish. In 1767, Michell was made Professor of Geology at Cambridge and then Rector of Thornhill in Yorkshire. He invented the first experiment to measure the forces of gravity in a laboratory, known as the Cavendish experiment and calculated the mass of the Earth and the values of the gravitational constant - all this for a man who was also famous for his work on geology and earthquakes[179] and identified the epicenter of the 1755 Lisbon earthquake (see chapter on theodicy in LE1). He also speculated on the existence of *dark stars* (in a 1783 paper) that were heavy enough to prevent light escaping – thus anticipating the twentieth century work on black holes! Michell regularly visited Erasmus Darwin's House in Lichfield and played his own key role with both Darwin and Boulton. Erasmus was a physician and something of a poet who had studied at Cambridge and Edinburgh. Like Michell, he opposed the slave trade and turned down an invitation from George 111 to be the Royal Physician. He shared with Boulton an interest in experiment, but had less practical experience. Their interests ranged from electricity and meteorology to geology. Boulton introduced Erasmus to the Derby based clockmaker **John Whitehurst** (1713-1788) another influential member of the

---

[179] *Conjectures concerning the Cause and Observations upon the Phaenomena of Earthquakes* in *Philosophical transactions* 1760.

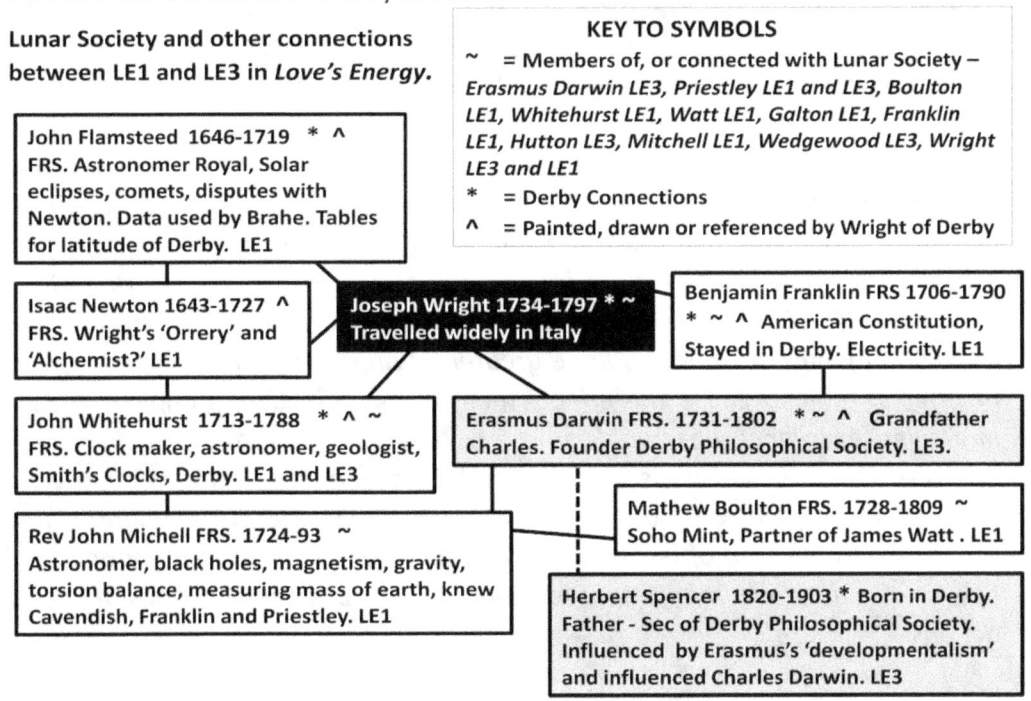

Lunar society, who had an even keener interest in experiments and was very creative in the way he used his knowledge of clocks to measure other things.[180]

Whitehurst, painted by Wright of Derby before 1798. Now in Derby Art Museum. P-D. Copyright expired.

John Michell introduced Erasmus Darwin, Boulton and Whitehurst to **Benjamin Franklin** when he visited Birmingham in 1758, searching out what he called 'Persons of Influence.' Franklin also spent time in Derby. Boulton and Franklin did experiments in electricity together. Franklin would return in 1760 to conduct experiments on electricity with Boulton. As we saw earlier, this was an age of societies and associations based on a developing excitement with new ideas, experiments and ways of understanding physical reality. Joseph Wright of Derby is well known for capturing these connections and referencing religious art in his painting of powerful scientific movements and moments. In Love's Energy, these natural philosophers connected the cosmology and physics of LE1 with the natural sciences of this book. Some of these connections can be seen in the following diagramme. I have probably missed out many figures who should be included, but my focus was on **Joseph Wright** and the wonderful collection in Derby Museum.[181]

---

[180] See his 1778 theory on geological strata in An Inquiry into the Original State and Formation of the Earth. In 1787 he published An Attempt towards obtaining invariable Measures of Length, Capacity, and Weight, from the Mensuration of Time based on studies of the shape of the earth measuring differences in gravitation (using the oscilation of pendulums!).

[181] I worked for some time in Derby as the Principal Social Responsibility Officer for the Diocese and return regularly to visit evidence of its regeneration, inward investment and increasingly creative vision for the future – much of it due to my friend John Forkin, Managing Director of 'Marketing Derby,' with whom I founded the Derby City Partnership.

The Power of Propensity and the Living Filament

Erasmushad often talked of the generation of species. He had formed the Litchfield Botanical Society and was responsible for the translation of the Swedish botanist, **Carolus Linnaeus** (1701-1778) from Latin into English – *A System of Vegetables* 1785 and *The Families of Plants* 1787. We still use many names of plants found in these works. Erasmus's long poem *The Love of Plants* made Linnaeus' work more popular. In his truly great work on medicine and physiology, **Zoonomia** (1794-96), he produced a chapter on *'Generation,'* producing ideas that would appear in the work of Jean-Baptiste Lamarck and even the later theory of evolution. His grandson must have read parts of this work several times. Sadly his grandfather died, before Charles was born, so they couldn't discuss this great work. Erasmus had been influenced by **David Hartley's** psychological theory of *Associationalism*. He comments extensively on other thinkers, including **Buffon** and **Linnaeus**, *'The late Mr. David Hume, in his posthumous works, places the powers of **generation** much above those of our boasted reason; and adds, that reason can only make a machine, as a clock or a ship, but the power of generation makes the maker of the machine; and probably from having observed, that the greatest part of the earth has been formed out of **organic recrements**; as the immense beds of limestone, chalk, marble, from the shells of fish; and the extensive provinces of clay, sandstone, ironstone, coals, from decomposed vegetables; all which have been first produced by **generation**, or by the **secretions of organic life**; he concludes that the world itself might have been generated, rather than created; that is, it might have been **gradually produced from very small beginnings**, increasing by the activity of its inherent principles, rather than by a sudden evolution of the whole by the Almighty fire. — What a magnificent idea of the infinite power of THE GREAT ARCHITECT! THE CAUSE OF CAUSES! PARENT OF PARENTS! ENS ENTIUM! For if we may compare infinities, it would seem to require a greater infinity of power to **cause the causes of effects**, than to cause the effects themselves. This idea is analogous to the improving excellence observable in every part of the creation; such as in the progressive increase of the solid or habitable parts of the earth from water; and in the progressive increase of the wisdom and happiness of its inhabitants; and is consonant to the idea of our present situation being a state of probation, which by our exertions **we may improve**, and are **consequently responsible** for our actions.*[182]

In LE1, I included a chapter on the physics of cause and effect. Now, we glimpse something of the theological and philosophical context inherited by Charles Darwin, as it impacts on the science of his times. Erasmus developed his own

---

[182] *Zoonomia* or SECT. XXXIX.' OF GENERATION.' Quoted by G.B Shaw in the preface to his Back to Methusaleh. My emphasis in bold to show the emergence of development language.

approach to the ideas of cause and effect in the conclusion to this section in Zoonomia, *'Cause and effect may be considered as the progression, or successive motions, of the parts of the great system of Nature. **The state of things at this moment is the effect of the state of things, which existed in the preceding moment; and the cause of the state of things, which shall exist in the next moment**. These causes and effects may be more easily comprehended, if **motion** be considered as a change of the figure of a group of bodies, as proposed in Sect. XIV.2.2. in as much as our ideas of visible or tangible objects are more distinct, than our abstracted ideas of their motions. Now the change of the configuration of the system of nature at this moment must be an effect its of the **preceding configuration,** for a change of configuration cannot exist without a previous configuration; and the **proximate cause of every effect must immediately precede that effect.** For example, a moving ivory ball could not proceed onwards, unless it had previously began to proceed; or unless an impulse had been previously given it; which previous motion or impulse constitutes a part of the last situation of things. As the effects produced in this moment of time become causes in the next, we may consider the **progressive motions** of objects as a chain of causes only; whose first link proceeded from the great Creator, and which have existed from the beginning of the created universe, and are **perpetually proceeding.***[183] So, he expresses a view, commonly held at the time, that *the progressive motion* (a key idea in the physics of the previous century) of living things leads to and constitutes a *chain of causes whose first link proceeded from the great Creator.* We note the use of the word *'link'* which Charles uses in *the Origin,* not least in connection with what he calls *'intermediate links'* or stages. We might pause to reflect that these ideas are found in the spirituality of many religions. Tibetan Buddhists, for example, believe in the *progressive motion* of life through many recycled phases of cause and effect, giving them the chance, in each linked rebirth phase, to transmute into something closer to the perfection of 'enlightenment.' They also believe that they have to reach out altruistically to benefit all creatures in the process of these *progressive motions in a chain of causes and effects.* Then Erasmus concludes by speculating in the following way *'Would it be too bold to imagine, that in the great length of time, since the earth began to exist, perhaps millions of ages before the commencement of the history of mankind, would it be too bold to imagine, that all warm-blooded animals have arisen from one **living filament,** which THE GREAT FIRST CAUSE endued with **animality,** with the **power** of **acquiring** new parts, attended with new **propensities,** directed by irritations, sensations, volitions, and associations; and thus possessing the faculty of continuing to improve by its own **inherent activity,** and of delivering down those improvements by generation to its posterity, world*

---

[183] Ibid VIII. 1.

without end!' [184] These are important passages for our understanding of Charles' later work on natural selection. They appear about sixty three years earlier, yet speak, controversially and prophetically, of *'millions of ages'* before humans appear and *'one living filament.. endued with animality...new propensities.'* They are also important concepts in our story. We found the idea of *propensity* in the physics of LE1 and *animality* as a living force or energy in the theology of LE2. In common with many in his times, he drew on the theology of *'The Great First Cause'* as the power which *endued* animals with their *animality* – a power capable of *'acquiring new parts!'* So, we are given a vivid picture of a *living filament* of power running through all creatures in ways that *enlivens* them and gives them the *propensity* to change and develop *'new parts'* in a *progressive motion of cause and effect*. It is hard for us to appreciate perhaps how radical this approach was and how formative these words were to become, especially the idea that the species themselves had the power to *'acquire new parts.'* The word *acquire* is pivotal and implies an internal process or *inherent* capacity that does not rely upon an external intervention except to do the *'enduing'* in the first place - *whose first link proceeded from the great Creator, and which have existed from the beginning of the created universe, and are perpetually proceeding*. Here in the idea of *progressive motion* and *perpetually proceeding* is adaption in its latent form. He is already very close to *The Origin*, especially when he goes on to assert that *'the strongest and most active animal should propagate the species, which should thence become improved and that this is motivated by 'lust, hunger and security.'*

The Life Force of Transformation and Spontaneous Generation
Erasmus was aware of the early 'evolutionary' thinking of **James Burnett, Lord Monboddo** (1724-99) and referred to him in his *Temple of Nature* 1803. Burnett had been a Scottish Judge, a deist, interested in language and philosophy who contemplated the idea of a common descent for all anthropoids. **Charles Neaves** (1800-76), one of his successors in the Scottish courts, [185] and also a theologian wrote

Though Darwin now proclaims the law
And spreads it far abroad, O!
The man that first the secret saw
Was honest old Monboddo.

---

[184] My emphasis in bold

[185] He was Solicitor General in 1852, judge of the Court of Session, the Supreme Court of Scotland in 1854 and Rector of St Andrew's University in 1872.

The architect precedence takes
Of him that bears the hood, O!
So up and at them, Land of Cakes,
We'll vindicate Monboddo."

Erasmus was certainly aware of Burnett, Lord Monboddo, and probably his reputation for being something of an eccentric - which probably had something to do with his controversial view that the anthropoidal ape is the 'brother of man.' He claimed that humans would once have had tails and that human language (and its organs) developed from the primates in response to changing and more complex, social structures. In this, he was far ahead of his times. As a horse breeder himself, he believed that selective breeding could improve quality. Charles Darwin was to spend years studying the effects of breeding processes in pigeons. Burnett understood, as Darwin understood, that natural processes were involved in changes that appeared in animals, and in this sense *artificial* selection was a good basis for later ideas of *natural* selection, though this was still too big a jump for many of their contemporaries. Erasmus Darwin and Burnett were deeply committed deists who believed that the universe was created by a divine First Cause who imbued it with some kind of inner *propensity* to develop and change. They struggled to understand the process of *development* from an animal to a human being and were aware of different cultures, diseases and climates. Erasmus saw early humans and apes as distinct from the rest of the 'lower' animals. It was only Lamarck's *Philosophie Zoologique* of 1809 that suggested a connection from the latter to humans, as we shall see. It is not known for sure whether Charles Darwin had come across Burnett's ideas, but he did comment on **Georges-Louis Leclerc, Comte de Buffon** (1707-1788) the French naturalist and cosmologist who was also the Director of the Jardin du Roi. Buffon's membership of the French Academy of Sciences in 1734 was due to his mathematics, and in particular his work on what became known as Buffon's *Law of probability theory* (see LE1). In the first three editions of his '*Historical Sketches,*' Charles Darwin said that he passed over Buffon, as he wasn't familiar

with his writings. From the first edition onwards he amended that to say '*the first author who in modern times has treated evolution in a scientific spirit was Buffon. But as his opinions fluctuated greatly at different periods, and as he does not enter on the causes or means of the transformation of species, I need not here enter on details'*.

Buffon 1753 by Francois-Hubert Drouais. Musée Buffon, Montbard P-D. Copyright expired

His *Histoire naturelle, générale et particulière* (1749–1788), consisted of no fewer than 36 volumes and was widely read all over Europe. His arguments bore little or no relationship to the Genesis account, and he proposed a theory of *reproduction* that countered the contemporary idea of 'pre-existence.' He also challenged **Linnaeus's** system of taxonomy. In his *Les époques de la nature* of 1778, he speculated that the solar system had been created by a comet's collision with the sun; that the earth was probably about 75,000 years old rather than the date of 4004 BC., given by Archbishop Ussher and that humans appeared about 6000 years ago. His work was condemned by the Theological Faculty at the Sorbonne and he published something of a retraction, but seemed to have paid little real attention in practice. He saw that different parts of the world had their own distinct species, and thought that species both improved and degenerated in their *development* from an original set of thirty eight quadrupeds. He rejected the idea of a common ancestor while accepting the similarity between humans and apes. He disagreed with Burnett when the latter argued for a close relationship between primates and humans. If it is the case that Darwin wasn't aware of Burnett, we are sure that he knew of **Buffon's** work and that Buffon had debated with Burnett. Buffon, who knew **Voltaire**, certainly influenced **Lamarck** and **Cuvier** who are about to figure in this work. Such are the patterns of inter-connection out of which ideas mutated and adapted influencing the environment out of which new species and varieties of thought would appear. The extraordinary German evolutionary biologist **Ernst Mayr** (1904-2005)[186] said of Buffon *'He was not an evolutionary biologist, yet he was the father of evolutionism. He was the first person to discuss a large number of evolutionary problems, problems that before Buffon, had not been raised by anybody.... he brought them to the attention of the scientific world...Except for Aristotle and Darwin, no other student of organisms has had as far-reaching an influence. He brought the idea of **evolution into the realm of science**. He developed a concept of the "unity of type", a precursor of comparative anatomy. More than anyone else, he was responsible for the acceptance of a long-time scale for the history of the earth. He was one of the first to imply that you get inheritance from your parents, in a description based on similarities between elephants and mammoths. And yet, he hindered evolution by his frequent endorsement of the **immutability** of species. He provided a criterion of species, fertility among members of a species that was thought impregnable.'* [187]

Another and the same

So, Erasmus Darwin was at the centre of very early speculation about the age of the earth, the development of different species, connections between them and

---

[186] It is worth noting in passing that Mayr knew that although Darwin had come up with the idea of multiple species developing from a common ancestor, he didn't know the mechanism. In Mayr's 1942 work *Systematics and the Origin of Species,* he redefined species as not just a group of morphologically similar individuals, but one that could breed only amongst itself. In isolation (on an island) one population of a species may differ from another population through genetic drift and through natural selection may even evolve into a new species. He rejected Dawkin's 'reductionist' gene centered view of evolution as being anything like a Darwinian approach, arguing that evolution pressures and processes affect the whole organism and not just particular genes.

[187] *The Growth of Biological Thought*. Cambridge: Harvard. 1981 p 330. My emphasis in bold

much more. In his poetry, he found another vehicle for expressing his many scientific and philosophical interests. In his poem of 1791, *The Botanic Garden, A Poem in Two Parts: Part 1, The Economy of Vegetation,* he connected cosmology (*Suns sink on suns, and systems systems crush*) and botany (*flowers of the* sky) and came up with a phrase that shows real imaginative insight as he hints at the way '*Immortal Nature*' emerges into '*another and the same.*'

Roll on, ye Stars! exult in youthful prime,
Mark with bright curves the printless steps of Time;
Near and more near your beamy cars approach,
And lessening orbs on lessening orbs encroach; —
Flowers of the sky! ye too to age must yield,
Frail as your silken sisters of the field!
Star after star from Heaven's high arch shall rush,
Suns sink on suns, and systems systems crush,
Headlong, extinct, to one dark center fall,
And Death and Night and Chaos mingle all!
— Till o'er the wreck, emerging from the storm,
Immortal Nature lifts her **changeful form,**
Mounts from her funeral pyre on wings of flame,
And soars and shines, **another and the same.**

He locates the new ideas of science within various classical themes - painting a picture of the mingled wreck of death, night and chaos, out of which (*Headlong, extinct, to one dark center fall*), emerges *Immortal Nature* in her *changeful form*, mounting from a funeral pyre, like a Phoenix, to be transformed into a state which is '*another and the same.*' This is a haunting phrase which anticipated the language of some dimensions of quantum theory. It could have meant many different things for him and it would be wrong to retrospectively apply to it a world of thought that only came later. It certainly underlines the pattern paradox which includes *connectivity* and *separation*. In LE1, we saw, in Quantum physics, how people wrestled with the strange, counter intuitive idea that a particle's position and speed cannot be measured at the same time and that a particle can act like a wave — its own kind of '*another and the same.*' Similar or other *pattern paradoxes* occur in logic too. Logic cannot create knowledge, but it helps to better understand the way we reason with, or handle, our processes of knowledge. We remember that the same Aristotle who had studied and begun to classify the details of animal and plant diversity, also gave us the principles of logic - particularly syllogisms[188] - to help us study and organize our reasoning processes. His work was influential for about two thousand years. Then, just five years before *The Origin* was published, **George Boole** (1815-1864) extended Aristotle by producing equations for the latter's work on logic. Boole's mathematical logic[189] worked on the principle that everything was either true or false and therefore could with *certainty* be reduced to symbols and numbers – even to something as simple as the numbers 1 and 0 – an idea which the brilliant and shamefully treated[190] **Alan Turing** (1913-1954) was to use to the full, nearly a century later. Boole could never have imagined its future use, but without his maths and Turing's 'algorithm' and 'computation' machine, our computers might not have developed (and we would not have decoded German communications in WW11).

Betrand Russell and **Alfred Whitehead** (both mentioned in LE2) wrote the three volume *'Principia Mathematica'* (first published in 1910) on mathematical logic to achieve what the German mathematician and philosopher, **Gottlob Frege** (1848-1925) had failed to do – to deal with the paradoxes of *'naive set theory'* using a *'theory of types.'* What was this about? In 1894, the University of Vienna commissioned a great painting for the ceiling of its main ceremonial hall. It was to celebrate its own contribution to science and the arts in an age of 'certainty' and the overcoming of light over darkness.[191] This was the theme given to the Autrian painter symbolist **Gustav Klimpt** (1862-1918). It was a theme that resonated well with Erasmus Darwin's poem considered above. But, what he produced was a surprise and it was rejected and hidden away.[192] In 1900, right at the turn of the century, he painted his depiction of 'Philosophy' as anything but certainty and conquest. Light wasn't simplistically separable from darkness, but merged in the shadows of naked men and women.[193] It was deeply unsettling for the aspiration that mathematics and science would give us provable certainty. The problem was how to prove that mathematical truth would never lead to a contradiction. The search was on for logic to provide an answer beyond Boole's certainties which may be the logic behind computers and the networked devices of our contemporary world, but doesn't answer the ontological problem of

---

[188] He identified 19 out of 256 kinds of syllogism as being logically valid – if the premise is true the conclusion must be true. If the premise is false the conclusion will be false. For example, all cats have 4 legs. All dogs have 4 legs, therefore all cats must be dogs

[189] See his 1854 *An Investigation of the Laws of Thought on Which are Founded the Mathematical Theories of Logic and Probabilities,*.

[190] He was prosecuted for the 'criminal' act of being a homosexual in 1952, and chose female hormone treatment as an alternative to prison. He committed suicide in 1954. He was posthumously pardoned in 2013 by the Queen.

[191] I am indebted to Professor Dave Cliff in his BBC Programme 'The Joy of Logic' February 2014 for many of these insights and descriptions and the research and skills he has developed.

[192] It was burnt by the Nazis. A copy was installed many years later

[193] His work at the Belvedere Palace, Wien Museum Klimt and Succession museums are a must see for visitors to Vienna

uncertainty. A decade or so earlier, Frege had believed that his new calculators could provide that certainty. Everything could be described or analysed by his logical quantifyers and Russell was doing similar kind of work. In 1902, Frege was about to publish his great work on logic, when he received a letter from Russell. Both relied, in their systems of logic, on consistently describing sets of things - numbers or groups of animals or species. The arguments were complicated - All 'sets' weren't in themselves members of themselves. If all these sets are combined, then they include things that don't contain themselves. But this is the set of all sets that don't contain themselves and they don't include themselves, so this set should include itself but if it does, then it's no longer the set of all sets that don't contain themselves so...and so on ad infinitum. Russell's reaction with Whitehead was nine years of work and the Principia Mathematica. The *Vienna Circle* (or the Ernst Mach Society) saw this work as the answer to their search for *certainty* and the group included **Albert Einstein** and was influenced by the ideas of **Wittgenstein**. The drift was from physics to mathematics and to logic with the purpose of prooving a logical basis for certainty. Everything else, including metaphysics, was rejected because it included statements that could not be verified internally. Their methods became increasingly pure, but unsuccessful in proving certainty. What would happen if *uncertainty* were built into the very structure of *certainty*? In this *pattern paradox*, was all *uncertainty* contained within *certainty* or was the latter a set that didn't contain itself or was that part of its inherent uncertainty?! Would the walls of logical defence prevent this *pattern paradox* from imploding? Would the optimism of absolute, provable truth be shattered and could that be allowed? What would the early appearance of fascism in the 1930's do to this search for certainty? It certainly dispersed the Vienna Circle.The Austrian **Kurt Gödel** (1906-1978) was a member of this Vienna Circle. He showed in 1931, that there was a problem with the first order logic in *'Principia Mathematica'* and its supporters. However useful Boole's equations were to prove, a century later, in developing computing science and software systems, could everything be reduced to the stark alternatives of true or false? I don't know if he ever considered Erasmus's poetic *'another and the same!'* Firstly, Gödel claimed that if the system is *consistent,* it cannot be *complete.* Secondly, the *consistency* of the axioms cannot be proven within the system. Like the position and the velocity of a particle, although for different reasons, *completion* and *consistency* were impossible to achieve at the same time. At the heart of this theory of *incompleteness*, is his idea that if something is provable, it would be false. This contradicts the idea that in a 'consistent' system, provable statements are always true. On this basis, there will always be at least one true, but unprovable statement, or that within any consistent, internal system there will be something that cannot be proved from within that system itself. Such a system cannot demonstrate its own consistency from within its own incompleteness!

Gödel's work was of course about mathematical logic or the philosophy of maths. I certainly don't understand the maths, and I am probably stretching his argument too far when I use it to reflect on what I have said about *'Life's longing for life'* in this book on evolution, or the way the nature of Love fulfils itself by being ek-statically outside of itself (see the early mystical writers in LE2). In Erasmus' poem, it is as if that which is one thing, can only be that one thing by being something else outside of itself – *another and the same*. So, internal systems are incomplete, if they are consistent. Theologically I cautiously apply this to what I have said about the nature of God as ekstatic Love. As an internal system, Love is incomplete, if it is only internally consistent. Its *consistency* and its *completeness* are found outside of its own consistent self referential 'system.' **Frege** and **Russell's** logical systems are transcended by **Gödel's** argument.[194] He was showing that, whatever our future computer power, a **complete** and **consistent**, finite list of axioms can *never* be created, nor even an infinite list that can be calculated by a computer program. This resonates with Heisenberg's claim that the indeterminacy principle is not affected by our capacity to observe/measure particles. However much we improve that capacity in the future, the ontology (my phrase) of indeterminacy will remain. So with Gödel - whatever the capacity of the computer, each time a new statement is added as an axiom, there are other true statements that still cannot be proved, even with the new axiom. If an axiom is ever added that makes the system complete, it does so at the cost of making the system inconsistent. Sometimes, in this connection, the 'liar paradox' is used as an illustration of Gödel's more complicated version.[195] It is based on the sentence – *'This sentence is false.'* It is one that no algorithm can solve. The history of the liar paradox is interesting. In the New Testament Epistle to Titus, the writer (Paul?) warns that *'a prophet of their own (referring to Jewish false teachers), said, the Cretans are always liars..'* This may refer to a very ancient story concerning Epimenides of Knossos in Crete. He lived in the 7th or 6th century B.C. well before Aristotle, and was associated with what became known as the Epimenides paradox. He may not have intended any irony when allegedly saying 'Cretans, always liars!' In the Middle Ages, using Aristotle's syllogism, this turned into the paradox statement 'all Cretans are liars.'

---

[194] Finally published in 1931 in a paper *On formally Undecidable Propositions in principia mathematica and related systems.* Von Neumann had independently verified Gödel's second incompleteness theorem in 1930 the year of the Konigsberg conference. The two theorems show that David Hilbert's attempt to find a complete and consistent set of axioms for all maths is impossible. At the conference, Hilbert had declared that there is no unsolvable problem in maths or in natural science.

[195] Gödel himself met a tragic end in America – when his reasoning led him to increasing self preoccupations with food and the fear of poisoning.

The paradox arises because the person saying this is himself a Cretan, so he is commenting from within the 'system' of being a Cretan. Therefore, if he was lying, then all Cretans weren't liars and he would be telling the truth.

When I first came across 'This sentence is false', I spent a sleepless night trying to analyse it from both directions, and internally, in its use of each word within its own structure or system. Basically, the argument explaining the paradox proceeds as follows – The sentence cannot be true because, if it were, then as it states, it would itself be false. But at the same time, it cannot be false, because then paradoxically, it would then be true. In my sleepless analysis, I decided that the problem was with the adjective 'This,' because it is the type of adjective that usually points to something specific within or outside of the sentence which contains it - so to 'this' or 'that' particular thing, or description which can then be considered. In this case, it is used in a way that creates or implies fulfillment, indeterminacy or incompletion. It is not referring, or pointing to any other thing in a sentence except to itself *as a sentence*. So we are given no additional external information, or point of reference outside of 'this,' which could then contain an axiom of meaning about a particular, but other object or thing. So the noun 'sentence' is grammatically correct as constituting the grammar – it is a structurally proper sentence, but it is emptied of any meaning because it has no external reference point which could be evaluated as either true or false, interesting or uninteresting etc. Its own veracity is therefore indeterminate because it cannot be revealed from within its own structural ontology. It uses the word 'sentence' to describe its own nature, in a self referential way that locks it into its own axiom of meaning - which cannot be proved either way and therefore is both true and false at the same time. However, when I showed the sentence to my wife, over coffee the next morning - and she is a trained teacher of language – she revealing said *'no one would write such a sentence in the first place!'*

It would take another book to attempt to apply Gödel's reformation of Frage and Russell to the processes of evolution itself. Again my disclaimer must be repeated - my inability with the maths and the science prevents me from doing this. Briefly and inadequately put, the argument might go as follows - that within the internal processes of natural selection, there is a fulfillment of completion in any living organism that can only be found in the movement from inside to 'outside.' That movement, which may take millennia, is the movement from internal consistency to outside completion – the latter not being any perfectible final end point, achieved through a straightforward, linear process, as Darwin implied. Rather, I am relating the idea of *outside completion* to what I said in LE1 about the energy transfer of proton gradients and in LE2 the energy of kenotic ekstasis. I saw both of these ideas as part of the physics of Love. I am using the idea of 'outside completion' as a way of relating the energy within an internal system, or living organism to its *potential* and *propensity* (rather than any predetermined intent or intervention) which can only be experienced or fulfilled outside of its present 'internal' nature. Love 'completes' itself outside of itself and that is the energy of *life longing for itself* and for growth and change. At any one particular moment in time, (as we learnt from Heisenberg) we cannot observe *completio*n and *consistency* side by side. The consistency within the processes of natural selection leads towards completion *only* by moving away from its self into diversity and into transmutations of adaptability within a wide *pattern paradox* of connectivity. This is not the same as arguing for the influence of the external environment on the processes of natural selection. It is saying that because the internal processes of consistency move in the direction of completion (never achieved) they lead *naturally* to adaption and change. This movement is part of their *consistency* and the only way it can fulfill its nature. They cannot be *completed* within their own internal system of *consistency* without this movement away from self. To pursue or test such a hunch, in any rigorous way, someone else would need to apply the maths of Gödel to the biochemistry of natural selection and to the latest genetics. In the absence of my ability to do that, I can only offer the thought and the possibility of potential connectivity. In our story of Love's Energy we owe much to Erasmus Darwin. In LE1, we saw that there is, below the surface of all things, a deeply embedded *ambiguity* and *unpredictability*. Here, in his work, over two hundred years ago, and well before any thought of Quantum physics, we find the language of **another and the same** in the *changeful form* of nature, emerging and then soaring from what he called the *storm.* True, this sounds like the language of poetry, but it comes from a great scientist and profound thinker who had studied the details of *generating processes* in a linked chain of *progressive motion,* and who had contemplated underlying connections within the *economy* of nature as it transformed itself and was transformed. This work must have made it easier for Charles to contemplate the idea of transmutation and to work on its implications for natural selection.

The idea of 'transmutation' was already in use, thanks to the work of the **Rev John Ray** (1627 – 1705), a Cambridge naturalist and clergyman and his early ideas of spontaneous or inner generation.

Ray Painter unknown but donated to NPG before 1939. P-D copyright expired

These individuals were all searching for an explanation of how A became B – of how things changed and what 'force' made this possible in the processes of nature and indeed how that related to the divine will, or design, as they sought for a way of picturing and defending the intervention of a creator God. With Erasmus, as with Ray, the images are already of an internal, if mysterious, process. These and many other scientists and thinkers used different language for this idea of an inner power or capacity for change and its cause and source. These thought experiments with language and natural theology, as much as with natural sciences, all contributed, in their different ways, to the changing pattern and context of theology within which the later work of Charles Darwin and his contemporaries developed. *Lamarck* used the idea of force in two different senses, as we shall see, and even used the term *alchemic force* which clearly drew upon centuries of earlier work and assumptions about the mysterious process of change in matter and living things. We remember that Newton (1642-1727) himself– the great experimenter with different ideas about *alchemic forces* - was a later contemporary of John Ray. Both would have naturally inherited a previous generation's assumptions about an inner force or

propensity for change – assumptions that had their own, long pre-history and by no means all based on mechanical metaphors! Charles Darwin had already picked up the notion of transmutation from his zoology instructor in Edinburgh **Robert Grant** (1793–1874), who was a physician and a biologist.

Grant, 1852, author unknown. See Archetypes and Ancestors, Desmond 1982. p117. P-D. Copyright expired

He was also, significantly for our story, a follower of **Lamarck**. Grant and Darwin both attended meetings of the Plinian Society for naturalists and the Wernerian Society for anatomists at Edinburgh University. When Grant became Professor of

Comparative Anatomy at University College London, he was attacked by the Tories for his '*blasphemous derision of the truths of Christianity.*' He thought that some kind of 'transformation' must affect all living forms and followed the evolutionary ideas of the French naturalist, **Étienne Geoffroy Saint-Hilaire** (1772 –1844) who, as a deist, insisted that there could be no divine interference in the details of the God given laws of existence, as they operated in natural

processes. *Geoffroy, anon, c 1842. US National Library of Medicine. P-D. Copyright expired.*

Geoffroy, a colleague of Lamarck, introduced the idea of a unity of underlying composition in all animals but also the possibility of the transmutation of species

over longer periods of time. He didn't believe in a common descent as such, but the emergence of an existing potential in a given species, triggered by the environment.

Cuvier. artist and date unknown. P-D. Copyright expired

His serious and famous, anatomical debates with **George Cuvier** in 1830[196] were based on a disagreement about how structure determines function (Cuvier) and function

---

[196] Cuvier insisted on the invariability of animals.

determines structure (Geoffroy), but the real tension between them may have been over religious and political ideas. Grant saw in geological strata, evidence of a natural succession of fossil animals. In 1826,[197] he declared that they must *'have evolved from a primitive model'* by *'external circumstances'* which appeared as a **Lamarckian** kind of view. Grant accepted the possibility of a common origin - a *monad* with the capacity for spontaneous generation. This was seen at the time as denying any divine intervention. Darwin visited Grant in 1831 for help with the storage of specimens immediately before his Beagle journey. Something happened to affect their relationship as Darwin turned down his support after the journey. This reminds us of the crucially human and subjective dimension of this story, with all its natural fragilities. It remains, after all, a human story, one of many, some overlapping, some developing within their own, isolated continents of conditioning, but all reflecting new light on the complex nature of our human condition, perceptions and understanding.

Charles's grandfather, Erasmus Darwin had hypothesized that warm-blooded animals sprang from simpler life forms and evolved by acquiring their parent's characteristics. So a paradigm was already appearing that the more complex came from the more simple. This anticipated the work of the French naturalist, biologist and soldier **Jean Baptiste De Lamarck** (1744 – 1829), who put forward his evolutionary theory of 'transformation' in 1809.

Lamarck 1893. Jules Pizzeta. Galerie des naturalistes de J. Pizzetta, Paris: Ed. Hennuyer, 1893. P-D. Copyright expired

Lamarck had introduced new terms in the natural sciences – chemistry, meteorology and geology, as well as botany. After injury on the battlefield in the Pomerian war with Prussia in 1766, he produced his three volume, *Flore Francaise* and gained membership of the French Academy of Sciences. He was made Professor of zoology at the Museum of Natural History in 1793 when it opened for the first time. His 1801 work on the classification of what he termed 'invertebrates' was a significant step forward. He is important for our story - in his work on what he called *soft inheritance* of acquired characteristics, or what became known as *Lamarckism*. This reflected what many naturalists were beginning to accept, but he produced the first cohesive theory of how an *alchemical force* drove organisms up a ladder of increasing complexity. He believed that a second 'force' caused their adaptation to their local environments through the *use and disuse* of their characteristics, so leading to differentiation from other animals. So, as an animal uses an organ, it becomes stronger and enlarged like a Giraffe's neck and this acquired characteristic is passed on to the next generation. Holding these two theories of forces together was his belief in *Le*

---

[197] See his *Observations on the Nature and Importance of Geology. Edinburgh New Philosophy Journal. **14**, p 270–84.*

pouvoir de la vie or *la force qui tend sans cesse à composer l'organisation.* In opposition to the French chemist and biologist, **Antoine Lavoisier's** (1743-94)[198] view that it was chemistry that explained this *complexifying force*, Lamarck insisted it could be explained by the traditional alchemy of earth, air, fire and water; in particular that it was the moving of fluids in living things that forced them into a greater stage of complexity.

Lavoisier. P-D. unknown date, Louis Jean Desire Delaistre, after Boilly, Courtesy of Chemical Achievers Copyright expired

With Laplace, Lavoisier had 'synthesised' water by burning oxygen and hydrogen in a bell jar over mercury, showing that water was not an 'element,' but a compound of gases. This was a very significant moment in the history of chemistry and of natural philosophy and theology. No longer could the traditional assumptions of early and medieval alchemy apply – that nature was based on earth, air, fire and water. This would have huge implications for how doctors understood and responded to the causes of illness. Millions died because the lack of balance between these elements was assumed to be the cause of disease and illness. It wasn't miasma, or bad air, but microbes – bacteria and viruses. It is worth pausing to remind ourselves briefly of some of the longer term, indirect effects of this early experiment which happened before later discoveries. **Louis Pasteur** (1822-1895) hadn't as yet looked down his microscope at bad wine to see microbes – the greatest killers of all, and one that must be part of our thesis that 'life longs for life,' even, or especially the life of those very ancient species of microbes. Without the microscope, it would have been impossible to see this 'universe' of life, which had its own story to tell about natural selection and the way microbes could adapt to benefit from the cells they 'infected.' If the telescope changed Galileo's view of the stars, then the microscope did similar things to revise our understanding of how the very small in nature really operated. The Franco Prussian war hadn't as yet happened and **Robert Heinrich Herman Koch** (1843 –1910), hadn't as yet identified germ theory as the cause of tuberculosis, cholera and anthrax. **Paul Ehrlich** (1854 – 1915) hadn't as yet discovered a drug against syphilis - which was killing millions – to replace the mercury treatment which was, itself, a killer.[199] He hadn't as yet experimented with dyes to discover a way of highlighting bacteria under a microscope,[200] and he hadn't, as yet, discovered that some compounds of toxic substances would 'poison' the bacteria. **Sir Alexander Fleming** hadn't as yet, by chance, noticed a fungus on a Petri Dish

---

[198] One of the greatest pioneers – along with his wife Marie Ann, who is too often written out of the story - in early chemistry and well known for his list of elements which replaced the Aristotlian earth, fire, water and air. He was the first to discover the role played by oxygen in combustion which so influenced Priestley when the latter met him in Paris. Lavoisier reversed Priestley's experiement to show it worked in both directions. He was the first to introduce the terms *oxygen* (1778) and later *hydrogen* (1783) – Greek for 'water former'. He was guillotined by the Revolution for being a tax collector.
[199] In 1909, he discovered Salvarsan.
[200] Published in his 1885 monograph *The Need of the Organism for Oxygen*

or, more importantly, the Oxford Group of the Australian, **Howard Walter Florey, Baron Florey of Adelaide** (1898 –1968), the German **Sir Ernst Boris Chaim** (1906-1979) and **Norman Heatley** (1911-2004)[201] hadn't as yet turned Penicillin into a life saving antibiotic. Nor had viruses as yet been discovered. They were at least a hundred times smaller than bacteria, and were capable of invading a cell and reprogramming it. In WWI, it is estimated that about 50 million people were killed by a Flu virus showing that it was at least as powerful as all the guns! Small Pox was the most deadly airborne virus in history with no cure, because the Englishman, **Edward Jenner** (1749-1823) hadn't as yet taken a milk maid's story seriously and confirmed that cow pox prevented small pox,[202] so saving more lives from a horrible disease than any other individual in history. The WHO hadn't, as yet, under the leadership[203] of **Donald Henderson,** taken the visionary commitment to eradicate the virus[204] from all populations. This it managed within just ten years – one of the greatest achievements of science ever. I view the capacity for natural adaption of bacteria and viruses as a natural part of Life's 'longing for life.' This is how I explain the theological problem of theodicy in killer diseases, as well as cosmological events (as we saw in LE1). If that is the case, then I also view the story of human intervention in anti-biotic and virus prevention and immunisation as part of the same story of Love's energy working through and in natural processes – in this case human courage, determination, patience and inventiveness. All of this is part of a long process of inner adaption where new knowledge freed us from the ancient and more recent assumptions about a divine intervention which either punished us with terrible plagues or intervened to miraculously cure us. The assumptions of alchemy had lasted for thousands of years and now following the work of **Lavoisier**, they were being dismantled, piece by piece. In the process, these extraordinary individuals were discovering, separately and together, sometimes cooperatively, sometimes competitively, something of how a new jigsaw puzzle of individual questions and discoveries might fit together, albeit still laid out on a basic template of beliefs. The map of our knowledge, as well as the world, was being redrawn in these centuries. We also note, as part of our developing story, Lamarck's idea that there was a power of life, a *life force* working ceaselessly to transform the nature of an organism – to *force* it into something more complicated that was capable of adapting to its local environment. True, he still held onto many of the assumptions of alchemy and the

---

[201] Who came up with a way of purifying penicillin and recorded the crucial trials, carried out on eight mice in May 1940 'After supper with some friends, I returned to the lab and met the professor to give a final dose of penicillin to two of the mice. The 'controls' were looking very sick, but the two treated mice seemed very well. I stayed at the lab until 3.45 a.m., by which time all four control animals were dead.' The team had to improvise with primitive equipment under war conditions so moved to the US where the drug could be produced in bulk in times for the D Day landings.

[202] The word 'vaccination' was named after the word for cow and enabled the immune system to prepare for and adapt to the virus – a human intervention in 'natural' processes.

[203] Many others were involved in the WHO and local teams and other attempts at been made before

[204] Controversially samples are still kept in secure locations

basic elements, even as he worked on these new ideas. Clearly, Lamarck's life and work were a crucial part of the changing context of ideas which, in turn, would influence Darwin's thought. Darwin acknowledged his debt to Lamarck, despite his disagreements with him. In his *'An Historical Sketch on recent progress of opinion on the Origin of Species,'* (added in later versions, right at the beginning of), he summarised a wide variety of alternative and critical views, and he pays special tribute to Lamarck. *'I WILL here attempt to give a brief, but imperfect sketch of the progress of opinion on the Origin of Species. The great majority of naturalists believe that species are immutable productions, and have been separately created. This view has been ably maintained by many authors. Some few naturalists, on the other hand, believe that species undergo modification, and that the existing forms of life have descended by true generation from pre-existing forms. Passing over authors from the classical period to that of Buffon, with whose writings I am not familiar, Lamarck was the first man whose conclusions on this subject excited much attention. This justly-celebrated naturalist first published his views in 1801, and he much enlarged them in 1809 in his 'Philosophie Zoologique,' and subsequently, in 1815, in his Introduction to his 'Hist, Nat. des Animaux sans Vertèbres.' In these works he upholds the doctrine that all species, including man, are descended from other species. He first did the eminent service of arousing attention to the probability of all change in the organic as well as in the inorganic world being the result of law, and not of miraculous interposition. Lamarck seems to have been chiefly led to his conclusion on the gradual change of species, by the difficulty of distinguishing species and varieties, by the almost perfect gradation of forms in certain organic groups, and by the analogy of domestic productions. With respect to the means of modification, he attributed something to the direct action of the physical conditions of life, something to the crossing of already existing forms, and much to use and disuse, that is, to the effects of habit. To this latter agency he seems to attribute all the beautiful adaptations in nature;—such as the long neck of the giraffe for browsing on the branches of trees. But he likewise believed in a law of progressive development; and as all the forms of life thus tended to progress, in order to account for the existence at the present day of very simple productions, he maintained that such forms were now spontaneously generated.'*[205] The relationship between Darwin's and Lamarck's views is a matter of considerable scholarly debate. Darwin seems to have accepted a variant of Lamarckism as a supplementary mechanism of natural selection and called it Pangenesis [206] - the idea that somatic cells, in response to the *use and disuse* of environmental stimulation throw off gemmules or pangenes which would travel round the body, although not necessarily in the blood stream. Pangenes were microscopic particles that supposedly contained information about their parent

---

[205] 3rd Edition. John Murray 1861

[206] As set out in the final chapter of Variation in Plants and Animals under Domestication.

cell which would eventually accumulate in the germ cells. Darwin helped his own half cousin, a polymath and very prolific inventor, anthropologist, traveler, proto-eugenicist[207] and geographer, **Sir Francis Galton** (1822-1911) to do experiments on

rabbits, transfusing blood from one to another, hoping that its offspring would show the characteristics of the parent.

Galton, date unknown, scanned from Karl Pearson's 'The Life, Letters, and Labors of Francis Galton.' P-D, Copyright expired

They did not, so Galton claimed that this disproved the theory of Pangenesis. Darwin objected that he had never claimed that pangenesis worked through the bloodstream of animals, but only in Protozoa and plants which have no blood. Many scientists followed the Lamarckian theory that acquired characteristics from the environment explained evolution better than natural selection.

From Malthus to Vestiges

Another important, perhaps crucial influence on Darwin's thinking was yet another clergyman, the **Rev Thomas Malthus** (1766-1834), the political economist and demographer who observed in *'An Essay on the Principles of Population'* (1798) that living things are enormously fertile, yet population sizes remain constant.

Thomas Malthus. P-D. Copyright expired.

Malthus proposed that the explanation was to be found in environmental pressures such as competition for food, predator - prey relationships and changes in the environment serve to limit increases in population. In 1838, we know that Darwin saw this as a significant insight. Malthus was arguing for limits to growth in the context of Enlightenment optimism. Darwin used this work as one explanation for thousands of species competing for space. In 1844, he placed Malthus at the heart of his theory of evolution - only *'favourable adaptions will survive.'* Darwin wrote, for example, *'In the preservation of favoured individuals and races during the constantly recurrent Struggle for Existence, we see the powerful and ever acting means of selection. The struggle for existence inevitably follows from the high geometrical ratio of increase which is common to all organic beings. This high rate of increase is proved by calculation, by the effects of a succession of peculiar season and by the results of naturalization as explained in the third chapter. More individuals are born than can possibly survive. A grain in the balance will determine which individual shall live and which shall die - which variety or species shall increase in*

---

[207] He invented the term eugenics in 1883. See his Inquiries into Human Faculty and its Development. He advocated encouraging eugenic marriages by supplying able couples with incentives to have children. He gave the second Huxley lecture at the Royal Anthropological Institute on eugenics in October1901. As Honorary President of the Eugenics society, he wrote the foreword for the first volume of the Eugenics Review. Winston Churchill attended the first International Congress of Eugenics in July 1912

number and which shall decrease or finally become extinct. As the individuals of the same species come in all respects into the closest competition with each other the struggle will generally be most severe between them; it will be almost equally severe between the varieties of the same species and next in severity between the species of the same genus.[208] Malthus observed that, sooner or later, rising population will be stopped by famine and disease in what became known as the Malthusian catastrophe. He was writing to challenge the popular view in 18th-century Europe that progress was inevitable and linear - always moving forward in a positive way. But what if *'the power of population is indefinitely greater than the power in the earth to produce subsistence for man?'*[209] He worked this out through his three principles that

> ➢ the increase of population is necessarily limited by the means of subsistence,
> ➢ population does invariably increase when the means of subsistence increase, and,
> ➢ the superior power of population is repressed, and the actual population kept equal to the means of subsistence, by misery and vice.[210]

As far as we can tell, Malthus thought that God was in some way intervening to teach moral behaviour. Many looked back at great catastrophes like the Black Death and other plagues believing that they were some kind of divine punishment. We saw how repugnant such views are theologically in LE1. If such events weren't a punishment from God, some writers to this day believe they are the punishing acts of the living organism of the planet as it cleanses itself of unsustainable population levels. Malthus's work remained controversial in political as well as scientific circles, but has influenced many subsequent thinkers and recent ecological movements. *Population* has remained the elephant in the room in many environmental discussions.[211] It is probable that even without Darwin's thesis, Malthus's ideas would have remained significant. He was certainly known by Wallace and other contemporary writers.

From Vestiges to the *self evolving* power of nature

In October 1844, Darwin heard of the publishing of *'Vestiges of natural history of creation'* authored anonymously by **Richard Chambers** (1802-71) a Scottish publisher and author[212] and member of the Geological Society and the Royal Society of Edinburgh.

---

[208] The Origin p 441-2. Randon Publishing 1979 edition

[209] *An Essay on the Principle of Population*. p 13 Oxford World's Classics reprint

[210] Ibid end of chapter VII

[211] When my national group on the environment in the Church of Wales organised a seminar on population we soon discovered how controversial that subject was with many environment NGOs.

[212] 1802 -1871.

Chambers, 1863 British Library Source. PD Copyright expired

He had travelled in Scandinavia and Canada looking at the geology. The book argued for a new vision of the *development* of the cosmos. It linked the development of stars with the progressive transmutation of species. The title alone is interesting, particularly the hesitancy hinted by the word 'Vestiges.' Chambers' original title '*The Natural History of Creation*' had been more confident. Friends referred him to the language of the Scottish geologist **James Hutton** who had written of the timeless horizons of geology, with 'no *vestige* of a beginning, no *prospect* of an end.' So 'Vestiges' it had to be, but it remained a natural history of creation, not just a natural history. It spoke convincingly of '*principles of development*' ordained by providence. The theological orientation is clear; there is no jump from these *principles* to a rejection of a Creator. But the challenge to creationist kinds of natural theology was nevertheless significant and in many ways cleared the way for what was to follow in the work of Wallace and Darwin. In particular 'Vestiges' focused on the implications of extinction, and this was to be pivotal for their work. The fossil record showed that some divine 'designs' were flawed. '*Some other idea must then come to with regard to the mode in which the Divine Author proceeded in the organic creation.*'[213] 'Vestiges' did not have in mind a natural mechanism of selection along Wallacian or Darwinian lines. But, clearly, that part of creationism which depended upon a God who was continually acting to separately create new species was at the very least questionable and unnecessary. '*..how can we suppose that the august Being who brought all these countless worlds into form by the simple establishment of a natural principle flowing from his mind, was to interfere personally and specially on every occasion when a new shell-fish or reptile was to be ushered into existence on one of these worlds? Surely this idea is too ridiculous to be for a moment entertained.*'[214] Yet, it still is *entertained* in many fundamentalist Christian circles and Islam as well. 'Vestiges' wanted to protect theology from such mistaken and untenable ideas. '*Thus, the scriptural objection quickly vanishes, and the prevalent ideas about the organic creation appear only as a mistaken inference from the text, formed at a time when man's ignorance prevented him from drawing therefrom a just conclusion.*'[215] More, the writer's own belief in a Creator had to be protected at all costs, both from the creationists of the time, and from their extreme critics who saw no space for any kind of Providence in nature. His theological, deistic language is of its own times when he argues, '*To a reasonable mind the Divine attributes must appear, not diminished or reduced in some way, by supposing a creation by law, but infinitely*

---

[213] '*Vestiges*' page 153
[214] Ibid page 154
[215] Ibid 156

exalted. It is the narrowest of all views of the Deity, and characteristic of a humble class of intellects, to suppose him acting constantly in particular ways for particular occasions. It, for one thing, greatly detracts from his foresight, the most undeniable of all the attributes of Omnipotence. It lowers him towards the level of our own humble intellects. Much more worthy of him it surely is, to suppose that all things have been commissioned by him from the first, though neither is he absent from a particle of the current of natural affairs in one sense, seeing that the whole system is continually supported by his providence.' [216]
'Vestiges' may well have facilitated different forms of theological and scientific coexistence for many in his generation, at least for believers willing to turn away from any miraculous intervention in species creation. But the vitriolic criticism it engendered was a warning to Darwin in particular about what was at stake. He took the scandal 'Vestiges' created very seriously and rethought what could be said and on what basis. (If the work of Wallace hadn't forced publication, Darwin's own work might have been postponed to even later). However, the author kept his name formally secret for 40 years – probably with religious critics in mind. Chambers tried to distance himself from Lamarck as many feared any talk of evolution *'Now it is possible that wants and the exercise of faculties have entered in some manner into the production of the phenomena which we have been considering; but certainly not in the way suggested by Lamarck, whose whole notion is obviously so inadequate to account for the rise of the organic kingdoms, that we only can place it with pity among the follies of the wise.'.*[217]
Lamarck's own ideas had been attacked by most contemporaries, except by those on the radical left and even **Charles Lyell** had criticised him in his own *'Principles of Geology.'* Some critics saw 'Vestiges' as a kind of Lamarckian hang over, and clearly Chambers himself was sensitively rejecting any connection with his work, just as Darwin was conscious that he had to protect himself from the critics of the 'Vestiges'. In its tenth edition, no less, Chambers added that he *'had heard of the hypothesis of Lamarck; but it seemed to him to proceed upon a vicious circle, and he dismissed it as wholly inadequate to account for the existence of animated species.'* Again we note the phrase *'animated'* as applied

to the process of change. In May of 1847, **Samuel Wilberforce,** Bishop of Oxford, who was to take sides against Darwin in the later Oxford debate, preached in a packed St. Mary's Cambridge on *'the wrong way of doing science.'*
Wilberforce by George Richmond 1868. P-D. Copyright expired
This was obviously an attack aimed at Chambers. He spoke of the dangers of seduction by the *'foul temptation'* of speculation looking for a self-sustaining universe in a *'mocking spirit of unbelief,'*

---

[216] Ibid 157
[217] Ibid page 231

showing a failure to understand the *'modes of the Creator's acting'* or to meet the responsibilities of a gentleman! By implying that God might not actively sustain the natural and social hierarchies, the book was seen as a threat to the social order, providing ammunition to Chartists and revolutionaries. Even Anglican clergy who were naturalists attacked the book. Sedgwick predicted *'ruin and confusion in such a creed'* which if taken up by the working classes *'will undermine the whole moral and social fabric'* bringing *'discord and deadly mischief in its train.'* The book was supported by many Quakers and Unitarians, but critics thanked God that the author began *'in ignorance and presumption,'* for the revised versions *'would have been much more dangerous.'* Nevertheless, the book quickly went through a number of new editions. *'Vestiges'* brought widespread discussion of evolution out of the streets and 'gutter presses' and into the drawing rooms of respectable men and women. **Rev Adam Sedgwick** called it a *'filthy abortion..if this book is true then religion is in vain..mutability is moonshine.'* This well illustrates the defensive 'take it or leave it' arguments used to protect established thinking from any new ideas. Darwin read these comments in fear and trembling. He shared his self doubt with **Hooker** - *'no one has the right to pronounce on species unless he has examined many species.'* So, Darwin began his eight years of work on barnacles and became a world expert on their ways of adaption. This is what Darwin himself wrote about Vestiges, in his *'An Historical Sketch'* at the beginning of the final edition of the 'Origin' in 1876.[218] *'The 'Vestiges of Creation' appeared in 1844. In the tenth and much improved edition (1853) the anonymous author says (p. 155):—The proposition determined on after much consideration is, that the several series of **animated** beings, from the simplest and oldest up to the highest and most recent, are, under the providence of God, the results, **first**, of an impulse which has been imparted to the **forms of life**, advancing them, in definite times, by generation through grades of organisation terminating in the highest dicotyledons and vertebrata, these grades being few in number, and generally marked by intervals of organic character, which we find to be a practical difficulty in ascertaining affinities; **second**, of another **impulse** connected with the **vital forces,** tending, in the course of generations, to modify organic structures in accordance with external circumstances, as food, the nature of the habitat, and the meteoric agencies, these being the 'adaptations' of the natural theologian." The author apparently believes that organisation progresses by sudden leaps, but that the effects produced by the conditions of life are gradual. He argues with much force on general grounds that species are not **immutable** productions. But I cannot see how the two supposed "impulses" account in a scientific sense for the numerous and **beautiful co-adaptations** which we see throughout nature; I cannot see that*

[218] 6th Edition. John Murray. Pages XVI and XVII. He comments on various authors in date order, beginning with Lamarck, Geoffroy Saint-Hilaire and Dr. W.C Wells.

we thus gain any insight how, for instance, a woodpecker has become adapted to its peculiar habits of life. The work, from its powerful and brilliant style, though displaying in the earlier editions little accurate knowledge and a great want of scientific caution, immediately had a very wide circulation. In my opinion it has done excellent service in this country in calling attention to the subject, in **removing prejudice***, and in thus preparing the ground for the reception of* **analogous views.**[219] We can be in little doubt that by *'analogous views'* he was also referring to his own work and we can sense the gratitude in such phrases as *'removing prejudice'* and *'preparing the ground.'* Again we note the (theological) beliefs and values indicated by phrases such as *'impulses which have been imparted to the forms of life'* and *'vital forces.'* Immediately after this section on the Vestiges, Darwin refers us to the prevalence of theological ideas of separate creation. *'In 1846 the veteran geologist M. J. d'Omalius d'Halloy published in an excellent though short paper ('Bulletins de l'Acad. Roy. Bruxelles, tom. xiii. p. 581), his opinion that it is more probable that new species have been produced by descent with modification than that they have been separately created: the author first promulgated this opinion in 1831.'* The subject of separate creation is clearly centre stage in Darwin's selection of contemporary or recent writers on the subject of how species are 'created' and change. He writes on page XVI, *'The Hon. and Rev. W. Herbert, afterwards Dean of Manchester, in the fourth volume of the 'Horticultural Transactions,' 1822, and in his work on the 'Amaryllidaceæ' (1837, p. 19, 339), declares that "horticultural experiments have established, beyond the possibility of refutation, that botanical species are only a higher and more permanent class of varieties." He extends the same view to animals. The Dean believes that single species of each genus were created in an originally highly* **plastic** *condition, and that these have produced, chiefly by intercrossing, but likewise by variation, all our existing species.'* It is an interesting quotation. It shows how one senior cleric was moving from separate creation towards modification and selection by use of the idea of an *original* genus being created in an originally **highly plastic condition**. So, the original version of plants and animals were created directly by a Creator but in such a condition that they themselves could then develop into different species. This belief in an 'original' appears in a related section, *'Professor Owen, in 1849 ('Nature of Limbs,' p. 86), wrote as follows:—"The archetypal idea was manifested in the flesh under diverse such modifications, upon this planet, long prior to the existence of those animal species that actually exemplify it. To what natural laws or secondary causes the orderly succession and progression of such organic phenomena may have been committed, we, as yet, are ignorant." In his Address to the British Association, in 1858, he speaks (p. li.) of "the axiom of the continuous operation of creative power, or of the ordained becoming of living things."*

---

[219] My emphasis in bold

We pause to note how belief in a created living force is operating even as the language changes – in this case to *'operation of creative power, or of the ordained **becoming** of living things.'* This use of the word *'becoming'* resonates with some of the vitalist ideas discussed in LE2 and clearly carries philosophical and theological associations. Darwin continues *Farther on (p. xc.), after referring to geographical distribution, he adds, "These phenomena shake our confidence in the conclusion that the Apteryx of New Zealand and the Red Grouse of England were distinct creations in and for those islands respectively. Always, also, it may be well to bear in mind that by the word 'creation' the zoologist means 'a **process he knows not what.'** He amplifies this idea by adding, that when such cases as that of the Red Grouse are "enumerated by the zoologist as evidence of **distinct creation** of the bird in and for such islands, he chiefly expresses that he knows not how the Red Grouse came to be there, and there exclusively; signifying also, by this mode of expressing such ignorance, his belief that both the bird and the islands owed their origin to a great first **Creative Cause**." If we interpret these sentences given in the same Address, one by the other, it appears that this eminent philosopher felt in 1858 his confidence shaken that the Apteryx and the Red Grouse first appeared in their respective homes, "he knew not how," or by some process "he knew not what." This Address was delivered after the papers, by Mr. Wallace and myself on the Origin of Species, presently to be referred to, had been read before the Linnean Society. When the first edition of this work was published, I was so completely deceived, as were many others, by such expressions as "the continuous operation of creative power," that I included Professor Owen with other palæontologists as being firmly convinced of the immutability of species; but it appears ('Anat. of Vertebrates,' vol. iii. p. 796) that this was on my part a preposterous error. In the last edition of this work I inferred, and the inference still seems to me perfectly just, from a passage beginning with the words "no doubt the type-form," &c. (Ibid. vol. i. p. xxxv.), that Professor Owen admitted that natural selection may have done something in the formation of new species; but this it appears (Ibid. vol. iii. p. 798) is inaccurate and without evidence. I also gave some extracts from a correspondence between Professor Owen and the Editor of the 'London Review,' from which it appeared manifest to the Editor as well as to myself, that Professor Owen claimed to have promulgated the theory of natural selection before I had done so; and I expressed my surprise and satisfaction at this announcement; but as far as it is possible to understand certain recently published passages (Ibid. vol. iii. p. 798), I have either partially or wholly again fallen into error. It is consolatory to me that others find Professor Owen's controversial writings as difficult to understand and to reconcile with each other, as I do. As far as the mere enunciation of the principle of natural selection is concerned, it is quite immaterial whether or not Professor Owen preceded me, for both of us, as shown in this* *historical sketch, were long ago preceded by Dr. Wells and Mr. Matthews.'*[220] The tension with Owen is palpable even though understated here.[221] Also, we note that Darwin attributes the principle of natural selection to **Wells** and **Mathews**.

From the self modifying power of nature to Adam Smith and Karl Marx

Professor Baden Powell was a priest and liberal theologian[222] who held the Savilian Chair of geometry at Oxford from 1827-1860. *Baden Powell, by Hartmann. P-D, Copyright expired*

He believed there were *uniform* and universal laws written into nature – so much so that miracles were a denial of the laws of God! He saw a conflict between the Word of God in the Bible and the Works of God in nature (hence 'natural' theology).

The Scottish polymath, **James Hutton** (1726-97) is often credited with these ideas of *uniformity* to explain slow changes in the earth's crust over long periods of time.

Hutton 1776. Artist Henry Raeburn. Scottish National Gallery P-D. Copyright expired.

The idea of 'geological time,' in contrast to Biblical time, was already raising new questions and dividing opinion. These questions were in Darwin's mind a few decades later. Hutton was a deist who saw divine design in nature, but he came to view the planet as a kind of super organism and in his own way made the science of geology respectable. He was not an evolutionist, but through his interest in agriculture, noted variations of adaption and survival and did experiments in plant and animal breeding. In his unpublished work, *Elements of Agricultu*re, he identified differences between *heritable variation* in animals as a result of breeding processes, and *non heritable variation* influenced by natural climatic, or soil and environmental conditions. He came to apply the ideas of uniformitarianism to natural things and **Charles Lyell** was certainly influenced by his work. It is possible that Darwin came across it[223] as a student in Edinburgh. In Hutton's 1794, three volume work, with a title that revealed the intellectual spirit of his age, *'Investigation of the Principles of Knowledge; from sense to science and philosophy,'* (Vol 2) he wrote *'...if an organised body is not in the situation and circumstances best adapted to its sustenance and propagation, then, in conceiving an indefinite variety among the individuals of that species, we must be assured, that, on the one hand, those which depart most from the best adapted constitution, will be the most liable to perish, while, on the other hand, those organised bodies, which most approach to the best*

---

[220] From Darwin's Edinburgh days

[221] His criciticism of Owen's writing style may be in reaction to Owen's comments in his Edinburgh Review of 1860. For detailed analysis see Professor Jeff Wallace in *Charles Dicken's Origin of Species*, Manchester University Press 1995. p 28 ff

[222] On of seven Anglicans who produced Essays and Reviews of 1860 to join in the evolutionary debate. They saw miracles as being irrational and this created a widespread debate which may have taken the focus of Darwin's work or in the minds of some been associated with it. 22,000 copies were sold in two years – many more than the Origin sold in twenty years.

[223] And the work on natural selection found in William Charles Wells and Patrick Mathew also in Edinburgh.

constitution for the present circumstances, will be best adapted to continue, in preserving themselves and multiplying the individuals of their race.[224]

Baden Powell became aware of the idea of *uniformity* from Charles Lyell's geology and he picked up evolutionary ideas from the 'Vestiges.' He supported both against what was then called 'catastrophism,' or the idea of a series of divine creations or interventions to create new species.[225] He believed that Christianity had to dismiss the 'science' in Genesis in favour of the ethical teaching and laws of the New Testament. He was widely read and **Joseph Hooker** commented in a letter to **Asa Gray** in March 1857 *"These parsons are so in the habit of dealing with the abstractions of doctrines as if there was no difficulty about them whatever, so confident, from the practice of having the talk all to themselves for an hour at least every week with no one to gainsay a syllable they utter, be it ever so loose or bad, that they gallop over the course when their field is Botany or Geology as if we were in the pews and they in the pulpit. Witness the self-confident style of Whewell and Baden Powell, Sedgwick and Buckland."* Whewell, Sedgwick and Buckland had of course opposed the idea of evolution. After reading or at least hearing about Wallace and Darwin's 1858 papers to the Linnaean Society, Powell believed that natural selection made the processes of creation at last seem rational. Powell commented on 'The Origin' in *Essays and Reviews* which included the significant presence of Frederick Temple later to become the Archbishop of Canterbury, *'Just a similar scepticism has been evinced by nearly all the first physiologists of the day, who have joined in rejecting the development theories of Lamarck and the Vestiges; and while they have strenuously maintained successive creations, have denied and denounced the alleged production of organic life by Messrs. Crosse and Weekes, and stoutly maintained the impossibility of spontaneous generation, on the alleged ground of contradiction to experience. Yet it is now acknowledged under the high sanction of the name of Owen (British Association Address 1858), that 'creation' is only another name for our ignorance of the mode of production; and it has been the unanswered and unanswerable argument of another reasoner that new species must have originated either out of their inorganic elements, or out of previously organized forms; either development or spontaneous generation must be true: while a work has now appeared by a naturalist of the most acknowledged authority, Mr. Darwin's masterly volume on The Origin of Species by the law of 'natural selection' – which now substantiates on undeniable grounds the very principle so long denounced by the first naturalist – the origination of new species by natural causes: a work which must soon bring about an entire revolution of opinion in favour of the grand principle of the **self-evolving powers of***

---

[224] See comments by Paul N. Pearson *'In Retrospect'*, Nature, V.425. 2003 p 665

[225] The debate between uniformists and catastrophists has continued in geology. It seems that most geologists now include dimensions of both positions.

nature.'[226]In the 'Historical Sketches,' Darwin commented *'The 'Philosophy of Creation' has been treated in a masterly manner by the Rev. Baden Powell, in his 'Essays on the Unity of Worlds,' 1855. Nothing can be more striking than the manner in which he shows that the introduction of new species is "a **regular, not a casual phenomenon**," or, as Sir John Herschel expresses it, "a **natural** in contradistinction to a **miraculous** process."* [227] It is interesting here that Darwin chose to use **Herschel's** rather than his work to make this point. Some theologians

in this ongoing and developing debate began to suggest that that the idea of separate creation was not in fact Biblical at all. One such, was the famous and prolific writer, **Thomas Henry Huxley** (1825-1895), who was to champion Darwin's position pubically, despite his private concerns, and to speak in the famous 1860 Oxford debate as 'Darwin's bulldog!'

Huxley. Print by Locke and Whitfield 1880.P-D, Copy right expired

In fact this nearly didn't happen at all, as Huxley was planning to leave Oxford the previous day, but was persuaded by **Robert Chambers**, of Vestiges fame, to stay and join the debate with Wilberforce who had been coached by Richard Owen – **Hooker's** old enemy. Huxley helped to campaign for the inclusion of more science in the British education system, and fought against some aspects of religious tradition and power. He coined the word 'agnostic' to describe his own position. Despite having little or no formal education, his work on comparative anatomy, firstly on invertebrates and vertebrates, helped develop and clarify differences between different groups. He is credited with the conclusion, in 1868, that birds evolved from small, carnivorous dinosaurs. Owen obstinately refused to accept this kind of adaption or transition as part of his opposition to Darwin.[228] One of the earliest known (160 million years ago) bird dinosaurs, *Anchiornis Huxleyi (Chen Yang museum)* was named after him.[229] Both legs and arms had 'simple' feathers, which helped it leap rather than being able to fly. The discovery in 1861, of a single feather of *Archaeopteryx*[230] in the lithographic limestone of Bavaria[231] helped convince Huxley of his position. Darwin commented in the Historical Sketches, *'In June, 1859, Professor Huxley gave a lecture before the Royal Institution on the 'Persistent Types of Animal life.' Referring to such cases, he remarks, "It is difficult*

---

[226] Essays and Reviews 1860 My emphasis in bold

[227] Page XXI ibid

[228] and despite or perhaps because of his own work on dinosaurs

[229] The first evidence of wings may be related to the fuzz over the skin of Tyrannosaurus Rex. Early wings had dinosaur claws that evolved on the skeleton. It is probable they evolved from both the flapping function which helped animals claw their way up trees and the gliding function which enabled them to fly to lower branches, but not flap their wings in the process.

[230] 12 fossils have been found. 'Pteryx' (πτέρυξ) means 'wing or feather so making the evolutionary point. In 1985 Fred Hoyle and others wrote a paper claiming that the feathers on the London and Berlin specimens were forged!

[231] The first evidence of wings may be related to the fuzz over the skin of Tyrannosaurus Rex. Early wings had dinosaur claws that evolved on the skeleton. It is probable that wings evolved from both the flapping function which helped animals claw their way up trees and the gliding function which enabled them to fly to lower branches, but not flap their wings in the process.

*to comprehend the meaning of such facts as these, if we suppose that each species of animal and plant, or each great type of organisation, was formed and placed upon the surface of the globe at long intervals by **a distinct act of creative power**; and it is well to recollect that such an assumption is as unsupported by tradition or revelation as it is opposed to the general analogy of nature. If, on the other hand, we view 'Persistent Types' in relation to that hypothesis which supposes the species living at any time to be the result of the **gradual modification** of pre-existing species—a hypothesis which, though unproven, and sadly damaged by some of its supporters, is yet the only one to which physiology lends any countenance; their existence would seem to show that the amount of modification which living beings have undergone during geological time is but very small in relation to the whole series of changes which they have suffered."*[232] At the end of the *'Historical Sketches'* he uses for the first time in *The Origin* the word *'doctrine'* - albeit of Hooker rather than himself - as applied to the modification of species *'In December, 1859, Dr. Hooker published his 'Introduction to the Australian Flora.' In the first part of this great work he admits the truth of the descent and modification of species, and supports this **doctrine** by many original observations.'*[233] The language of 'doctrine' reminds us of many theological disputes through the ages, and at the heart of the conflict between Owen and Darwin there was indeed a 'doctrinal' dispute of sorts, however indirectly articulated. It was one that went very deep, and often became vitriolic and hurtful on both sides. If anyone doubts the depth of division generated by *The Origin*, they need look no further than this dispute, which was probably much more serious in the scientific community than even the Oxford debate itself.

The Doctrinal Dispute between Darwin and Owen

Richard Owen (1804-1892), as we have seen, played a significant role in the story of Darwin's life and work.

Owen in the 1870s. Author unknown. P-D Copyright expired

If we step back for a moment, before looking at some of the detail, we might see this dispute as part of its own *'pattern paradox.'* Within its dialectical nature, and only from an historical perspective, we might look upon this as being, in itself, a process of evolution which brought new intellectual adaptions in the face of their competition and struggle. At the time, of course, it was very painful for both of them. Of all the portraits or photos I have used in this 'picture gallery' of famous scientists, I admit to seeing in the face of Owen a vivid expression of hurt and bitterness, or similar negative emotions. So, it is worth saying something about the deep seated conflict that developed between them, before leaving this

---

[232] My emphases in bold
[233] 6th Edition John Murray Page XXII

section on the 'Historical Sketches.' It is probable that Owen's own public and vitriolic disputes with Huxley and with Hooker[234] were involved in this tension, as Owen saw them both as Darwin's chief supporters, disciples and champions. The conflict with Hooker was focused on the politics of museum collections, prestige and power. With Huxley, it was even more serious. As we shall see, it came to a head over their different interpretation of the comparisons between ape and human brains, but it was clearly more fundamental even than that. It is surprising that Darwin himself didn't say even more about Owen and his position in *The Origin* or even speak about it more publically. Perhaps Emma was a restraining influence or he was aware of the risks of more open conflict. Owen had been developing his own theories, influenced by the German physiologist, **Johannes Peter Muller** (1801-1858). According to Muller, all living matter had an organizing energy. This was a life force that directed growth processes and determined the life span of an individual species. In December 1838, Darwin was secretary of the Geological Society in London. He was still feeling his own way and noticed how Owen and his supporters ridiculed what became known as the *Lamarckian heresy* espoused by **Robert Grant** – one of Darwin's tutors. Darwin's relationship with Owen is complicated. When Owen became ill in 1841, Owen was one of his few visitors from the scientific community. Despite this, Darwin was very cautious, sensing how strongly Owen opposed any thought of the transmutation of species. In fact, Owen came to believe that species do change as a result of some kind of evolutionary process which had, he thought, six mechanisms - parthenogenesis, prolonged development, premature birth, congenital malformations, Lamarckian atrophy, Lamarckian hypertrophy and transmutation.[235] He considered the last the least likely. Some believe that Owen himself became hesitant about sharing these ideas as he watched the critics' treatment of Vestiges in 1844. In Owen's own 1849 work *Nature of the Limbs,* he suggested that there must be natural laws explaining how humans had evolved from fish. The media immediately criticised him for denying that humans were created by God.[236] Owen remained open to the possibility that humans had evolved from intermediate, extinct and therefore unknowable species in a process of *'ordained continuous becoming,'* but in 1854, gave a talk to the British Association dismissing the idea that 'bestial apes' could be transmuted into humans. Owen was nervous about the political implications of Darwin. In 1861, Karl Marx had written *'Darwin's work is most important and suits my purpose in that it provides a basis in natural science for the historical class*

---

[234] In 1871, Owen tried to prevent further Government funding for Hooker's botanical collection at Kew probably so that he could bring to the British Museum. Darwin was horrified and allegedly commented that he would cherish his hatred and contempt for Owen for the rest of his life. See Joseph Dalton Hooker. W.B. Turrill 1963. Nelson, London. P 90.

[235] Nicolaas Rupke, *Richard Owen: Victorian Naturalist.* New Haven: Yale University Press.1994, p. 226

[236] See Evellen Richards, *A Question of Property Rights: Richard Owen's Evolutionism Reassessed, British Journal of the History of Science*, 20: 129–171.1987

struggle.' [237] Owen was about to become President of the Royal Association, and something of a famous establishment figure, not least for his work on Dinosaurs,[238] which was of significance in the Great Exhibition of 1851. Clearly, with all his faults, he was still a very influential and dominant figure in Victorian society who gave biology lessons to the Queen's children and was Professor of Comparative Anatomy and Physiology (1836). He was certainly capable of becoming in 1856 the first Superintendent of the Britism Museum's Natural History sections and then the founding visionary behind the huge Natural History Museum in South Kensington (1873-1880). However, he chose this moment, in the dispute with Darwin, to announce that he could show that the human brain had structures which apes lacked. To him, this prooved that humans were a separate sub-class or species. Darwin disagreed and Huxley gave his 1858 lecture to the Royal Institution to deny Owen's structural claim for human uniqueness. During and following the publishing of the *Origin*, Huxley's arguments with Owen continued. In April 1860, Owen wrote an anonymous review of the *Origin* for the *Edinburgh Review.* He wrote to decry what he saw as Darwin's caricature of the creationist position and in particular Owen's own *'axiom of the continuous operation of the ordained becoming of living things'*. He included Hooker and Huxley in his criticism of Darwin and described the *Origin* as an *'abuse of science... to which a neighbouring nation, some seventy years since, owed its temporary degradation.'* With this clever phrase, he was implicating Darwin and his supporters in the horrors of the French Revolution, which is ironic as Darwin had been influenced by at least one French scientist (Lavoisier) who had been executed by the Revolution. Darwin and Huxley were clearly hurt by this article. Darwin described it as *'spiteful, extremely malignant, clever, and... damaging.'* Darwin's friends tried to reassure him that many now saw Owen as being mad with envy[239] because of the popularity of the *Origin*, but it was clearly painful for Darwin to feel Owen's intense hatred. Again this shows how complicated and duplicitous the relationship was, given Owen's private remarks and letters to Darwin. Apparently Owen was among the first to respond when he received his copy of the *Origin* directly from Darwin. In his discussions with Darwin, Owen claimed that his own work had shown that 'existing influences' were partly responsible for the *'ordained birth of species,'*- an expression he kept using - and that the *Origin* offered the best explanation for the manner of the formation of species. But, he retained the strongest concerns about the implications of transmutation for the relationship between humans and 'the beasts.' Darwin tried to enter into the theological world

---

[237] Letter to Ferdinand Lassalle, 1861

[238] His own research should not be dismissed lightly, particularly his work on vertebrates. See his *Comparative Anatomy and Physiology of Vertebrates* (3 vols. London 1866–1868). This was in the same league and style of vertebrate paleontology as George Cuvier's *Leçons d'anatomie comparée.*

[239] See *The Life and Letters of Charles Darwin: Including an Autobiographical Chapter* (7th Edition) London Francis Darwin 1887. Page 149

view of Owen and admitted there was, in nature, a 'designed' law. Owen may have interpreted this, at least for a moment, as a shared belief in a Creator. The problem was that Owen never showed in public any sense of shared agreement between them. He tried to discredit Huxley by portraying him as an *'advocate of man's origins from a transmuted ape.'* Owen was using this as a weapon against Darwin and gave several talks on the subject. In 1862, Huxley arranged various public dissections of the ape's brain anatomy to demonstrate the presence of the very structural parts alleged by Owen to be missing. This was tantamount to calling Owen a lier and discrediting his reputation. It also undermined one of Owen's main objections to the Origin. Huxley began to win the argument over brain size and structure, persuading many other scientists that they should ignore Owen and take Darwin more seriously. This angered and frustrated Owen even more. In a paper directly criticizing Owen, Huxley wrote, *'if we place A, the European brain, B, the Bosjesman brain, and C, the orang brain, in a series, the differences between A and B, so far as they have been ascertained, are of the same nature as the chief of those between B and C.'*[240] Owen continued to argue that the brains of all human races were of similar size, but in any case patently superior to apes and gorillas. The ground had shifted from size and structure to intellectual capacity, and the debate continued to rage. It must have been embarrassing for all uninvolved observers. If Huxley entered the room, Owen would often leave. When Huxley joined the Zoological Society Council, in 1861, Owen left. In 1862, Huxley opposed Owen's election to the Royal Society Council, on the grounds of his *'wilful and deliberate falsehood.'* It has to be admitted that although Owen was an unpopular person and could be vitriolically critical, he was in a difficult position as the Head of the Natural History Collections at the British Museum. In that role, he was the recipient of many complaints against the *Origin*. When arguing before a Parliamentary committee for a new, Natural History Museum he used Darwin's popularity, or infamy, as evidence of the need for more space – to show all the relevant varieties of species mentioned by Darwin. The new building in London was designed on the assumptions of the time - that such a place would demonstrate natural theology as well as nature itself. It was a kind of cathedral in which God could be proclaimed through the study of nature, and Owen was its

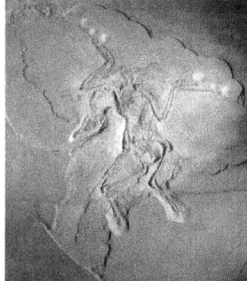

first Director. In January 1863, Owen bought the Archaeopteryx fossil for the British Museum.

There are many specimens, this from the Museum für Naturkunde in Berlin. Photo by H. Raab 2009. Creative Commons Attribution–Share Alike 3.0 Unported Licence.

It was a crucial bit of evidence in the Darwin debate, supporting his prediction of an intermediary link between the dinosaurs for which Owen was himself famous and birds

---

[240] Huxley, 'On the Zoological Relations of Man with the Lower Animals', *Natural History Review* 1: 67–84.1861,

- a proto-bird with unfused, wing fingers. Owen continued to insist it was just a bird. Owen's statue was erected in the main hall of the Natural History Museum and stood there until 2009, when it was replaced by a statue of Darwin as part of the special exhibition on Darwin – named 'The Big Idea.'[241] The history of the Museum is almost as fascinating as its 7 million specimens, and it was a tangible expression of the issues in the Big Debate behind the Big Idea.

I have taken time to look at some of the quotations, used or made by Darwin in his 'Historical Sketches' to illustrate the way he helped create a new approach to the assumptions of a *separate creation*; to build on what had gone before and to address or challenge the concerns of that belief's supporters. In 2014, it is possible to still find such ideas. They may not be openly articulated and many of their followers may not have fully recognised the significance of their own positions. I often wonder whether Darwin, writing the notes in his historical sketch, might have wondered how long it would take for such ideas to disappear in the light of new ideas about species development and change. In his work on barnacles, he continued to ask questions about how they evolved and wondered whether natural selection was a good enough theory to explain the principle of divergence itself. He wrote of modified offspring, diversified in the *economy*[242] of nature. The more divergence there was, the easier to adapt. The Glaswegian Professor of Moral Philosophy, **Adam Smith** (1723-1790) had already written on the natural division of labour through specialization.

Adam Smith. Etching created by Cadell and Davies (1811), John Horsburgh (1828) or R.C. Bell (1872).P-D. Copyright expired

In his great work '*An enquiry into the Nature and Causes of The Wealth of Nations*'[243] in 1776 – ironically the year of American Independence – he had written of the 'invisible hand' which enabled the separate contribution of people in different trades to contribute to the wider common good, without any direct or controlling plan to make this so – '*every individual necessarily labours to render the annual revenue of the society as great as he can. He generally, indeed, neither intends to promote the public interest, nor knows how much he is promoting it... he intends only his own security; and by directing that industry in such a manner as its produce may be of the greatest value, he intends only his own gain, and he is in this, as in many other eases, led by an invisible hand to promote an end which was no part of his intention. Nor is it always the worse for the society that it was no part of it. By pursuing his own interest he frequently promotes that of the society more effectually than when he really intends to promote it.. It is not from the*

---

[241] In 1909 fifty years after The Origin, the National history museum celebrated his work with exhibits, but no reference to the theory of natural selection – it was still too controversial and a scientific consensus had shifted against the theory.

[242] A term used by Erasmus Darwin as we have seen

[243] His earlier, important work *The Theory of Moral Sentiments* was written 100 years before *The Origin*.

benevolence of the butcher, the brewer, or the baker, that we expect our dinner, but from their regard to their own interest. We address ourselves, not to their humanity but to their self-love, and never talk to them of our own necessities but of their advantages.'[244] This was a time when the Dutch naturalist, **Bernand de Mandeville's** influential paper *The Fable of the Bees*[245] on private vice and public benefit was being discussed across Europe. It was seen as offensive and dangerous in its criticism of British society. **Adam Smith** disagreed with much of what it said, but introduced his own significant idea of 'natural' benefit from individual self-interest and specialised function. I spent some time in my professional life studying the implications of Smith's two great works in the context of the religious and political assumptions of his times. He contributed something of interest to our story of how the ideas of previous decades came to influence Darwin. For Smith, there was something *natural* about a process in society whereby individuals contributed to something greater. By functioning as individuals, they were connected to the whole of which they were unconsciously a part. While Smith was emphasizing the economics of this natural process, in the American colonies, they were fighting for its revolutionary politics. On the 4th July 1776, the same date as his more famous book, the Continental Congress announced that thirteen American Colonies, then at war with Great Britain, were *independent* States forming a *Union* of States of their own that would become a new nation, thanks to the work amongst others of John Adams and of course the drafting of Thomas Jefferson and the vote on July 2nd. The greater the recognition of the individual parts and their role and significance within a *connected* whole, the greater their (species) *development*. The greater the specialisation, the greater (common) wealth was created. Darwin realised that more life could be sustained on a plot of land if organisms diverged (specialised) with their different use of limited space. If all were of the same kind, using the same space in the same way for the same purpose, they couldn't survive. He experimented with up to 20 different species in a small plot in his grounds. This was a time when many were thinking through the symbiotic relationship between different individuals, species and varieties; between *diversity* and *connectivity*; between *independence* and *union*; between divergence and natural selection. If there was divergence there was adaption and in the adaption process there was natural rather than divinely intervened selection and change, however complicated and painful its processes.

Karl Marx (1818-1883), another polymath, commented on Darwin's discovery of Malthusian struggle for existence and applied it to his own class struggle theories. *Marx 1875. Author unknown. International Institute of Social History, Amsterdam. P-D. Copyright expired.*

---

[244] *The Glasgow edition*, pp. 26–7.
[245] 1724. London. Tonson

He was more of a social determinist than Darwin, in his constructs of the inevitability of class struggle (mutability) based on Hegelian dialectics. Darwin may not have agreed with the assumption of automatic linear progression through species adaption, but the thought patterns were close. If something was true for nature why not for human societies? Darwin had stopped well short of discussing detailed, human, social behaviours, even though much of this is implicit in his later work. Marx observed the movement through different stages of society from feudal to capitalist industrialisation, mechanisation and competition to socialism and finally communism as the functioning of a law of class struggle. Each stage would lead to the next, not because of violent revolution, but through a process of social evolution and diversification – the seeds of change being sown in each phase. In Marx, there was an inevitability about the movement to mutate and adapt from within. No forced revolution was needed to bring this about. Marx had learnt this dialectical materialism from the neo – Hegelians, particularly **Ludwig Feuerbach** (1804-1872)[246]. If only the Bolsheviks had taken this part of Marxism more seriously, the world of the twentieth century might have been a very different place. Darwin saw differentiation taking place over much longer periods of time and natural selection working in many different directions, rather than one predictable and straight, linear even dialectical line. He hesitated to discuss any theories of cultural differentiation amongst humans as a form of differentiation in evolutionary change. All of this and the sciences of neurology and evolutionary psychology were to come much later. Having spent so long with barnacles, Darwin then chose pigeons, because of their popularity with Victorian readers and the common experience of domestic cross breeding. He continued to ask the question – how do all species relate to their original ancestor. He saw that nature 'naturally selected' over longer periods of time in the way that pigeon fanciers domestically bred to create new strains of bird.

Wilberforce and the Oxford Debate

In the famous 30 June 1860 Oxford debate, Darwin himself wasn't present. Huxley defended the corner of evolution with Bishop Samuel Wilberforce speaking for the opposition. The cartoons of the time made much of the 'ape like' inheritance of human beings. Bishop Wilberforce's earlier review of the *Origin* had been condescendingly critical. The young Huxley responded with an opportunistic attack on Wilberforce and the Church. The supporters of Darwin used this event as a paradigm of success over the forces of reaction. Wilberforce attacked Huxley with words to the effect – *'was he descended from monkeys on his father's or mother's side?'* and Huxley's reply was that *'he'd rather be descended from apes than a cultivated bigot like the Bishop'*. However, the debate wasn't typical of the wider

---

[246] See *'From Hegel to Nietzsche, the evolution in 19th century thought.'* Karl Lowith 1964. Holt Rhinehart and Winston.Chapter 11 p 53 ff

discussion in the country, but some church and establishment opponents continued in their attacks. Fourteen years after the debate, Wilberforce was still taking the issue of natural selection very seriously. In *Essays Contributed to the Quarterly Review*, Wilberforce wrote, *'First, then, he (Darwin) not obscurely declares that he applies his scheme of the action of the principle of natural selection to man himself, as well as to the animals around him. Now, we must say at once, and openly, that such a notion is absolutely incompatible not only with single expressions in the word of God on that subject of natural science with which it is not immediately concerned, but, which in our judgment is of far more importance, with the whole representation of that moral and spiritual condition of man which is its proper subject matter. Man's derived supremacy over the earth; man's power of articulate speech; man's gift of reason; man's free will and responsibility; man's fall and man's redemption; the incarnation of the Eternal Son; the indwelling of the Eternal Spirit---all are equally and utterly irreconcilable with the degrading notion of the brute origin of him who was created in the image of God, and redeemed by the Eternal Son assuming to himself His nature. Equally inconsistent, too, not with any passing expressions, but with the whole scheme of God's dealings with man as recorded in His word, is Mr. Darwin's daring notion of man's further development into some unknown extent of powers and shape, and size, through natural selection acting through that long vista of ages which He casts mistily over the earth upon the most favored individuals of His species....Nor can we doubt, secondly, that this view, which thus contradicts the revealed relation of creation to its Creator, is equally inconsistent with the fullness of His glory. It is, in truth, an ingenious theory for diffusing throughout creation the working and so the personality of the Creator. And thus, however unconsciously to him who holds them, such views really tend inevitably to banish from the mind most of the peculiar attributes of the Almighty.'*

Wallace

But he was not alone in his thinking. In 1858, the explorer **Alfred Russell Wallace** (1823-1913)[247] published his *'Malay Archipelago'* and humbly, perhaps innocently, dedicated it to Darwin with whom he had been corresponding.

Alfred Wallace, Photo, Singapore, 1862. Unknown author.P-D. Copyright expired

In Indonesia, he had been wrestling with similar ideas. Struck down by malaria and stuggling financially, it was perhaps when faced with his own mortality that he asked himself why some die and others live. In his struggle to adapt to his own hostile environment he concluded that animals better able to adapt would survive. Wallace was extremely aware of the importance of different environments, particularly when

---

[247] Born in Usk, Monmouthshire, Wales.

humans interfered with the balance of nature. In this, he was decades ahead of his time – and warned of the human impact on forests and the destruction of species.[248] It is worth pausing in our story of Wallace to note how both he and Darwin would later influence the study of the environment in the light of natural selection. Darwin had developed interdependent linkages in nature and their implications for survival. Later, aged 72, he published 'The Formation of vegetable mould by actions of worms with observations on habits '(1871) which outsold even the Origin during his lifetime.[249] He came to believe that a significant loss of lowly organisms could make a large part of the planet hostile to human survival. The environment in Britain was changing with the affects of a steam powered, industrial revolution. By the 1980s, **William Morris** (1834-1896) was expressing his concern, appalled by the working and living conditions of industrial classes. There was a growing awareness of the need to preserve the beauty of the countryside against the 'unutterable grime threatening nature.' The English entomologist, **Jame W Tutt** (1858-1911), a supporter of Darwin, studied the evolution of the peppered moth in the face of changing environmental conditions. Its pale coloured wings camouflaged it against the pale lichen on trees. When the lichen was darkened in Industrial areas, the moth evolved darker colour wings and increased in number. The pale varieties didn't stand out and were increasingly spotted and eaten by birds. Humanity was directly affecting natural selection. By 1896, when he published his work on British moths, 86% of peppered moths in the north now had black wings.[250]

There are many practical examples of the challenging implications of evolutionary adaption to changing environments. The ecologist, **Charles Sutherland Elton** (1900-1991) was asked for help by the Government, during WW11, as huge food storage depots were being over run by mice. [251] There were many examples of the scale of human impact on animal environments and he issued early warnings that the post war increase in mechanised farming and pesticides were causing huge dislocations in nature. Meanwhile in the US by 1956, DDT was being used indiscriminately on farming land. Bee colonies were lost, horses killed, and local wildlife devastated. The biologist **Rachel Carson** (1907-1964) was told by a friend that the birds in her garden had died. Carson discovered that pesticides were killing animals beneficial to nature and affecting the food chain. She foresaw an ecological catastrophe[252] and in 1962 published '**Silent spring**' one of most influential, ecological books of the century. She too used Darwin to explain the

---

[248] Haeckel, of course, was inspired by Darwin to define a whole new science of ecology.

[249] 'humble worm played more significant impact on the history of the world than usually supposed.'

[250] Following the 1956 clean air act the darkening of tree bark and lichen was reduced and the moth had to lighten its wings again or become a new target for birds. See later reference to Kettlewell.

[251] He had to set thousands of traps

[252] For her, personally, there was some urgency in her work, as she had breast cancer.

negative effects of DDT. Pests were adjusting to pesticides, evolving immune 'super races,' which forced the production of ever more deadly pesticides. Government scientists accused her of getting the facts wrong, and the attacks on her compared with the earlier criticism of Darwin. **President Kennedy** supported her and finally the Government confirmed her findings. Again Darwin's interconnections in nature were confirmed and her book became an inspiration to environmental movements. This was the age of the space race and James Lovelock was working for NASA, studying the possibility of life on Mars. Given how hostile that planet's environment seemed, he began to wonder why the conditions on Earth were so right for sustaining life. His answer brought Darwin's ideas into the space age. He developed his theory that our whole planet was a self-regulating organism, a global eco-system. **William Golding** (1911-1993) of *Lord of the Flies* fame suggested 'Geia' as name for this. Earth herself was a living thing – an idea that fitted with the values of the sixties.

Wallace wrote of the tendency of varieties to differ in the same type, based on his studies in the Indonesian archipelago. He too had been affected by Malthus. Because he was in awe of Darwin, a member of the Royal Society, he had been hesitant to publish straight away. Wallace's life shows an extraordinary mixture of interests from spiritualism (*'On miracles and modern spiritualism'* in 1875) to politics (*'The revolt of democracy'* in 1913). His main work on natural selection was published in 1889, and he named it with characteristic humility 'Darwinism.' He had travelled much further than Darwin and many of his contemporaries – at least 14,000 miles alone within the Malay archipelago collecting 125,600 specimens. He had produced the Wallace geographical 'line' dividing the appearance of certain species in what he called the Australian and the Indian zones -arguably one of the first major studies of bio-geography. His observations of variation within species were enhanced by the commercial self interest of his own poverty, at least in relation to Darwin's wealth and support. He had negotiated a deal with **Samuel Stevens** to provide specimens of new species to fuel the new demands for such 'trade,' back in the museums and universities of Britain. Patterns of species variation appeared in these collections. In 1852, he published *'On the monkeys of the Amazon'* describing how different species of monkey were localised on different sides of the Amazon basin. He asked himself the crucial question - If God had created all species from the beginning and placed them in their appropriate locations and environments, why hadn't God put these monkeys on both sides of a given river? Three years later in Borneo, Wallace reflected on **Swainson's** *'Treatise on the geography and classification of animals,'* **Humboldt's** *'Travels,'* Darwin's *'Beagle Journal'* and of course the *''Vestiges' of the Natural History of Creation'* (1844) Why are hummingbirds native only to the Americas? Why do hornbills locate in the same types of areas as Toucans, but only in tropical Africa, Asia and the Eastern Islands? The implication of these facts had never been fully appreciated before. Wallace saw them as indications of the way species had come into existence, and it led him to design an early draft of a possible family tree of species development and mutation. Both Darwin and Wallace had studied the raw data of species variation. While Darwin had concentrated for many years on only a few species e.g. barnacles and pigeons, Wallace had reflected across a much wider experience of observation, reading, species diversity and natural environment. For him, it was also a question of moving on from what was commonly known in his times, and before, as 'natural theology' turning it into just 'nature' and then nature's natural selection. The church had interpreted natural theology in such a way as to imply a creationist position in broad terms – with each species having a *divine* design and origin, rather than a *natural* one. The idea of nature could now stand alone, separate from any theological edifice. It is difficult for people in the twenty first century to appreciate how large a shift this was and how difficult it must have been for many to accept its implications. Nature could now be studied in its own right, with natural processes as its only context. Of course, there were many theologians who came to accept the idea and find their own ways of fitting it into a belief in a Creator. The widespread existence of creationism in our own times reminds us how hard it was to let go., I am offering my own way of putting natural selection centre stage with all its implications, but setting it within a theology of nature that sustains belief in a Creator alongside all that Wallace and Darwin and so many others have taught us.

Wallace combined geographical and geological (very much influenced by **Charles Lyell's** work on the fossil record) evidence with the prompts from Malthus on the relationship between population growth and food and habitat limits. Reflecting on the evidence that most offspring cannot survive, he commented *'Vaguely thinking over the enormous and constant destruction which this implied, it occurred to me to ask the question, why do some die and some live? An antelope with shorter or weaker legs must necessarily suffer more from the attacks of the feline carnivore.'* The idea of nature selecting the fittest, or as Darwin insisted,the most adaptable

now had its own momentum.

Painting of Wallace by J W Beaufort Installed in January 2013 by Bill Bailey next to the statue of Darwin in the Grand Hall of the Natural History Museum

Darwin was distraught by the letter from Wallace. Hooker and Lyle came up with their own solution - to organize the presentation of separate papers by Darwin and Wallace to the Linneum society, where, in fact, little interest was shown in either. Wallace meanwhile was stuck on the coast of New Guinea, suffering hunger, fever and wet weather. Darwin resolved to publish quickly and the *Origin* appeared in November of 1859, with only two brief references to the origin of humans. Darwin sent a courtesy copy to Wallace who allegedly read it several times. He commented *'it is the "Principia" of Natural History. Mr Darwin has given the world a new science and his name should in my opinion stand above that of every philosopher of ancient or modern times. The force of admiration can no further go!'* Wallace didn't allow any personal jealousy or lost reputation to undermine the significance of either Darwin's work or his respect for him – quite the opposite. He was not driven by a ego or fame, but a delight in the natural world in situations of personal discomfort, disease and domestic difficulty. He showed no bitterness towards Darwin, whatever Darwin felt the other way round! Wallace continued to send papers to the Linnean society journal on subjects like *'the zoological geography of the Malay Archipelago.'* He continued to work in ways that contributed to the Darwinian revolution. The fact that he did so at such a distance from the debates in England perhaps made his support less significant to those who opposed Darwin. Many later scientists[253] acknowledged their debt to Wallace and this can be overlooked in the dominance of Darwin's name and work in the popular imagination. To commemorate the centenary of Wallace's birth, the 1923 portrait by J. W. Beaufort has been re-erected in the Natural History Museum in London, on the wall a few steps higher than the statue of Darwin. [254]

If Wallace was a distant but great supporter, Sedgwick was painfully disappointed, seeing the *Origin* as claiming a break between God and nature. Like Emma, he saw the fabric of faith and 'salvation' tottering before the impact of Darwin's work. Darwin was clearly all too painfully aware of the sensitivity of his work and did his best to acknowledge what was at stake. He went out of his way to affirm the *enobling* implications if creationism is removed by the new science of natural selection – *'Authors of the highest eminence seem to be fully satisfied with the view that each species has been **independently created.** To my mind it accords better with what we know of the laws impressed on matter by the Creator, that the production and extinction of the past and present inhabitants of the world should have been due to **secondary cause**s, like those determining the birth and death of the individual. When I view all beings not as **special creations**, but as the **lineal descendants** of some few beings which lived long before the first bed of the Silurian system was deposited, they seem to me to become **ennobled**. Judging from the past, we may safely **infer** that not one living species will transmit its unaltered likeness to a distant futurity. And of the species now living very few will transmit progeny of any kind to a far distant futurity; for the manner in which all organic beings are grouped, shows that the greater number of species*

---

[253] **Alfred Wegener's** continental drift theory in the early twentieth century is a good example.

[254] Mainly due to the work of Bill Bailey, Patron of the Wallace memorial fund and much encouraged by David Attenborough. **Dr George Beccaloni**, Director of the Wallace Correspondence Project and a curator at the Museum says, *'Wallace's correspondence (over 4000 pieces) is second only to Darwin's in terms of its importance in the history of biology. It is the major primary source of information about his life and work.'*

of each genus, and all the species of many genera, have left no descendants, but have become utterly extinct.'[255]

Evolutionary Teleology

We sense the Darwin's anxiety about the transition he was suggesting and many parts of the *Origin* are expressed with a certain tentativeness. There is little that is arrogantly triumphalistic in his style. He chooses to use phrases such as *'we may safely infer'* as a gentle entry into a new insight, rather than a more bombastic statement of certainty. There is humility in his style which speaks of a finely attuned, spiritual sensitivity and honesty about the intellectual challenge. This honesty is mixed with caution in the way he unpacks his theory in the *Origin*, beginning with Chapter 1's appeal to what is already common practice in breeding techniques.[256] It also shows a developed imagination to anticipate reactions to the dilemmas and questions raised by his Big Idea, as he builds his argument for struggle and natural selection from Chapter 1V onwards. His inclusion of what he calls 'difficulties' may be both evidence of his honesty and humility and also his rhetorical ability to engage critical readers – both specialists and disbelievers in the general public. The shadow of reactions to the *Vestiges* must have constantly crossed his mind and his desk, not least that of Owen's. By admitting *'difficulties,'* he was clearly acknowledging the unresolved *dilemmas*, not least in the geological record, and those he couldn't have uncovered in the genetic 'mechanism' of natural selection. *Difficulty* also implies its own potential. If there is a difficulty, it opens the door to sympathetic engagement with the theory as well as future research. By admitting *difficulty*, Darwin is drawing back from any final, fixed position without undermining his own confidence in the theory – frequently he seems to draw attention to a difficulty, even *grave* ones, only to dismiss them as being 'apparent,' rather than 'real.' The way he selects and then deals with some *difficulties* indicates a clever plan as well as intellectual rigour. He sometimes turns difficulties and ignorance into proof, or at least the next stage of his argument for natural selection *'We are far too ignorant in almost every case to be enabled to assert that any part or organ is so unimportant for the welfare of a species that modifications in its structure could not have been slowly accumulated by means of natural selection.'[257]* It is as if the book contains its own teleology, as it moves from hesitancy to confidence in the final chapter. In the process of building to that state, he is crearly reaching out empathetically to those who are, like him, still struggling with the *dilemmas*. He is giving permission to explore futher the difficult questions, albeit taking the

---

[255] *Complete works of Charles Darwin online*, editor John Van Whye, p 576.
[256] Though he knew breeders themselves wouldn't automatically accept natural selection theory – see his comments in Ist Edition p 88-89
[257] *Origin* Ist edition p 231 ff

theory seriously, rather than dismissing it entirely, as he knew so many would, despite the fact that it was an idea whose time had come, given the widespread debate triggered by the Vestiges, Wallace and the ideas of Lamarck and others. How could he move the debate forward without adopting the *certainty* of previous (metaphysical) positions, especially when he wanted to introduce a note of *uncertainty* about them? He was caught in a genuine dilemma – if he didn't show his *difficulties* or those of various critics, he would rightly be accused of the very certainty he was challenging in others. He was positioning himself in the *pattern paradox* of *knowing* and not *knowing* that would commend his theory to any open minded reader.It would be implication suggest that the only way forward is to proceed by questioning previous knowledge. By admitting certain difficulties, he was also trying to distract readers with closed minds from the real import of the theory - beyond any technical details or difficulties - with its implications for current beliefs and assumptions about *how things really came to be as they are* and how and why they had been created like this. Compared to those difficult teleological questions, many of the difficulties he admits have less radical implications.

One significant example of difficulty is relevant to my list of *pattern paradoxes*. It is the paradox of *perfection* and *imperfection,* and it arose most acutely in his understanding of the evolution of the eye. It is worth pausing to look at this in some detail. Put very briefly the dilemma was this – if evolution adapts or improves organs by natural selection, how do we explain the eye's ability to see. Does this mean that, in its many stages of evolution over millennia, its sight was imperfect, or did the eye appear as perfectly adapted to its task at a much earlier stage than natural selection normally requires of its internal processes? Underneath this dilemma was the issue of teleology – the question of (final) purpose and direction in the internal processes of change. In one letter to **Asa Gray** (1810-1888) he wrote *'I remember well the time when the thought of the*

eye made me cold all over.'[258] He also saw the Cambrian explosion as a major challenge to natural selection.

Gray 1864, John Whipple, University South Carolina, P-D, Copyright expired.

We now know from the Cambrian period discoveries[259] that, in Trilobite eyes, there were already many lenses well before more complex lenses evolved the ability to see shape and depth. Prior to an image appearing on the retina, early eyes could only see light and dark. Improved sight benefitted both predators and prey. In the Cambrian seas, the explosion of species development was probably triggered by the much earlier end of 'snowball earth,' increased levels of

[258] Life and Letters 11, 296. April 3 1860
[259] Not least from the Burgess Shale. See also the work of Stephen Gould

oxygen,[260] calcium build up in oceans[261], ozone layer protection, changes in the genome, increased Plankton and changes to the eye.[262]

The Eye

Clearly there are many different types of eyes, in different species, and they have evolved to do different things.[263] Given our present knowledge, we take much for granted, but it is worth pausing, briefly, to remind ourselves of how its functioning had been understood, scientifically and theologically at certain points in history. As we shall see, there is a connection here with early cosmologies (LE1) as well as theology (LE2). From the Greeks and the Egyptians on, people had discussed how light either came from (extramission) or went into the eye (intromission) – which was seen by the ancients as the *window into the soul*. Plato believed that light came from the 'cone' of the eye, as if it were comparable to the rays of the sun. The Greek atomists began to consider the small parts of the structure of the eye. They believed light came into the eye, but puzzled over how, for example, something as large as a mountain could be seen by something as small as the structure of the eye. By contrast with Plato, Aristotle believed in intromission, but that some 'crystalise lens' reacted with incoming light in a strange way. This was influential for the science of the next centuries. The Islamic world, influenced by the Persians and their translations from the Greek texts, began to experiment with light's rays, well before Newton. The polymath and philosopher, Abū ʿAlī al-Ḥasan ibn al-Ḥasan ibn al-Haytham (965-1040), or **Al Haythan** for short, was born in Basra, but lived in Cairo for most of his life.[264]   He was a famous physicist and known as *Ptolemaeus Secundus*, as well as being the greatest, optical experimenter of his times, seeking to understand the correspondence between the inner and outside worlds of sight. The eye, with its symbolic importance, became a focus of special interest in medieval, Islamic medicine and philosophy, much influenced by Galen's writings. Many, if not the majority of Islamic scholars seemed to have supported the extramission theory, with the exception of **Avicenna**, who tended to follow Aristotle. It became common to think of a '*visible spirit*' moving from the inside of the brain through the eye into the outside world. Al Haythan had improved the Aristotelian theory and also speculated on the metaphysics on light.

---

[260] Due to photosynthesis

[261] Helping skeletal development

[262] See work of Andrew Parker but disputed by some because other pre Cambrian senses as effective under water.

[263] In the 1990s we discovered how the PAX 6 gene controls other genes in the eye and mutations. Human and fly eyes evolved independently but appearance of this gene proved they have common evolutionary histroy. PAX 6 is present in all eyes despite their differences. All eyes can trace their origin back to one simple creature living 1 billion years ago.

[264] Where the Caliph ordered him to build a dam to control the flooding of the Nile. He saw the task as impractical and pretended to be mad to escape the Caliph's anger. While under house arrest for ten years he wrote his *Book of Optics*

Roger Bacon (c. 1214–1294), was a Franciscan Friar who also placed great emphasis on experimentation. He tried to integrate **Al Haythan's** work and other theories, including extramission – the soul of the body projecting an *enobling force* outwards to see things as they really are. Both lived during the East West tensions of the different crusades, as did **Robert Grosseteste** (1175 –1253) who was probably a teacher of Bacon, and the latter acknowledged his debt to him. Grosseteste was a scientist and statesman and also a famous **Bishop of Lincoln.** He is credited with the founding of scientific thought in medieval Oxford, where he may, at one time, have been Chancellor of the University. Grosseteste wrote on astronomy, as well as the the metaphysics of light (*De Luce*). In *De Iride* (on rainbows) he wrote '*This part of optics, when well understood, shows us how we may make things a very long distance off appear as if placed very close, and large near things appear very small, and how we may make small things placed at a distance appear any size we want, so that it may be possible for us to read the smallest letters at incredible distances, or to count sand, or seed, or any sort of minute objects.*' He is credited as being the first of the Scholastics to understand the significance of Aristotle's way of working from the particular to a general law, and back again to predictions about particulars – for example using observations of the moon to prove a universal law about planetary motion. This was to influence the later work of **Galileo Galilei** (see LE1). The Belgian anatomist, **Andreas Vesalius** (1514 –1564) became a Professor at the University of Padua and published his influential, seven volume *De humani corporis fabrica* in 1543.[265] This included his work on the anatomy of the eye. He was building on the work of **Galen's** dissection of animal eyes and the theory that it was the retina that explained sight. Leonardo da Vinci's drawings of the eye and the ventricles of the brain[266] were also based on Galen, and have been proved entirely wrong. He initially supported the extramission theory but ten years later, in the 1490s changed his mind. The theological assumption was that as God created light, the eye was the most spiritual of organs, as well as being that most at risk from the evils of lust. Natural theology and Aristotelianism went together in these times. God had created the eye so we can see things as they really are. The wondrous construction of the eye witnessed to a divine creator. The German astronomer and astrologer, **Johannes Kepler** (1571-1630)[267] inherited this type of natural theology. His *Mysterium Cosmographicum* of 1596 was the first published defense of the Copernican system (*De Revolutionibus*).[268] He understood it as revealing something of God's geometric plan for the universe and he was fascinated with patterns in natural things.

---

[265] The same year as Copernicus's discoveries (see LE1).

[266] As vehicles of memory. This was a time when the seat of the soul was being relocated from the heart to the brain. Christopher Wren was the first to do a correct drawing of the brain, correcting Leonardo.

[267] famous for his laws of planetary motion

[268] Removing some of Copernicus's reliance on Ptolomaic explanations of eccentric orbits.

Kepler 1610 P-D Copyright expired
His idea of bringing Copernicus and theology together sprung from his conviction that the physical and the spiritual were part of the same *connectivity* or even, we might say, the same *pattern paradox*. The universe itself was an image of God, with the Sun corresponding to the Father, the stellar sphere to the Son and the intervening space to the Holy Spirit. His first draft of the *Mysterium* included a chapter reconciling heliocentrism with contradictory biblical passages. Working with the astronomer, **Tycho Brahe** (see LE1), and just a few years before Galileo's telescopic discovery of the four moons of Jupiter in 1610, he began to work on lenses for astronomy, and these proved useful in optics as well. In these times, optics was part of astronomy, because light was seen as the same everywhere. In his great 1604 work, *Astronomiae Pars Optic,* he included a study of the eye, perhaps being the first to discover that images are inverted by the lens onto the retina. '*Therefore vision occurs through a picture of the visible things on the white, concave surface of the retina.*' The enigmatic playright and scientist, **Giambattista della Porta** (1535-1615) was also part of the scientific revolution, and in 1558 wrote his *Magiae Naturalis.* He used the camera obscura – important for Renaissance painters – and Kepler deduced from this that the image was refracted upside down as if 'painted' on the retina. He assumed that the image was corrected '*in the hollows of the brain*' due to the '*activity of the Soul.*' **Vesalius** had recognised the importance of the crystalline lens, but disagreed that the optic nerves were hollow.

René Descartes (1596-1650) used the Aristotelian idea of light as pressure, but Newton (1642-1727) focused on colour. Descartes had dissected animal eyes as well as producing theories based on his assumption that mind and body are separate. He used Kepler's anatomy of the eye, but Newton did experiments on his own eye to investigate the problem of how sight actually worked. By sticking a bodkin into his own eye and looking at the sun repeatedly through a coloured prism[269] he worked on the problem of how colour is produced. In his times, it was thought to be come from the soul, the mind, the body or light itself.[270] Newton and later **Goethe** and **Turner** worked on colour theory. Goethe saw it as part of human perception. Newton saw it as light broken down in a prism. These were hard philosophical, as well as scientific problems to solve - do colours exist in the real world, or are they wave lengths compared by our brain on our eye receptors? I return now to Darwin's direct dilemma over evolutionary teleology that he shared with Grey. Hooker had introduced Darwin to Grey at Kew and

---

[269] Self experiment was not unusual in these times.

[270] Pseudo Dionysius (Part Three) had speculated on colour in sight as a function of the divine, particularly the colour blue.

they became firm friends, exchanging letters on many occasions. In 1868, Gray visited Darwin during a year of absence from Harvard, where he was Professor of Botany. Gray helped Hooker with research on Darwin's behalf when Hooker joined Gray on a Rocky Mountain expedition in 1877. In the Linnaean Society meeting of July 1858, when papers by Darwin and Wallace were read in their absence[271] (arranged by Lyell and Hooker) Darwin's position included the reading of one of his letters to Gray written in 1857. Gray had arranged the first American publication of the *Origin* and Darwin dedicated his *Forms of Flowers* in 1877 to Gray. Gray was anxious to convince Darwin that design was present in all forms of life and yet, despite his concerns, remained a close friend and supporter. Darwin wrote '*I feel most deeply that the whole subject is too profound for the human intellect. A dog might as well speculate on the mind of Newton.*' [272] Gray's own book '*Darwiniana*' attempted a reconciliation between natural selection and theism, but Darwin found that hard to accept. At the heart of this challenge of Gray's, was the issue of teleology and the really hard question about Darwin's continuing relationship with Paley and the question of design. There was a haunting optimism in parts of the *Origin* and notably in its final paragraphs which showed Darwin's belief that something '*exalted*' ie the *higher* animals could evolve, or, as he put it be *produced* '*from the war of nature, from famine and death.*' For him, there was still this haunting '*grandeur*' which he saw as having been '*originally breathed by the Creator into a few forms or into one.*' Was there an incipient anthropomorphism of a pre-Copernican type in this approach? Did all the processes of natural selection lead to the higher good and eventually the *superiority* of the human species?[273] Teleology was, by its very nature, optimistic. Darwin had observed nature's hard struggles and experienced the misery of personal loss himself. He had confided to Hooker in 1856 that nature '*could be clumsy, wasteful, blundering, low and horribly cruel.*'[274] Yet, in the *Origin* and certainly in its moving conclusion, there is a strong sense of final and higher purpose, emerging from the process. **Mary Midgley's** thinking is again

---

[271] Their tittle was revealing in its language and timings just a year before publication of the *Origin - On the tendency of Species to form varieties; and on the Perpetuation of Varieties and Species by Natural Means of Selection.* Darwin was absent because children in his village were dying of scarlet fever and his own child Charles Warring had just died from it in June 1858.

[272] Letter 2814 Darwin to Gray May 1860

[273] See Midgley's discussion of utopian teleology and power fantasies which imply their own quasi religious versions of immortality in 'science' fiction writers where Dyson (*Time without End* 1979), Bernal (*The world the flesh and the devil* 1929), Day (Omega Man), Barrow and Tipler and others imagine evolutions into superior kinds of creatures that transform and control their stellar environment - *Science as Salvation* Chapter 14 Evolution and the Apotheosis of Man p 147. This departs hugely from Darwin's evolutionary realism. As Midgley says 'possibility is not limitless.' P 157. She sees these writers as being caught in their own literalistic interpretation of science's potential without allowing for organic or other limits. Other genre of utopias e.g. Plato and More can be useful as moral comments on the world as it is and its need for improvement and new direction. Dyson et al assume all social and scientific problems have already been solved as a basis for new direction. See her comments on Paul Davies's '*Superforce*' 1984 and J.B. S. Haldane's '*Possible Worlds*'1927.

[274] More Letters, I, 94

helpful here. She challenges any idea of dropping teleology altogether as *'purpose centred thinking is woven into all our serious attempts to understand anything, and above all into those of science.'* [275] She notes that discerning a pattern in any natural process is not to infer design, but simply to note the meaning and presence of its content and relationships.

Matter and our thinking about matter cannot be reduced to a series of inert, unconnected packets of information, as the atomists thought in the seventeenth century. The 'individual' bits are part of a complicated, connected and interactive system over time and space. Connection implies some kind of pattern which in turn implies some kind of order. This 'order' may contain its own *pattern paradoxes* of disorder and is part of nature's chance selections and changes. She sees teleology as an explanation of a thing by mentioning its *function,* not its cause, or design, arguing that it is a misreading of Aristotle to see his use of purpose as implying cause. From this, she argues that all discussion of function implies some presence of teleology, and therefore of design, but this can be separated from assumptions about a designer. The legitimate concern to dismiss the argument from design needn't lead to a dogmatic avoidance of any discussion of design, pattern, meaning, order, purpose in the functioning of organisms and organisations, and can allow for the presence of chance and indeterminacy within those processes. The mind seems to be programmed to find patterned meaning, in order to manage its perceptions of reality, and we continue to use metaphors and myths of meaning in our descriptions of natural processes. She quotes science's use of the words 'information' and 'communication' to describe unconscious processes such as the transfer of DNA. She claims that both Science and Religion have grown out of the human mind's capacity and propensity to see a unifying, connecting and ordering vision, rather than just disorderly and purposeless confusion and chaos.[276] Theologians and philosophers had long searched for ways of understanding *order* and *disorder* paradoxes and the *good* in its relationship with *evil,* particularly in Augustine and Aquinas. Could the good ever come out of evil; was the latter's existence, in its natural and free will forms, a necessity for finding the good? Did Darwin confirm this view and was nature's cruelty part of a *designed* process with a higher, if not

---

[275] See her section *Questions about Teleology* in *Science as Salvation* p 9 ff. She quotes C.S.Lewis to make her point *'We find that matter always obeys the same laws which our logic obeys..A great many people think that it is due to the fact that Nature produces the mind. But on the assumption that Nature is herself mindless, this provides no explanation. To be the result of a series of mindless events is one thing; to be a kind of plan or true account of the laws according to which these mindless events happen is quite another.. It is as if cabbages, in addition to resulting from the laws of botany, also gave lectures in that subject... We must seek the real explanation elsewhere. I want to put this other explanation in the broadest possible terms and am anxious that you should not imagine I am trying to prove anything more, or more definite, than I really am..Unless all that we take to be knowledge is an illusion, we must hold that in thinking we are not reading rationality into an irrational universe, but responding to a rationality with which the universe has always been saturated.'* Christian Reflections, Collins, Fount 1967 p 89 cited in Midgley p 14.

[276] See reference to debate between Einstein and Bohr and Heisenberg in 'Does God play Dice' chapter in LE1.

final purpose? Was it a stage in the evolutionary purpose with natural selection as its means to a higher end?

Asa Grey wrote in his *Darwiniana* that nature had *'received at its first formation the **impress** of the will of its Author, **forseeing** the varied yet necessary laws of its action throughout the whole of its existence, **ordaining** when and how each particular part of the stupendous plan should be realised in effect.'* [277] This went much too far for Darwin, not that he didn't struggle with the dilemma. Many of his contemporary readers rejoiced in finding traces of design or teleology in t*he Origin* and, because of that, came to accept selection as the process through which the *impress* worked itself out as a *secondary law*. There was much at stake. Should Darwin tactically admit he was still using teleological assumptions from a design belief he could no longer accept, or should he reject teleology altogether and stick to random selection, despite its radical, anti-teleological implications for many of his contemporaries. Or was he in something of a muddle, and unconsciously still drawing upon the left-overs of a teleology he had largely rejected? If each variation was pre-determined to lead its *futurity* to a final purpose, how was selection random in any meaningful sense? In which case, was the evolutionary pattern part of some greater design *impressed* upon it and if so who was doing the designing, if not nature itself? At times he came close to personifying nature as a force outside of itself, capable of impressing a pattern and purpose on its own processes. It was as if he was replacing a creator God with a creator Nature. If he was caught in the vortex between an intellectual Sylla and Charybdis, his friend Hooker seemed to oscillate back and forth in his own presentations. Sometimes, he was avidly opposed to any kind of teleology. At others, he appeared to collude with the theological assumptions of the times - to show that there was an ordained purpose in the processes of selection - not least to show that if there had to be a providential process, it was better Darwin's that anyone else's! When it came to selecting quotes for the frontispiece of *The Origin*, Darwin chose the highly esteemed and prolific theologian and scientist, **Rev Willliam Whewell**, Master of Trinity College Cambridge whose significance for natural theologians we considered earlier. *'But with regard to the material world, we can at least go so far as this—we can perceive that events are brought about not by insulated interpositions of Divine power, exerted in each particular case, but by the establishment of general laws.'* So, again we note that Darwin is using this to insist on general laws rather than Divine interposition. He placed this quotation alongside **Bishop Butler's** (1692-1752)[278] *'The only distinct meaning of the word 'natural' is stated, fixed, or*

---

[277] *Darwinia* p 57

[278] A highly influential English Bishop, known for his critique of Thomas Hobbes and some of Locke's theorires. His work influenced David Hume and Adam Smith

settled; since what is natural as much requires and presupposes an intelligent agent to render it so, i.e., to effect it continually or at stated times, as what is supernatural or miraculous does to effect it for once.'[279] So, we note, that Darwin is distinguishing between a supernatural agent affecting a one off event, and a natural agency producing a fixed or continuous process. This was placed alongside **Francis Bacon's** *'To conclude, therefore, let no man out of a weak conceit of sobriety, or an ill-applied moderation, think or maintain, that a man can search too far or be too well studied in the book of God's word, or in the book of God's works; divinity or philosophy; but rather let men endeavour an endless progress or proficiency in both.'*[280] So, we note Darwin seems to be balancing out his position by saying that God's 'Word' and 'Works' go together. The territory for what I call the *teleological dilemma* in Darwin had already been set out, even before the title page. He was using the first two of the three pillar quotes[281] to indicate the transition he was making and then, by including the third, to reestablish a connection back to an inclusive position.

Gray genuinely believed that Darwin had restored teleology to natural science and in *The Descent* Darwin more or less admitted that the *Origin* had put too much emphasis on random selection as a natural process to compensate for his continuing teleological dilemmas. He does this in a passage we shall use again, but here quote at greater length – *'but I now admit..... that in the earlier editions of my 'Origin of Species' I probably attributed too much to the action of natural selection or the survival of the fittest. I have altered the fifth edition of the Origin so as to confine my remarks to adaptive changes of structure. I had not formerly sufficiently considered the existence of many structures which appear to be, as far as we can judge, neither **beneficial nor injurious**; and this I believe to be one of the greatest oversights as yet detected in my work. I may be permitted to say as some excuse, that I had two distinct objects in view, firstly, to show that species had not been **separately created**, and secondly, that natural selection had been the **chief agent of change**, though largely aided by the inherited effects of habit, and slightly by the direct action of the surrounding conditions. Nevertheless I was not able to annul the influence of **my former belief**, then widely prevalent, that each species had been **purposely** created; and this led to my tacitly assuming that every detail of structure, excepting rudiments, was of some special, though unrecognised, service. Any one with this assumption in his mind would naturally extend the action of natural selection, either during past or present times, **too far**. Some of those who admit the principle of evolution, but reject natural selection, seem to forget, when criticising my book, that I had the*

---

[279] From his *Analogy of Revealed Religion*.
[280] Advancement of learning
[281] Three in the later editions and only the first two in the earlier ones.

*above two objects in view; hence if I have erred in giving to natural selection great power, which I am far from admitting, or in having **exaggerated** its power, which is in itself probable, I have at least, as I hope, done good service in aiding to overthrow the dogma of separate creations. That all organic beings, including man, present many modifications of structure which are of no service to them at present, nor have been formerly, is, as I can now see, probable. We know not what produces the numberless slight differences between the individuals of each species, for reversion only carries the problem a few steps backwards; but each peculiarity must have had its own **efficient cause. If these causes, whatever they may be**, were to act more uniformly and energetically during a lengthened period (and no reason can be assigned why this should not sometimes occur), the result would probably be not mere slight individual differences, but well-marked, constant modifications. Modifications which are in no way beneficial cannot have been kept uniform through natural selection, though any which were injurious would have been thus eliminated. Uniformity of character would, however, naturally follow from the assumed uniformity of the **exciting causes,** and likewise from the free intercrossing of many individuals. The same organism might acquire in this manner during successive periods successive modifications, and these would be transmitted in a nearly uniform state as long as the **exciting causes** remained the same and there was free intercrossing. With respect to the **exciting causes** we can only say, as when speaking of so-called **spontaneous** variations, that they relate much more closely to the constitution of the varying organism, than to the nature of the conditions to which it has been subjected.*

Conclusion.*—In this chapter we have seen that as man at the present day is liable, like every other animal, to multiform individual differences or slight variations, so no doubt were the early progenitors of man; the variations being then as now induced by the same **general causes**, and governed by the same **general and complex laws.** As all animals tend to multiply beyond their means of subsistence, so it must have been with the progenitors of man; and this will inevitably have led to a struggle for existence and to natural selection. This latter process will have been greatly aided by the inherited effects of the increased use of parts; these **two processes** incessantly reacting on each other. It appears, also, as we shall hereafter see, that various unimportant characters have been acquired by man through sexual selection. An unexplained residuum of change, perhaps a large one, must be left to the assumed **uniform action of those unknown agencies**, which occasionally induce strongly-marked and abrupt deviations of structure in our domestic productions. Judging from the habits of savages.. and even the ape-like progenitors of man, probably lived in **society**. With strictly **social animals,** natural selection sometimes acts **indirectly** on the individual, through the preservation of variations which are beneficial only to the* *community. A community including a large number of well-endowed individuals increases in number and is victorious over other and less well-endowed communities; although each separate member may gain no advantage over the other members of the same community.* [282]

We note how he places the much used phrase *'exciting causes'* alongside *unkown agencies* and *general and complex laws* in this passage. The idea of *causes* and *laws* inevitably qualifies the freedom of any more radical and indeterminate process. They hint at what he calls *the influence of my former belief.* Laws were, as we have seen, his way of rescuing nature from any supernatural agency, but his *former belief* may have produced a certain implicit teleological, or patterned process in his understanding of evolution, reminding us how hard it was for him to escape from the horns of this dilemma. Let's return to that conversation between Darwin and Gray about the eye and its place in this teleological question. **Getrude Himmelfarb** asks the question *'How can selection, knowing nothing of the end or final purpose of this process, function when the only test is precisely that end or final purpose?'* [283] If sight is the final purpose of selection how is an organ, like the eye, of use in survival adaption in its different stages, if it wasn't perfect from the start? In one sense, this dilemma could be applied to all organs, including the heart, but obviously it is acute in the case of the eye. Himmelfarb accuses Darwin of dealing with the problem by admitting it, rather than facing its implications, and by implying that if the theory was right in other cases it must be in this. Darwin has been proved right by later scholarship. It seems that the process of evolution doesn't need to *know the final purpose* of the process and the only *test* isn't that *final purpose,* but the way each stage learns to adapt. The first, eye like structures were light sensitive pigmentations on the skin which evolved into different and more complicated types of eyes – some far more effective than the human eye. So the eye did evolve through many *intermediate* stages and has *different* structures in *different* animals. Here is how the zoologist, **Dan Erick Nillson** describes the process in simple terms. *'Random changes then created a depression in the light-sensitive patch, a deepening pit that made "vision" a little sharper. At the same time, the pit's opening gradually narrowed, so light entered through a small aperture, like a pinhole camera.* **Every change had to confer a survival advantage, no matter how slight. Eventually,** *the light-sensitive spot* **evolved** *into a retina, the layer of cells and pigment at the back of the human eye. Over time a lens formed at the front of the eye. It could have arisen as a double-layered transparent tissue containing increasing amounts of liquid that gave it the convex curvature of the human eye. In fact, eyes corresponding* **to every stage** *in this sequence have been*

---

[282] *The Descent 1*, p 152-155. My emphasis in bold
[283] *Darwin and the Darwinian Revolution*, Getrude Himmelfarb, Chatto and Windus 1959, p 277

found in existing living species. The existence of this range of less complex light-sensitive structures supports scientists' hypotheses about how complex eyes like ours could evolve. The first animals with anything resembling an eye lived about 550 million years ago.' [284] Is Himmelfarb's use of the polarities *perfection* and *imperfection* helpful here? If we start with the idea that the human eye is now perfect, then we work backwards to assume, in previous stages, it was imperfect and by her implication, therefore useless in those stages. This has often been used as an argument against evolution on the basis that if God was the designer or creator of the eye he would have given all animals a perfect eye from the beginning. All eyes seem to be 'perfect' therefore God must have been the designer! In fact eyes are very vulnerable to all sorts of malfunctions and *imperfections* at every stage of their development – for example, blood vessels run across the surface of the retina rather than beneath it and easily proliferate or leak. From one perspective, the eye in any species is certainly an extraordinary thing, but is it perfect? Perhaps not! The polarity of *imperfect* and *perfect* may itself be the problem here in this *pattern paradox*, at least as used by Himmelfarb and by some anti-evolutionists. She offers the following sharp criticism *'Credited with such marvels as the origin of new species natural selection might have been expected to accomplish such lesser miracles as the elimination of their grosser imperfections.'* [285] Not only is she assuming perfection in the origin of a new species but imperfection as a problem in adapted species. She goes on to criticise the theory of selection by asking *'if it is intended to account for the development of species from the simple to the complex and from a low to a higher order of organisation how can it also account for the persistence of the simple and low? ...why have not the superior or higher forms supplanted the inferior or lower?'*[286] To uphold this argument, she is making the most of the polarites *low* and *inferior* and *high* and *superior* in relation to the correlates of *simple* and *complex* forms. There are certainly ways of reading the *Origin* and some Darwinian supporters which imply such polarities, but I am not sure that is correct. It may well be that, from within her own cultural and moral assumptions, it was natural to see simple as inferior, but is this always the case? I am no Darwin expert, but sense something in his use of the term *diversity* which may help here. While he often implies selective adaption as leading to *higher* forms of organisation, his *entangled bank* contains many forms of *diversity* which must have included differences that should not and could not be described within this comparative polarity of (moral) absolutes.

---

[284] Evolution; Darwin's dangerous idea.
[285] Ibid p 280
[286] Ibid p 280

Natural selection doesn't always move from *'imperfection'* to *'perfection,'* (let alone the other way round) in parallel with its movement from *simple* to more *complex or inferior* to *superior* forms. We should not assume that all imperfect forms disappear because of the linear movement of selection forward in time. This is to impose a view of final *purpose* as something *perfect* onto the processes of evolution. It implies that selection always dictates a linear movement from A to B as from *imperfection* to *perfection*. The truth is more complicated and the intermediary stages, within this pattern paradox, may be more about *different* presentations and movements of change than from one absolute to another absolute. We cannot judge earlier stages of life as imperfect from a later stage of life's apparent, teleological perfection. As we saw in LE1, we have to allow for *indeterminacy* in the process at every stage. Admittedly Darwin cannot be blamed for not being aware of Heisenberg's Indeterminacy Principle - which of course came later. We also have to allow for the cultural assumptions of Victorian society about optimistic, teleological process which probably affected his thinking. What some later commentators, like Himmelfarb, saw (wrongly perhaps) as *perfection* and *final purpose* in adaption, surely needs qualifying for a more realistic assessment of nature's own *indeterminacy* and *finitude*; its potential for further *propensities* in a process that is not pre-planned and never perfect, however wondrous. Nowhere does this hit home harder than in our human experience of suffering in those we love, as well as ourselves, and we now turn, briefly, to that subject. If he did justice to the presence of suffering and extinctions in his understanding of nature and species evolution, the least we can do is to honour that by including in Love's Energy a reference to his own experience of love and loss in the human condition.

Chapter 6
Love and loss. Their influence on Darwin

The thesis of Love's energy, as we saw in LE1, doesn't rigidly separate the facts of knowledge from our human subjectivity and experience. Most biographies of Darwin (and this book doesn't pretend to be that) comment on the influence and importance of his family life, with its private tensions and tragedies. It is worth saying something about these dimensions of his life to understand something of their influence on his work. I particular any talk of *love* has to take seriously the risk and reality of *loss*. He was part of a very significant merging of two great families – the Wedgwoods and the Darwins and some measure of intermarrying. Charles was born in 1809, as the fifth of six children, named after his father's brother Charles who had died early, while studying medicine at Edinburgh. His father, Robert Darwin, was a wealthy, society doctor and financier; his mother Susannah Darwin was the daughter of **Josiah and Sarah Wedgwood** – the great pottery family with their influence on early Victorian industrial life and the anti slavery struggle. Charles was brought up under the influence of his father and mother's Unitarian free thinking and the Wedgwood's growing interest in Anglicanism. Josiah was part of the Lunar Society, discussed earlier. On his father's side, as we have seen, was the great **Erasmus Darwin**. Charles knew of his poetry, his science and significant role in the Lunar society. Erasmus was a leading light in his family, and his regional and national reputation must have influenced Charles' own early life.

When thinking about Darwin's personality, I sense a likeness with certain aspects of **Edward Elgar**, born two years before the *Origin*. Beneath the well cultivated self-image of Edwardian *'pomp and circumstance,'*[287] beat a very sensitive and trembling heart, full of enigmatic ambiguity, overlapping between the *pattern paradoxes* of the major and minor keys of the transitions in his music. A good illustration is his *'Sospiri'* written in 1914, at the beginning of the war. At one stage, he called it the *'sighing of love'* and spoke of a *reaching out*, tentatively and hesitantly, to the world around him. In his life, as in Darwin's, we see a *reaching out* across the *pattern paradoxes* of *separation* and *isolation* to connectivity and relationships from one who lived from within the inside as well as the outside of his personality and work. Darwin was brought up in Shrewsbury. His mother died, probably of some kind of stomach cancer when he was just nine, in 1817. He later wrote with great pathos, *'It is odd I can remember nothing about her except her death bed.'* He remembers being locked in rooms

---

[287] He was always aware of being from a lower class that his wife Alice. Even when knighted he longed for nobility and even higher social status.

on many occasions. His father accused him of being good for nothing except for hunting, and his headmaster saw him as a failure, unable to pay attention to anything. The introspective personality was emerging from the mist memory these early experiences. From an early age, it became his habit to go on long, solitary walks, and to be a collector of various items - from coins to minerals. The passion for collection which *'leads a man to be a systematic naturalist, a virtuoso, or a miser is very strong in me,'* he wrote much later. Both families believed in the importance of 'freedom' - free trade, free religion, free individual enterprise, free thinking, and free democratic choosing within the limits of the day. Charles supported the freedom of slaves and the French and American revolutions, at least on the surface. His natural friends at Cambridge were Quaker reformers. His belief in freedom drove him to pursue his own work, in his own way and to develop his own, life long interests, which included various changes of direction and times of hesitation. His father, a Doctor, encouraged him to do medicine, with his brother Erasmus, and they went to the best medical school in the country – Edinburgh in 1825, where he came across **Robert Grant** and his support of **Lamarck's** ideas – which he had also come across in his grandfather's journals and from reading *Zoonomia*. It was perhaps Grant who began to ask him leading questions that were to challenge his acceptance of Paley's theology. He found medicine unpalatable, distressing or boring, but became interested in plant classification and enjoyed the Plinian society. His father continued to be anxious about his son's attitudes and lack of application, so he decided he should go to Christ Church Cambridge with a view to priesthood. He didn't exactly shine, and seems to have spent time riding and shooting (confirming his father's fears that he was a bit of a wastrel), but he did meet and form an important relationship, as we have seen, with the botanist **Professor Henslow**. He took up beetle collecting on the suggestion of his cousin, **William Darwin Fox**. He did meet famous naturalists – most of them ordained - and he was influenced by the natural theology assumptions that dominated Cambridge culture. He even lived in Paley's old rooms at one point and greatly respecting his work began to think about how God might act through the laws of nature.In his exams, his best results came in the paper on Paley's *'Evidences for Christianity.'* He attended **Sedgwick's** geology lectures and was persuaded to join him on a field trip. He read **Herschel** and **Humboldt**. This was where he came across **Milton's** *Paradise Lost* which he took with him to South America. It is hard to know how Milton's theology influenced his later thinking.

His older sister Caroline played a crucial role in his upbringing. He was part of a larger, extended family and seems to have relished their company. He became romantically involved with **Fanny Owen**, the sister of his school friend **William Owen** and a friend of his own sisters, but when it came to the long Beagle voyage neither asked the other to wait, and before that there had already been a drawing apart. She wrote to him in September 1831, just before he left in October, *'My dear Charles, I have this evening heard from Caroline that you leave home the end of this week—and that you wish to have a good bye from me before you go. I had not the **least idea** you were to go so soon, for they told me it was the end of October you sailed, so I **hoped** and fully expected I should have been at home in time to see you— I **cannot** tell you how disappointed & vexed I am that that cannot be. Little did I think the last time I saw you at the poor old Forest, that it would be **so long** before we should meet again!! ....My dear Charles I do hope you will enjoy yourself & be the happiest of the happy, I would give any thing to see you once more before you go, for it does make me melancholy to think the time you are to be away—& Heaven knows what may have become of all of us by this time two years. at all events we **must** be grown **old** & steady— the pleasant days, and fun we have had at the Forest can never come over again— how I wish I was there this week to have one last chat with you I cannot bear to think you are really going clear away, without my saying one good bye!!....how vexed I am you are going it is too selfish of me to say so, for I am sure it will be the very thing to suit you— did you throw yourself on the Governor's mercy, & confess your creditors, or what have you done? What a capital way of escaping un-gentlemanlike Tailors &c— When you are far from the Land they may whistle for their cash for what you care! Well, don't be surprised if you hear I have taken Ship too and fled my duns ..Pray write to me one last Farewell my dear Charles & tell me all your plans & prospects—where you are to go to—& all about it? And tell me too if I shall look out for a nice little Wife for the Parsonage by the time you return. tell me what you require and I will look about and get one in my eye by the time you want her—a proper knowledge of the Beetle tribe of course you require—As for all your Sisters I think they are gone crazy or sulky or sleepy or something for not one line have I had from any of them these two months—they treat me with the most marked contempt.— I went on board the Adelaide and all over it—so can fancy you in your little Cabin—and I assure you, you will not be forgotten, I shall often long to have you to laugh with and scold out of the Painting room— I wish I had made your Pincushions they might have been useful—and occasionally in taking out an instrument of death for a Beetle you would have called to mind the Manufacturer of the useful article —but it can't be helped now— this letter is most prosy, & duller than letter ever was before—but I cant help it you must take the will for the deed..I must now conclude—can only add—I most sincerely wish you every amusement & happiness possible— but only wish most heartily you were not going quite so soon that we might have one more talk & laugh first—but it is not to be— so good bye my dear Charles. Burn this before you sail for pity's sake'* The style is not exactly Jane Austin, but it reveals something of the passion she felt, as well as some tensions and her perception of his future needs in the 'parsonage' with his beetle collecting. Fanny married **Robert Myddelton Biddulph** of Chirk Castle while Darwin was away, and then wrote. *'Believe me Charles that no change of name or condition can ever alter or diminish the feelings of sincere regard and affection I have for years had for you. And as soon as you return from your wanderings I shall be much offended if one of your first rides is not to see me at Chirk Castle and find out what curious beetles the place produces.'* She had been his first love and he always remembered her, keeping the letters she had sent to him at University. Seven years later, working on his notes, Darwin realised how important sexual connection and reproduction was. This was the fundamental unity of animals and plants. He revealingly drew up a balance sheet in 1838 to decide whether to marry, or not to marry. He took a cataloguer's view of the structure of his life. The story of his relationship with **Emma Wedgwood** (1808 –1896), his first cousin is well known. In 1838, he

proposed to her while she was nursing her sick mother.

Based on Water-colour of Emma Darwin (1840) by George Richmond. P-D. Copyright expiredUS P-D

She was musical (having had piano lessons with Chopin) and had turned down six other marriage proposals! She saw Charles as being transparent and affectionate. They were to create a marriage that was rich in warmth and honesty. She was indeed a significant part of his life and his work and deserves a higher profile in his story. They had ten children, but three died in their early years. **George** an astronomer, **Francis** a botanist and **Horace** a civil engineer became Fellows of the Royal Society. **Leonard** became a soldier, politician, economist and eugenicist. Charles's family were free thinking members of the Church of England. Emma's grandfather, the non conformist, Unitarian, **Josiah Wedgwood** was the famous pottery scientist and entrepreneur – one of the great industrialists and inventors of his age and an anti slavery campaigner. It was probably his wealth, handed down via Emma, which enabled Darwin to spend his life on research and science. Darwin's working life and his undiagnosed illness, with its many debilitating symptoms was protected by the constant support and love of Emma, the peace of Down House and its natural environment, a special closeness with his children and a private income – without which his years of study and reflection would have been impossible. Unlike Einstein, he did not have to fit his life's work and thoughts into the rhythm and pressures of paid employment. He could walk round the famous 'sand walk' for his regular reflections, whenever he wanted. The wall of protection was not, however, impregnable. It was porous to the negative influence of certain, academic clergy at Cambridge, and in particular the ever present shadow of his teacher the **Revd Adam Sedgwick**. The effect of the highly dismissive letter Sedgwick sent to Darwin after the former had read Chamber's *Vestiges,* demonstrated Darwin's vulnerability to their criticism. Sedgwick described the

Vestiges as mischievous and dangerous. Darwin knew only too well that his own work questioned the religious pillars of morality and culture and that the threat to their edifice risked evoking anxiety and anger from many. Emma's own attitudes and beliefs were a daily reminder of this and the tensions they faced over this issue helps explain his caution and hesitation over publishing. Perhaps these tensions also contributed to his poor health. The protective wall quickly came down at times like this. Sedgwick's' criticism only increased Darwin's appreciation of the potential and actual opposition which had prevented him from writing his theory of transmutation of species ever since 1844. His study was full of correspondence as he shared and tested his ideas with colleagues and friends. If Sedgwick reacted so vehemently to Chambers, what would he say to his old student, if Darwin were to publish something similar? Perhaps the letter confirmed to Darwin the hopelessness of trying to convince Sedgwick, freeing or motivating him to put his work in the public domain. Of course, Wallace's ideas and Hooker's encouragement also prompted him to set about the ten months work in 1858 which would lead to publication. Emma was concerned to see the effect of his critics on Darwin's health and wellbeing. She was also concerned about the implications of his thinking for his own faith and their relationship. Apparently this was there from the beginning. He had already shared with her some of his new ideas on the transmutation of species before they were married, in particular on a visit to her in July 1838. He proposed in November of that year. After the proposal visit, she wrote to him *'When I am with you I think all melancholy thoughts keep out of my head but since you are gone some sad ones have forced themselves in, of fear that our opinions on the most important subject should differ widely. My reason tells me that honest & conscientious doubts cannot be a sin, but I feel it would be a painful void between us. I thank you from my heart for your openness with me & I should dread the feeling that you were concealing your opinions from the fear of giving me pain. It is perhaps foolish of me to say this much but my own dear Charley we now do belong to each other & I cannot help being open with you. Will you do me a favour? Yes I am sure you will, it is to read our Saviour's farewell discourse to his disciples which begins at the end of the 13th Chap of John. It is so full of love to them & devotion & every beautiful feeling. It is the part of the New Testament I love best. This is a whim of mine it would give me great pleasure, though I can hardly tell why I don't wish you to give me your opinion about it.'*[288] There is something ominously revealing in that last sentence.

In 1851, while Emma was pregnant with Horace, he wrote of the love and joy he felt for his beloved daughter Annie, who then tragically died at Easter time 1851, aged just 10 years old. Darwin asked himself *'does God punish ten year olds?'* He

---

[288] Letter 441 in *Darwin Correspondence Project* (21-22 November 1838).

faced, in the most personal terms possible, the ultimate question about species adaption and survival.

Annie Darwin 1849. Lukas Fenner, Matthias Egger, Sebastien Gagneux. P-D copyright expired

Perhaps, this was a more critical time for his faith than all his observations of natural selection. From Malthus he might have learnt of warring, competitive nature, but this personal experience of death, in the struggle for existence, remained with him in all his life and work. Seven years after her death, in Chapter three of the *Origin,* Darwin talked of the *'smiling face of nature but under the surface the young are dying young in the struggle for existence.'* The death of **Anne** (2 March 1841-23 April 1851), **Mary Eleanor** (23 September 1842-16 October 1842) and **Charles** (6 December 1856 -28 June 1858) must have been on his mind as well as his reflections on many observations of different species. The face of nature was yielding to the *'blows of wedges'* (a phrase from Malthus) being driven in with the crowded and struggling space for survival. The *'smiling face of nature'* might well have been a memory of Annie's 'brilliant face' and her fate. Darwin's great work was touched by the realities of species survival, suffering and death. However much contemporary theologians and he himself spoke of the wonders and beauty of nature, he knew its dark side and it gave him pause when it came to speaking of a benign God in relationship to this kind of reality. The protective wall around him was shattered by the death of Anne and the loss of the joy she gave him in their relationship together. It was probably her death that destroyed his faith in a traditional view of salvation and a divine purpose and plan in creation. It was her death that probably reinforced his view that Sedgwick was wrong.

Scholars still argue about the influence of this loss and illness in Darwin's life on his theory of natural selection. His own memory of the loss of his mother, and all the darkness in that episode, the death of three of his children and the draining symptoms of his own illnesses all played their part in influencing his emotions and views. He feared that consanguinity in his marriage to Emma might be partly to blame for the death of his children. His own search for exotic cures, including water treatment for himself and for Anne had failed. These emotions of guilt, helplessness and anxiety were surely a significant part of his big idea and his predilection for collecting, gathering things and reflecting on the key role of species extinction, suffering and survival instincts in all species. Perhaps the faces of those he had loved and lost merged into his reflections on the millions of individual creatures involved in a daily struggle for survival. Perhaps, without this amount of personal suffering and loss, the collecting, evidence gathering and intellectual life of Darwin would have been less creative.

Somewhere within the symbiotic relationship of these and other parts of his life the theory of transmutation of species appears. Any great achievement has its cost and in Darwin, it emerged out of considerable personal struggle and loss. At the heart of human suffering and loss there is a shadow of the image of love and its embedded place within the 'entangled bank' of creation. This shadow is part of love's powerful reality which entangles us in daily complexity, struggles, pain, opportunities and wonders. The experience of wonder cannot be separated from the experience of loss and suffering. At its deepest level, wonder emerges from the same place as the hard and dark realities. A romantic view of nature, so popular at the time, was only a superficial cover to what Darwin saw and studied below the surface. So, in his painstaking study of barnacles, pigeons, finches and many other species, he was confronted by a profound emotional challenge to the worldview and assumptions of so many of his contemporaries. It must have been hard for him to attend his local church with Emma, where the **Rev John Innes** was a good and critical friend. In that church, Darwin must have heard things being preached or sung about God's plan in creation. In 1848, **Mrs Cecil Alexander** wrote the hymn *'All things bright and beautiful..the Lord God made them all,'* popular ever since in childhood spirituality, and influential in wider Christian views about creation. The words may have been inspired by a verse from **S.T.Coleridge** (1772-1834) in his *Rime of the Ancient Mariner 'He prayeth best, who loveth best; All things great and small; For the dear God who loveth us; He made and loveth all.'* Verse two of the hymn refers to *'wings'* and verse seven to *'eyes,'* both cited by **Paley** as examples of the intricate complexity of God's direct design. It must have been agony for Darwin to watch his dear Emma sing or play such hymns, knowing how far they were from his own understanding of creation. It is hard for us, who now take natural selection for granted, to retrospectively appreciate the depth of such pain, and the sense of isolation and criticism he must have felt from the widespread, religious attitudes and assumptions of his times. The face of nature is always pierced by the thorns of reality. Of course the clergy of his times had wide experience of the pastoral implications of pain and loss. For many of them, this was all part of the mystery of God's plan, but Darwin, helped by Malthus, had begun to see a symbiotic relationship between species struggle and species survival and transmutation. Darwin's courage was to learn to accept, emotionally, what he observed and thought rationally. It is one thing to do this privately. To publish was quite another matter. To love is to become more embedded in *entangled* vulnerability. But alongside these thorns, roses can sometimes grow. Any anthropomorphic projection of human love onto non-human, species experience is questionable and dangerous. If anyone had the right to use the word 'grandeur' and 'wonder' at the end of his observations of an *entangled bank,* full of loss and suffering, destruction, violence and vulnerability, then it was Darwin. In this sense, his spiritual legacy is crucial for our modernity and our religious views are hardly credible, unless they take his work and life experience seriously. In the spirit of Darwin's huge contribution to science and humanity, my own thesis of Love and its energy moves us far from any romantic content and context for the Big Bang or the Big Birth. It is a search for a new way of embedding the wonder and reality of Love's energy in the awesome and awful realities of both cosmology and evolution. The debt theology, spiritual attitudes and religions owe to modern science is profound and far reaching. One of the great tragedies of the story of religion is that it has, with notable exceptions, made itself incredible and irrelevant in the eyes of many, by ignoring or rejecting science. Some argue that religion in Britain and beyond has never recovered from the 1860 Oxford debate between Wilberforce and Huxley. Certainly a gap was opened then and indeed before, which was never completely closed. In the perception of many, Darwin's success led to religion's failure. It can no longer credibly talk about belief in a dualistic or isolated way. As we saw in LE2 the early mystics did their best to oppose dualisms which separated the sacred from the profane, the spiritual from the material, not to say the religious from the secular. These dualisms are widespread still, and they do us much harm not least in the reaction of many thinking atheists. Many commentators, including **Lord Richard Harris,** an ex Bishop of Oxford, are convinced that Darwin never became an atheist, in any strict sense. However, it's clear that his experience of great love and loss, particularly of Annie, shook his belief – and many of the religious and cultural assumptions of his times - to its very foundations. We should not underestimate the personal cost of those years of delay in publishing the *Origin*, nor his surprise and frustration when he learnt of Wallace's independent work on the same theme. So, at the heart of the *Origin*, we find two, very simple and challenging convictions.

1. All the theological assumptions, traditionally made and shared by far too many people, that belief in creation meant belief in the individual, separate creation of every species, could not in fact be true. This was probably not as widespread amongst Anglican clergy as is often assumed, though many, as now, would have accepted it. However appealing it was in popular religion, it could no longer be defended as any kind of scientific truth. Truth is never the enemy of God or belief in God. Belief in the direct 'separate' creation of everything could never again be used credibly in any religious faith.

2. Every articulation of the wonders of Creation has to come face to face, not just with finitude and mortality, but horror stories about species suffering, loss and extinction on a tragic scale – all at the heart of the mainstream processes of creation. It is one thing to be in awe of the

wonder and beauty of life; it is quite another to romanticise, ignore or deny the horrors of non-human and human behaviours and processes. Darwin refused to do this. It was because he constructed his formative theory out of the reality of these horrors, that we are forced to make sense of them, or at least accept their place in the same *entangled bank* as everything else in Creation.

We cannot overestimate the significance of the emotional implications of both these positions and their symbiotic relationship. Both imply a huge sense of loss for those who've based their lives on different assumptions. While many Anglicans, of the time, did not all believe everything was independently created, many lived within a world view of popular religion which continued to make these assumptions. Many hymns and prayers, still used today, do the same. Since Darwin, such belief was impossible. Then, more than now, revising or giving up those assumptions led to considerable emotional upheaval and mourning. We have hundreds of Darwin's letters and we know much about his life, but we cannot know what it was like to be Darwin and live with these thoughts, and his fears about the repercussion of publishing. We know that Emma read the text and, in the end, supported its publishing – a tribute to her love for him and its capacity to respect the growing 'faith' difference between them, even where she was most concerned for his 'eternal salvation.' It is one of the subtexts of this story that her love continued through the tragedies they both experienced and the tensions she felt about his big idea. His love for her drove him to act sensitively towards her beliefs and to anticipate her concerns. In exploring the implications of Love's energy, we are constantly reminded to remember the human face and cost of Love. This compels us to face the hard implications of physical reality in daily living, as well as the discoveries of science and to recognise and admit the scars and wounds we receive and make in the process. We handle holy things when we talk of Love.Living in its shadow and calling, we have to face the ways in which we marr its image in ourselves and others. Ironically and tragically, the churches, as conduits of practical, Christian love, have often undermined it, either by their actions or inactions. Religion should not be blamed for all the wars and violence in the world,[289] but has been responsible for many, and provided implicit support for the negative side of values such as nationalism and patriotism. The inquisition was its own role model and Protestants and Catholics learnt to torture and kill each other through the centuries, as did their Kings and Queens. When the EU constitution was first debated, many churches said it inadequately reflected the Christian origins of the European project; but the history of 'Christian' Europe was far from being a universally positive story. 'Christendom' was always divided, often violently and tragically in the history of Europe.

---

[289] See the magisterial work of Karen Armstong *'Fields of Blood, Religion and the History of Violence'* Bodley Head 2014

Love and loss are not opposite polarities, but part of the same continuum. Every individual who has ever lived has known some kind of loss in love, in life and death. A neurological, genetic or social science study of love and its presenting behaviours may teach us more than we now know. Love, it seems, is ubiquitous in human experience. There seem to be certain common experiences which transcend cultural diversity of expression. Loss and the tragedies of human experience can blunt Love's appeal or credibility. It is fragile as well as tough. It is as vulnerable as our most difficult experiences and while it may change lives, if not the world, it rarely stops wars, domestic violence or terrorism. Yet, it survives the worst of our inhumanity, and may be the source of human, evolutionary development – if humans choose to trust and invest in that strategic direction rather than others. It is most itself when it doesn't force, but compels our transformation. It is most itself when given away. This is the process of its own renewal. The wellspring of its wellbeing seems to fill itself up, even as it is emptied. Its stretch can reach beyond other human boundaries – social, political and cultural. It inspires and facilitates the truly, transformative acts of forgiveness, understanding communication, redemption, compassion, care-fullness, empathy, justice, and solidarity. Finally, it is the only gift which outlasts all matter and yet its meaning can be expressed through material processes and in the flesh. It hints at a mystical presence, but remains available to those who block or deny such a presence. It is its own kind of divinity and expression of practical divinity. It holds us in an embrace in which we are free to discover our own freedom, even as we wriggle around, searching for new ways of understanding ourselves and others; new attitudes and new actions to achieve self- actualisation and communion with other *others*. It is quietly present in that search for a greater union with others and all things, and its destination is often present in its journey and diverse pathways of seeking as well as finding. What has been called, in various contexts, the 'making of love' is perhaps the ultimate and most down to earth form of knowledge of the other, as the holy other, that we can have; it takes many forms, only some of them sexual. Love is the attitude and action towards the other that allows the other to be who and what they are, beyond and through our desires and needs. In the entangled bank of human relationships, and our relationships with non-human creatures, the ultimate possibilities of wonder and communion, connection, value and trust, learning and sharing are inspired by the attitudes and actions of love. When we trust it in our seeking, it seems to find us. When we ignore it, it waits patiently for us. When we betray and deny it, it suffers with our suffering. It has chosen to make its home in our relational *entangled bank,* as well as everyone else's on this, and presumably every other possible world.

Chapter 7
The Importance and Influence of Darwin

Sub Headings
The theological umbrella; The Entangled Bank; The Descent of Man; Evolution, Nature and Morality; Science and culture - the intellectual debate about difference and division; The Leadership debate - Adaptablity and innovation; Darwinian Economics; Darwin, religion and liberal democracies; Scientific developments.

The theological umbrella
Interest in Darwin's work and influence continues to be extensive. In the popular mind, he represents for evolution what Newton and Einstein symbolised in cosmology and physics. His face is on at least as many tee-shirts as their's. His influence continues globally and his book has never been out of publication. No religious person should lightly ignore or dismiss this hugely significant step forward in human understanding of the natural world. It comes from a time when theology and the study of nature were intimately connected and many leaders from one discipline were leaders in the other. These were different times from our own. Very few theologians today are natural scientists. The cognitive gap which developed between science and theology after Darwin reflects the problems confronted and initiated by his work. However much the detailed science has changed, no religious belief which denies Darwin's basic achievement can be taken seriously by educated parts of the modern world. Of course, some parts, of some religions, still use their own educational and cultural influences to defend that kind of denial.

Why was it that in Darwin's time universities and 'science' were conducted under such a wide, theological umbrella and not now? Could such an umbrella have been expected to cope with the storm of new Enlightenment ideas? Of course, there were other social and political realities which contributed to the changing perception of religion. Yes, there have been great revivals and exceptions, but generally those parts of Christianity and Islam which are growing in strength largely reject Darwin's idea and opt for various kinds of creationist position instead. The effect of Darwin on religion has been well documented and discussed and yet there is still much to consider. Did the Church retreat more into its own self-referential concerns, or had this begun much earlier? If Darwin helped to create or rather develop the intellectual framework of modernity, then fundamentalist religion grew in response to this development. This is a crucial question for Islam as well as Christianity. Islam in the golden age before 1492, and particularly in the translation movement, was inclusive of the best of science, philosophy, medicine and mathematics. Believers naturally engaged in these disciplines as part of the expression of their faith, as did Newton and the theologians of Darwin's time. Their faith motivated their studies of the natural world. Now, it seems that religion has left that world to the scientists and retreated into its own narrow areas, nourished by its own discourse, unchallenged by new science, philosophy or even some social and economic questions. Did Darwin so destroy the confidence of Sedgwick's successors? Was the Darwin thesis a mortal blow to the old way of doing theology from which it apparently still hasn't recovered? If so, then it is not Darwin who is to blame, but those who retreated with their theological paradigms untouched, withdrawing from further scientific exploration of the real world. Many religious thinkers have been profoundly grateful to Darwin for opening up new ways of understanding the natural world in its own right. There is no way of turning back the clock, either for scientists, or for faith communities. There is much more to be explored and considered. Creationism should not be seen or taught - as it still is in some places - as an alternative and therefore 'equal' explanation to evolution. Evolution fitted the facts as then understood in ways that creationism couldn't. Ironically the ghost of Sedgwick haunts us still. He was the Anglican clergyman who said truth is never the enemy, and yet used his own beliefs to turn mutability through natural selection into the enemy. Some might say the battles of Darwin's day are over. In this work, I have tried to explore something of their continuing importance in their own right and because of their influence on other, related concerns. Support for the 2009 anniversary of his birth show something of a continuing interest as do the many books still being published to defend or attack his ideas. I was recently surprise when speaking to a senior clergyman in the Church in Wales. We were discussing a talk I was to give to a group of clergy on this subject. He warned me that he was a creationist, preferring to believe the Bible rather than Darwin. The real surprise was that he didn't consider this to be a strange position to defend in any way at all.

The Entangled Bank

We have already seen the obvious influence of family and close relationships on Darwin's work. We know about the influence of political as well as scientific and theological ideas on thinkers of his times, for example the anti slavery movement. We know that he was horrified by what he saw of slave owners on the plantations in Brazil. Slaves were treated as a separate species and he was nearly thrown off the Beagle because of his views. This experience taught him something of the diversity of how humans could be treated, and this remorseless competition and struggle for survival in a human context contrasted with the ideas of liberty and fraternity which was his Enlightenment heritage. He arrived back from the voyage to see a rising population and social turmoil in London. Conditions were only barely tolerable for the poor, consumed by misery and the harsh nature of their conditions. This was the world so graphically described by **Charles Dicken's** *Hard Times* 1854, *Bleak House* 1853 and *Oliver Twist* 1838, **Benjamin Disraeli's** *Coningsby* 1844 and *Sybil*, 1945[290] **Elizabeth Gaskell's** *Mary Barton* 1848 and *North and South* 1855, **Rev Charles Kingsley's**[291] *Alton Locke* 1849 and *Yeast* 1848, Trollope, Tonna, Martineau and others including D.H.Lawrence and Thomas Carlyle. In France we mention only **Victor Hugo's** *Les Miserables* 1862, **Emile Zola's** *L'Assommoir* 1877 and *Germinal* 1885; in America, **Harriet Beecher Stowe's** anti-slavery novel *Uncle Tom's Cabin* 1852 and **Mark Twain's** social protest, *Huckleberry Finn* 1884. [292]

People had to compete for work or face the workhouse. No wonder Malthus influenced him and helped him find a key to the harsh and diverse nature of life on earth in the idea of a perpetual struggle, where survival really mattered. How much he imported from this rich experience, with all its light and dark shades, into his observation of other species is hard to say, and it would be wrong to project too much. He did study his own children's behaviours and physical characteristics in relation to those of primates and other animals. We do know that he did this without the luxury of modern genetics and DNA coding. Writer and poet, **Ruth Padel** comments on the qualities of her great-great-grandfather. *'It all fits together – from rocks to mammals through an experience and belief in connected relationships. Family was crucial to Darwin. His letters from the Beagle illustrate this as does his pride in his family.'* It is highly likely that his famous last paragraph description of *'an entangled bank,'* also reflected his experience of family life, and his wider observations of other people. Perhaps he was thinking of them when he said of nature that it was *'clothed with plants of many kinds.'* As I reread this passage and think of his large, extended family and its history, I sense connections he might himself have been making – *'and to reflect that these elaborately constructed forms, so different from each other, and dependent on each other in so complex a manner, have all been produced by laws acting around us. These laws, taken in the largest sense, being Growth with Reproduction; inheritance which is almost implied by reproduction; Variability from the indirect and direct action of the external conditions of life, and from use and disuse.'* As he contemplated the entangled bank of the human species over

---

[290] Published at the same time as Friedrick's Engel's *The condition of the Working Class in England* 1844 and of course the early writings of Marx

[291] One of the first novelists to praise Darwin's 'Origin' wehn sent a review copy before publication. In later editions of the Origin, Darwin wrote *'A celebrated author and divine has written to me that 'he has gradually learnt to see that it is just as noble a conception of the Deity to believe that He created a few original forms capable of self-development into other and needful forms, as to believe that He required a fresh act of creation to supply the voids caused by the action of His laws.'*

[292] See also Thomas Hardy's *A Pair of Blue Eyes* referring to fossils, published a year after the *Origin*

time, he followed with these words *'a Ratio of Increase so high as to lead to a Struggle for Life, and as a consequence to Natural Selection, entailing Divergence of Character and the Extinction of less-improved forms.'* He includes in this powerful image, the ideas of species development over time, as well as species variation over space, on this space-time 'entangled bank' which he observed so carefully and reflectively. He knew about the co-evolving, connected relationships between for example a flower and its pollinator, or the cooperative relationships that evolved between the Madagascan, ten inch tongued moth and certain local flowers. He gave us an insight into biodiversity as a system of competitive relationships, causes and effects. Yet, *'Thus, from the war of nature, from famine and death, the most exalted object which we are capable of conceiving, namely, the production of the higher animals, directly follows.'* Despite the horrors and tragedies of the real world of which he and his family are a part, we should perhaps be surprised that he could still conclude *'There is grandeur in this view of life, with its several powers, having been originally breathed into a few forms or into one; and that, whilst this planet has gone cycling on according to the fixed law of gravity, from so simple a beginning endless forms most beautiful and most wonderful have been, and are being, evolved.'* Darwin developed his theory of the laws of natural selection by reflecting back in time as far as he could go. Cosmologists share a similar exploration about the Big Bang. Physics and biology can only get so far, albeit very close to a cosmological and a biological common ancestor. We know roughly what the earliest, single cell animals were like as they appeared out of the sludge of bacteria, amino acids and then proteins, butut we are still trying to work out the conditions for those bacteria to exist and then the crucial catalysts for their development. We have tried to reproduce the conditions for the 'Big Birth,' and understand far more than even a generation ago, let alone the thinkers of Darwin's time. Similarly, we keep pushing back the age of the universe now we know it was not a steady state continuity. Perhaps Hoyle found it as difficult to accept the theory he labeled as the Big Bang, as Sedgwick and Wilberforce did with natural selection and Darwin. **Edwin Powell Hubble** (1889-1953) profoundly changed astronomers' understanding of the nature of the universe by demonstrating the existence of other galaxies besides the Milky Way. He also discovered that the degree of redshift observed in light coming from a galaxy increased in proportion to the distance of that galaxy from the Milky Way (Hubble's law). This proved Hoyle wrong and helped establish that the universe is expanding. Hoyle's critics had to battle to get beyond his paradigm that the universe was fixed and immutable, rather than expanding from a single point. This was as controversial as the implications of Darwin's claim that all species could be traced back to a common ancestor and were therefore not immutable or fixed. Since Darwin's time, embryologists, geologists, evolutionary biologists, geneticists, zoologists and others have worked more closely together within his paradigm of evolution, as well as from within their increasingly specialised areas. There is convergence as well as differentiation. We now understand much more about how coding triggers evolutionary steps and we have discovered more evidence of mid point mutations in the fossil record. How does that coding get laid down? How does the environment produce a biological mutation or selection at the coding level? Such questions include the best of our theories, observations, instincts, knowledge and equations, but also transcend them. You can only go back so far using observation based theories of the past. Even the mathematics seems to stop at the door of those two big questions of Big Bang and Big Birth. Any natural theology for this century has to face these questions with people from different disciplines, as they try to find the key to that door, or at least to look through the right key hole. The final frontier lies behind us as well as ahead of us in the work of people like Darwin. His influence on subsequent ways of thinking about our world was subtle and indirect, beyond just the study of biological processes. We make assumptions now about historical, human behaviour and culture which draw upon his influence and how he has been used, even without us consciously realising it. We might, for example, look back at wars as patterns of natural selection, whatever their immediate causes. We might see politics and culture itself as an adaption of the effects of wars, or competitive struggles between different groups over ideas, values, resources, or territory. Sometimes, we might see both war and culture overlapping as they evolve new ways of developing human ideas. For example, the Greek city states showed considerable cultural diversty and they fought amongst themselves. Ideas of democracy and theatre coalesced as spaces within which ideas could be tested and the tragedy of contemporary events explored in what Aristotle called the purging of *catharsis*. After the Peloponnesian wars, the culture of both democracy and theatre changed when Athens was defeated by Sparta. The culture of tragedians and actors moved out from Athens to cross the boundaries of enemy, city state territories. The great Phillip of Macedonia, father of Alexander the Great, sought cultural as well as military superiority. He used actors as diplomats or cultural mercenaries to help people adapt to new ideas. We might jump from the 5th and 4th centuries BC to the age of Frankish chivalry in the 10th-13th centuries AD. This part of human history should not be dismissed too lightly, despite the horrors of the Crusades. Chivalry gave Christian knights in Europe a sense of responsibility to *reach out* to each other across whatever differences divided them, bringing values and laws into the guidance of their conduct. I do not want to underestimate the violence involved in these conflicts, but rather underline this aspect of reaching out across *divisive* boundaries, in the name of a sustainable *connectivity*. Although there were economic reasons for some of the practices of chivalry - for example taking knights and their expensive horses for ransom in tournaments and in war - they also sprang from cultural and religious values.[293] The patterns of competition and adaption to changing power relationships were held together by at least some shared codes and laws. This may have held true only at certain times, and for certain knights, but it did appear to transcend various 'species' varieties of conflicting allegiances to different Barons and Kings.[294] In chivalry, there was at least a shared set of cultural values which influenced behaviour even in and across conflict between divided groups. If only that reaching out quality could have been sustained! The new weapons of guns and artillery were to increase the geographical distance and undermine the code. The responsibility for conflict was extended from the knights of landed families to the officer class of trained soldiers and mercenaries. Over time, the unthinkable began to happen. Much of this had been exemplified in the crusades and the crusade mentality used by the Pope to suppress theological diversity (including the Cathars in Southern France) where non-combatant 'civilians' were slaughtered. It was only a small jump to the excesses of genocide and the bombing of cities, although in WW1 there were occasionally examples of a reaching out across the dividing lines of no-man's land before returning to killing each other. It seems that our human capacity to reach out towards the other, for the sake of reconciliation across threatening divisions is very limited, despite notable exceptions. In domestic or national conflict resolution, we depend upon the inspiration of special individuals to find a way for each side to reach out beyond the boundaries of their divisions to a sense of underlying, connected and shared humanity. It is understandable, that we associate Darwin with conflict and competition in natural selection. This assumption would be later misused, but not always without justification. It was perhaps inevitable that the less palatable realities of evolutionary process would

 be transferred metaphorically and ideologically to human experience. Much of the attributed influence of Darwin came in fact from the phrase *'survival of the fittest,'*[295] coined not by him, but **Herbert Spencer** (1820-1903), who applied Darwinianism to the social realm, and we shall consider some of its effects later.

Spencer, author and date unknown. Portraits from the Dibner Library of the History of Science and Technology. P-D. Copyright expired

---

293 The Indian Mahabarrata (5th century BC?) discussed just cause (never out of rage and always after every attempt at reconciliation), proportionality (chariots cannot attack cavalry) and just means (no poisoned arrows) well before Augustine and Aquinas. In ancient Rome war seen as wrong or forbidden without a just cause – in theory to be declared by the priests. See Cicero's De Officiis, Bk 1, sections 1.11.33–1.13.41

294 The idea of acting as a 'gentleman' may date back to the gentle man knight and the laws of chivary across different kinds of rivalry.

295 *'This survival of the fittest, which I have here sought to express in mechanical terms, is that which Mr. Darwin has called 'natural selection', or the preservation of favoured races in the struggle for life.'* Principles of Biology of 1864, vol. 1, p. 444,

Darwin and Wallace had probably meant, in their use of this phrase, the capacity of species adaption, rather than any sense of physical fitness. *'In the future,'* Darwin wrote in *The Origin, 'I see open fields for far more important researches... Psychology will be securely based on the foundation already well laid by Mr. Herbert Spencer, that of the necessary acquirement of each mental power and capacity by gradation. Much light will be thrown on the origin of man and his history.'*[296] There were enough references and hints to the position and significance of the human species in the *Origin* to excite conflicted interests on all sides. We can only imagine the speculations about what was to come next and perhaps the expectations weren't to be fully satisfied. So, we need to take into account *The Descent* as well as *The Origin* and understand something of its place in the story.

The Descent of Man

In 1867, while waiting for the proofs of what he called his big book on *'The Variation of Animals and Plants under domestication,'* Darwin began work on what began as a small piece and then developed into a two volume work entitled *'The Descent of Man and Selection in Relation to Sex'* – not completed until 1870 and published in February 1871. There was obviously huge hesitancy in this process; perhaps because he thought the work had already been done. Wallace, Huxley and Lyell had already covered much of what he wanted to say, as had the German biologist and philosopher, **Ernst Heinrich Philipp August Haeckel** (1834 – 1919) in his *'History of Creation'* 1868. Haeckel's contribution was an important challenge to critics like **Richard Owen** who had focused mostly on the deficiency of the fossil record – with its little or no evidence of middle stage mutations of natural selection.[297]

Haeckel date and author unknown. The National Library of Medicine believes this item to be in the public domain.

Haeckel was a brilliant anatomist, artist and German Romantic[298] who was besotted with his cousin (a parallel with Darwin), Anna Sethe, who became his first wife, but died a few months after their marriage, leaving him feeling suicidal. He found a kind of religious purpose for his life in following Darwin and worked on reconstructing the evolutionary past via the embryo, rather the fossil record. He noticed that most vertebrates looked similar before their limbs and head appeared. As he projected backwards, he concluded all animals must have come from a gastrulation in a ball of cells that moved through the pre-Cambrian seas.

---

[296] *'Complete works of Charles Darwin online,'* editor John Van Whye, p 576.

[297] Since 1950s, fossils from pre Cambrian rocks have been found showing transition from land to sea, with evidence of how front limbs become flippers and rear ones withered away (whales have traces of ankle bones similar to cows and pigs and the DNA proves and link between whales and hippos as evolving from another a common ancestor about 55 million years ago.

[298] Credited with first calling the European War the First World War September 1914.

On this basis, he attempted to put every stage of development in order on his first evolutionary family trees. He was certain of a connectivity under the surface of the diversity of 'endless forms' and behaviours. He couldn't have known the internal processes, but put the first gastrula at the base of that tree. In our times, individual cells and some proto-creatures, not that different from Haeckel's gastrula, can be seen in computer imaging. Through **Crick** and **Watson's** work we can use DNA to date the past and this confirms animal life in the pre-Cambrian. Without fossil evidence of a link between apes and humans, Haeckel 'invented' an intermediate form (Darwin's term) he called 'ape man'- Pithecanthropus, and imaginatively challenged his students to find it. The Dutch geologist **Eugene Dubois** (1858-1940), greatly influenced by Haeckel, dug up the remains of 'Java Man' in the East Indies and this was later reclassified as Homo erectus. Haeckel believed that all embryos are similar and then diversify, based not just on a struggle for survival, but a law of systems which were the engine of organised life forms. We now know this is triggered by molecular signals from the HOX genes which determine the type of segment structures that will develop. Haeckel beleived natural selection rationally explained why things looked as if they had been designed, even if it didn't explain predicatively what might have evolved – for which there were too many possibilities. One of the limits of natural selection was that the changes made over time in any one form could only be reconstructed retrospectively. Unlike Newton's mathematical laws of planetary motion, the 'law' of evolution was not fixed, but ruled by apparent chaos and contingency. There may be, in the near future, ways of understanding more about the internal geometry of living things and the laws of evolution which limit if not predict the pathway patterns of development and change. As we think ourselves back to Haeckel and Darwin, we wonder at their ability to intuit and observe as much as they did, and have the confidence to theorise about the laws of nature in a process that seemed so unpredictable. In writing *The Descent*, Darwin had been full of self doubts and even wondered whether the book was worth publishing – a concern shared apparently by his publishers. Looking forward he could not have predicted the outcome. It sold out and Hooker commented that he regularly dined out – at least three days a week – on the *Descent of Man* issues, which had become a 'normal' part of conversation amongst the chattering classes of the day. The major scientific reviewers were surprisingly un-shocked, but also unconvinced. The Edinburgh Review said that the *Descent 'had raised a storm of mingled wrath, wonder and admiration'*, but in fact the storm was little more than a drizzle. The Times reacted more critically – partly because it hadn't covered the *Origin*. Darwin's ideas were seen as mischievous as they were unscientific. The biggest fear was probably close to **Gertrude Himmelfarb**[299] commentary that *'morality would lose all elements of*

---

[299] *Darwin and the Darwinian Revolution* 1959 Chatto and Windus London. An American historian well known for her

stable authority'. Of course, the 1870s were explosive times, politically in Paris and Europe. The Times was outraged that Darwin's arguments seemed to be unscientific and, therefore, reckless at a time of political unrest and volatility. Again, according to the reflections of Himmelfarb, *'no man will ever develop religion out of a dog or Christianity out of a cat'*. The main concern at the time was to protect the distinctiveness of humans from non-human species by discussing the former's legal and moral conscience. During these years the variety of reactions to the implications of *The Origin,* and later *The Descent,* ranged across many different interests and emotions. Criticism was often tempered with praise; skepticism with discussion; outrage with indifference.

The Roman Catholic biologist, **St. George Jackson Mivart** (1827-1900) offered the

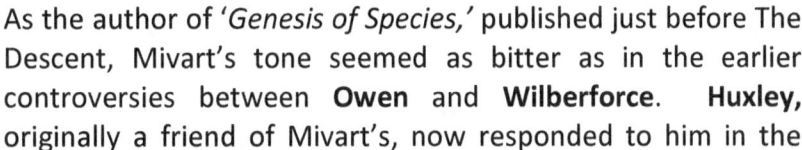

strongest criticism in any of the review articles.
Mivart date and author unknown. P-D copyright expired
As the author of *'Genesis of Species,'* published just before The Descent, Mivart's tone seemed as bitter as in the earlier controversies between **Owen** and **Wilberforce**. **Huxley,** originally a friend of Mivart's, now responded to him in the strongest terms, pointing out that public reaction to Darwin had changed, as evidenced by a comparison of the Quarterly Review's treatment of *The Descent*, compared with *The Origin* a decade earlier. Mivart originally agreed that, as a physical creature, man had evolved from the animals. Then in 1873, Darwin's son George published a short article in *The Contemporary Review* proposing that divorce should be made easier in cases of cruelty, abuse or mental disorder. Mivart was horrified and called it 'hideous sexual criminality' and 'unrestrained licentiousness.' Huxley and Darwin both broke off their relationship with him after this. Darwin had taken Mivart seriously in his later editions of *The Origin.* Mivart had questioned the inability of natural selection to explain the emerging stages of useful structures. Darwin responded by showing the different stages of light sensitivty in the eyes of different animals arguing for intermediate development. Mivart was also concerned that natural selection seemed helpless to explain examples of parallel evolution. Huxley replied that this would happen in places with the same environment. Mivart then wrote his *Nature and thought* in 1882, and the *Origin of human reason* in 1889, arguing for the *soul* as the ultimate and radical distinction between human and non human species. Mivart, quoting Augustine and Aquinas, was glad to emphasis the distinguishing presence of a divinely imparted soul, as indeed the Catholic Church would have been at that time, and probably ever since. The fact that Mivart was reinforcing a commonly held dualism between the physical nature of the person and their soul didn't seem to trouble the Church at the time. Ironically, the Catholic Church was

studies of Victorian Britain, including Adam Smith and Thomas Malthus,

later to excommunicate Mivart for his 'liberalism,' though not over this particular issue. In Darwin's autobiography, *Recollections of the Development of my Mind and Character,* published five years after his death in 1887,[300] he made his peace with all his critics, with the exception of Mivart. It was written in 1876, and probably with his extended family and close friends in mind. It was edited before publication by his son Francis who removed some passages concerning Darwin's critical views of Christianity. Francis was obviously sensitive to the debates that still continued and the risk that many would dismiss Darwin's work because of his views on Christianity. The passages were restored by Darwin's granddaughter, **Nora Barlow** (1885-1989)[301] in a 1958 edition.[302] *The Descent,* in its original edition, was in two volumes – the first covering the descent of man and the second, sexual selection – as applied mostly to insects, fishes, birds and mammals. Ironically, sexual selection, with its far more important role in *The Descent* than in *The Origin,* referred to the argument of *The Origin* as not so much natural selection but *'the principle of evolution'. 'For my own part I conclude that of all the causes which have led to the differences in external appearance between the races of man and to a certain extent between man and the lower animals, sexual selection has been by far the most efficient.'* [303] Wallace had argued in *The Development of Human Races under the Law of Natural Selection of* 1864 that races must have developed at an early point as humans moved and reacted to different physical contexts. He believed that the differences between races became fixed at this time because there was no longer any need for further evolution of man's physical structure - clothes replaced hairiness; technology replaced physical adaption. Darwin explained the differences more in terms of sexual selection – each tribe developing in its own way *'a slightly different standard of beauty.' These different standards of beauty can be deduced from the elaborate practices of self decoration and self mutilation, all encouraging the selection of variations to exaggerate difference from neighbours.'* [304] Darwin saw speech, music and cultural patterns as being derived from the courtship patterns of our ape like ancestors. Both Darwin and Wallace produced ambiguous arguments for differentiated, social development. Darwin claimed that the 'sweeter voices' of women, for example, were acquired in order to attract the male. Gertrude Himmelfarb, in her chapter on *The Origin of Man* in *'Darwin and the Darwinian Revolution',*[305] points out the thinness of

---

[300] By John Murray as part of *The Life and letters of Charles Darwin including an autobiographical chapter.*

[301] She was perhaps one of the first Darwin scholars. Daughter of Sir Horace Darwin and Lady Ida Darwin (née Farrer), she studied plant genetics under William Bateson at Cambridge and published two papers on plant genetics 1913 and 1923 drawing on her Grandfather's *The Different Forms of Flowers on Plants of the same Species.* 1877, John Murray, London

[302] The Autobiography of Charles Darwin 1809-1882. With the original omissions restored. Edited and with appendix and notes by his granddaughter Nora Barlow.

[303] *Descent* Part 2 pages 389-90

[304] *Descent* Part 2 page 370.

[305] P 290

the argument as it developed in The Descent. She argues that Darwin has practically abandoned natural selection by this time, when faced with the development of human from non-human species or for the distinctions of race and sex. Natural selection had assumed that beneficial variations alone would be reserved. Amongst different human races and ethnic groups, variety and difference did not have any simple, natural selection cause. According to Himmelfarb, Darwin slowly realized that sexual selection could work at cross purposes with natural selection, if 'sexual admiration' interfered with the more elementary struggle for existence. Himmelfarb argues that the '*Lamarckian principle of the inherited effects of use and disuse came to replace natural selection'*. She cited the '*smaller tail in some monkeys and its absence in man, the development of the vocal organs and power of speech, the thin legs and thick arms of Indians who spent most of their lives in canoes, the larger hands of English labourers compared with those of the gentry, the hardened skin on the soles of the feet, the inferiority of Europeans compared with savages in sight and other senses, customs such as the liberate eradication of hair and other mutations.'*.[306] So Darwin himself was able to finally confess, as we saw earlier, in The Descent Part 1[307] '*a very large yet undefined extension may safely be given to the direct and indirect results of natural selection; but I now admit.......that in the earlier editions of my Origin of Species I probably **attributed too much to the action of natural selection or the survival of the fittest**.....I had not formally sufficiently considered the existence of many structures which appear to be, as far as we can judge, neither beneficial nor injurious; and this I believe to be one of the greatest oversights as yet detected in my work. I may be permitted to say as some excuse, that I had two distinct objects in view, firstly to show that **species had not been separately c**reated, and secondly, that **natural selection had been the chief agent of change,** though largely aided by the inherited effects of habit, and slightly by the direct action of the surrounding conditions. Nevertheless, I was not able to annul the influence of my form of belief, then widely prevalent, that each species had been **purposely created**; and this led to my tacitly assuming that every detail of structure, excepting rudiments, was of some special, though unrecognized, service. Anyone with this assumption in his mind would naturally extend the action of natural selection, either during past or present times, too far..If I have erred in giving **to natural selection great powe**r, which I am far from admitting, or in **having exaggerated its power**, which is in itself probable, I have at least, as I hope, done good service in aiding to overthrow the **dogma of separate creations.'* Himmelfarb maybe one of the few commentators who quote this passage and in the 2009 Darwinian centenary celebrations, it was largely, absent from the debate. What it does show is his strength of purpose in

---

[306] Himmelfarb page 301.
[307] *The Descent*, 1871, John Murral. pp 152-153. My emphasis in bold

addressing what he described *'as the dogma of separate creations'*. It is understandable that in addressing the strength of that position, which he saw as no longer tenable because of natural selection, he needed to admit much later that there were some things that natural selection in isolation could not explain, but not because separate creation had the answer! Himmelfarb believes that Darwin had moved some distance from the doctrine of natural selection at this point and in his own words admitted that some variations arise 'spontaneously' in ways that weren't subject to any selective process, either natural or sexual, or due to the effects of *use and disuse* or from the environment. Darwin admitted the nature of the cause of this kind of variation was unknown. Given that it acted 'uniformly and energetically' over a long period, the result would be the production not of *'mere slight individual differences but well marked constant modifications.'* [308] Darwin attempted to modify this concession and minimize its effects in the second edition of The Descent – fearful that people like Mivart would see this as opening the way for a stop-gap God to appear. Himmelfarb quotes the Scottish **Rev Henry Drummond** as an example of this[309] and also believes that a case can be made for theological theories of development having always been part of Christian teaching, so preparing the ground for Darwin, not least in the work of Cardinal Newman. *'In the Catholic Church, Darwinism had a history and literature of its own, As had already been observed Catholics were, in some respects, better situated than Protestant to accept Darwinism. Depending rather on the authority of the Church than on Scripture they were not bound by the literal Biblical texts; ecclesiastical authority could and did interpret the Biblical account of creation as it saw fit. It was indeed the purpose of Newman's theory of development which anticipated the Origin by over a decade to provide a rationale for the evolution and progressive reinterpretation of dogma. Liberal Catholics wanting to exploit this scriptural freedom and at the same time wanted to endear themselves to other liberals besought the church not to intervene in scientific quarrels. On the eve of the publication of the Origin, Newman himself urged this course upon the hierarchy, not only because Catholics deserved to be spared any addition to the heavy load of controversy which they already bore but also because by copying 'the divine wisdom in not making formally binding the old accepted cosmology' they would demonstrate to the world 'the divinity of their religion.'* All very well, but this view from this great thinker somewhat underplays the negative hierachical reaction to most earlier science, to Newman's theory of development itself and, as Himmelfarb admits, in 1864, Pope Pius IX issued the Syllabus of Errors condemning 'progress, liberalism and modern civilization,' including Erasmus Darwin's Zoonomia, though not the *Origin* by name, though that was implied. As late as 1950, the Encyclical Humani

---

[308] Descent Part 1 p 153
[309] ibid p 325

Generis *'condemned the idea that evolution could account for the origin of all things on the grounds that such an extension of the theory was conducive to a variety of heresies – monism, pantheism, dialectical materialism, idealism, immanentism, pragmatism and existentialism – in short all those 'false opinions' as the subtitle put it 'which threaten to sap the foundation of Christian teaching.'* [310] It is interesting to see how this encyclical lists together many conflicting intellectual and scientific ideas as being false, and to reflect on the assumptions behind such a church statement, dated 1950 not 1850! Himmelfarb continues *'But it conceded that a more modest theory of evolution concerning itself only with the physical development of the human body would avoid this litany of errors and be a legitimate subject of research...'* provided of course it didn't question the existence of the soul. If Darwin had specifically brought the soul into his idea of natural selection, then the statement would have read very differently! With the soul, as it were, secure in their back pocket, many Catholics embraced the theory of evolution. These decades demonstrated complicated alliances of what Himmelfarb calls reconcilers and irreconcilables[311] on all sides. **Huxley**, heavily engaged in debate with **Gladstone**, was desperate to find some way of clarifying and healing the debate –*'There must be some position from which the reconcilers of science and Genesis will not retreat'*[312] Many conservative people retreated behind the negative religious response to protect traditional values, against what they saw as the materialistic implications of Darwin. Many adopted evolution as a quasi theology of nature to protect them against the conservative and worn out ideas of religion. If the religious conservatives feared that evolution was code and cover for other new ideas, including atheism, undermining everything that mattered, the supporters rejoiced in the idea that an impoverished view of a monarchial and hierarchical Deity intervening and controlling human affairs had at last been unveiled as a sham. Many believers privately struggled to bring their theological assumptions up to date, or live with the tensions created. Many found it hard to reconcile an omniscient God with the slow progress of clumsy nature, groping her way forward. They feared that a new kind of God of Nature was being created in ways that would naturally lead to its own demise. Was the new evolutionary 'theology' vulnerable to its own laws of evolution and would that lead to more Darwins and a less confident, less relevant and diminishing church? The process may not have begun with Darwin, and can be glimpsed at many points in the previous century, but publication of the *Origin* and *Descent* had brought it well into the popular, public domain. Himmelfarb quotes the English historian, **James Froude's** (1818-

---

[310] Idid p 327

[311] Ibid p 328

[312] Ibid p 328 citing Huxley *The interpreters of Genesis and the Interpreters of Nature* 1885 p 89

1894) work *'Carlyle; his life in London'*[313] to depict the atmosphere in the 1840s – one that Darwin must have sensed, well before he plucked up courage to publish the *Origin*, which both reflected and extended this atmosphere. *All round us, the intellectual lightships had broken from their moorings and it was then a new and trying experience. The present generation which has grown up in an open spiritual ocean, which has got used to it and has learned to swim for itself, will never know what it was to find the lights all drifting, the compasses all awry and nothing left to steer by except the stars.* Himmelfarb believed this atmosphere was part of the age even without Darwin's *Origin* and refers to the writings of **Tennyson**, **Ruskin**, **J. S. Mill** and **Arnold** as illustrations of this growing sense of loss and despair.[314] Students of the Victorian age will have their own views of how this can be understood alongside the great achievements and confidence of the times, including the Reform Act of 1832, Victoria's crowning in 1837 and epitomized by all that led up to the Great Exhibition of 1851. The Pax Britannica would be dented by the Crimean war of 1854 and the spirit of the previous decade was still affected by many political and social changes including the Irish Famine of 1845, the Repeal of the Corn Laws in 1846 and the Cholera epidemic of 1848. Theologically, the perceived shift was from a God who had omnisciently made everything as part of the providential and pre-determined laws of the universe, to a universe that was based on its own laws. What if those laws led to chaos and the overturning of order and authority? Within those laws, theologians had to find a a place for God, fitted into an ever smaller religious corner, until even that too might be locked away in a dusty room that no one cared about. It was as if those who framed and defended this original view of God would always be fighting rear guard actions on increasingly less credible ground for belief. The real sense of loss in a divinely controlled providence was a gap that had to be filled or redefined. With or without a conscious connection with the effects of Darwin on religious beliefs, the Victorians turned to spiritualism, burial and mourning rites as a way of dealing with an after world without formal religious belief. They could manage a global Empire and still go to church; they could talk of reforming the causes of poverty, but something underneath the edifice was crumbling away.

Evolution, Nature and Morality

Moral behaviour[315] had depended upon belief and obedience to religious commandments and teaching. As the ground under that began to show signs of sinking, a new basis for morality was needed. Herbert Spencer had used the basic

---

[313] 1, p 290-291. He had thought about taking Anglican orders, but then published his controversial novel *Nemesis of faith* in 1849 revealing his doubts about doctrine.

[314] Ibid p 330

[315] See the fuller discussion in the later chapter on Sociability and Altruism

principles of evolution to create a philosophy and science of ethics. This needed its own justification to motivate support. Moral conduct was redefined as behaviour leading to higher values in adaption and selection. Utilitarian morality and happiness were brought together as *higher stages* of evolution. The moral person combined the promptings of his nature and his individual needs with the needs of wider society. The morality of individual happiness was a question of social utility in the evolutionary process. But others warned that neither *'survival'* nor *evolutionary process,* in themselves, would necessarily promote the greatest happiness of society or the individual. Others feared that the implied determinism of evolution undermined any sense of free choice in questions of moral agency (see section of Freedom in LE2). Theories of evolution were not capable of producing any moral code, in themselves, but only described 'facts' rather than values. Paley's Natural Theology had appeared optimistic about the benevolence of nature. Darwin's *Origin* described something altogether different. Others before Darwin had warned not just of its neutrality but its 'cruelty.' The **Marquis de Sade** (1740-1814), famous for his attacks on the Catholic Church and as an advocate of complete individual freedom had written, as only he could perhaps write, and very much in contrast to the romantic poetry of **William Wordsworth** (1770-1850) and **Samuel Coleridge** (1772-1834) *'I tell you nature lives and breathes by crime; hungers at all her pores for bloodshed, aches in all her nerves for the help of sin, yearns with all her heart for the furtherance of cruelty..Unnatural is it? Good friend. It is by criminal things and deeds unnatural that nature works and moves and has her being..If we would be at one with nature, let us continue to do evil with all our might.'*[316] These contrasting views of nature could be found in classical literature. The point is that Darwin didn't invent this view of nature. He confirmed it scientifically. That which de Sade had called cruelty, Darwin saw as part of the governing force of nature and part of natural process. **Huxley** did more work than Darwin on the moral implications of the *Origin* and drew upon the ideas of Spencer, as well as his own experience – including the sudden death of his four year old son in 1860. The theological questions and the moral dilemmas involved in this kind of loss, as in Darwin's experience, were very real. **Huxley**, like many of his contemporaries, came to believe that the processes of nature, rather than a divine judge, imposed their own rewards and punishments. It was as if the less people believed in a providential creator and after life, the more they had to turn to nature as the source of meaning and purpose in its processes. Darwin never produced, even in the *Descent,* a developed science of morality. Huxley was to write in 1893 *'a real and living belief in that fixed order of nature which sends social disorganization upon the track of immorality, as surely as physical disease after physical*

---

[316] Ibid p 332 citing Swinbourne's *Blake* p 158

*trespass.'[317]*Huxley came to reject the ideas of **Spencer** and the two clearly fell out over this. Huxley felt that the ethics of evolution were indeed a real threat to the morality of society. He became convinced that Spencer's *struggle of the fittest* for existence shouldn't be projected onto human society. Nature was neither benevolent, nor did it improve morally with evolution. Even if, over longer periods of time, there were signs of linear progression, they were achieved at great cost of suffering and death. However realistic this might be, it was no moral template. Huxley came to see evolution and ethics as the great opposites, albeit caught in a paradoxical relationship. Humans were capable of moral progress but were trapped in the struggle for survival as part of nature's processes. So, it seemed, we had to develop beyond the higher primates only by acting like them! In *Evolution and Ethics* he said *'The optimistic dogma (of progression), that this is the best of all possible worlds, will seem little better than a libel upon possibility.... From the point of view of the moralist the animal world is on about the same level as a gladiator's show.'[318]* *'The best of all possible worlds'* was probably a reference to Leibnitz's use of this phrase in his theodicy. In 1859, significantly in the same year as the *Origin's* appearance, **Dickens** published his *Tale of Two Cities,* set in Paris and London at the time of the French Revolution. It opens with the well known and haunting description which Victorian readers would have recognised as applying to their own, as well as the end of the 18th century.*'It was the best of times, it was the worst of times, it was the age of wisdom, it was the age of foolishness, it was the epoch of belief, it was the epoch of incredulity, it was the season of Light, it was the season of Darkness, it was the spring of hope, it was the winter of despair, we had everything before us, we had nothing before us, we were all going direct to Heaven, we were all going direct the other way...'* If this was the beginning of the novel, with its pattern paradoxes of moral and social certainty and uncertainty, the novel ended with the altruistic, self-sacrifice of its 'anti-hero.'

By the 1890s, **Huxley** had concluded there could be no ethics based on evolution (he revealingly called this the 'the cosmic process'). In that famous lecture *Evolution and Ethics* he revealed the reach of the wider moral debate, triggered by Evolutionary theory and the Victorian capacity to set it in a deeper, philosophical context.[319] *'As I have already urged, the practice of that which is*

---

[317] *Evolution and ethics* 1893 p 146

[318] Ibid p 196

[319] The prestigious *Romanes* (famous biologist) lecture, given in the Sheldonian Theatre Oxford. Huxley in 1893 followed Gladstone, the first lecturer in 1892. Huxley's was extraordinary in its breadth and depth covering ancient religions and philosophies (with detailed scientific notes) as context for discussing justice and ethics in the face of suffering e.g *Theories of the universe, in which the conception of evolution plays a leading part, were extant at least six centuries before our era. Certain knowledge of them, in the fifth century, reaches us from localities as distant as the valley of the Ganges and the Asiatic coasts of the Ægean. To the early philosophers of Hindostan, no less than to those of Ionia, the salient and characteristic feature of the phenomenal world was its change[54]fulness; the unresting flow of all things, through birth to*

ethically best–what we call goodness or virtue–involves a course of conduct which, in all respects, is opposed to that which leads to success in the cosmic struggle for existence. In place of ruthless self-assertion it demands self-restraint; in place of thrusting aside, or treading down, all competitors, it requires that the individual shall not merely respect, but shall help his fellows; its influence is directed, not so much to the survival of the fittest, as to the fitting of as many as possible to survive. It repudiates the gladiatorial theory of existence. It demands that each man who enters into the enjoyment of the advantages of a polity shall be mindful of his debt to those who have laboriously constructed it; and shall take heed that no act of his weakens the fabric in which he has been permitted to live. Laws and moral precepts are directed to the end of curbing the cosmic process and reminding the individual of his duty to the community, to the protection and influence of which he owes, if not existence itself, at least the life of something better than a brutal savage' and elsewhere *'social progress means a checking of the cosmic process at every step and the substitution for it of another, which may be called the ethical process; the end of which is not the survival of those who may happened to be the fittest...but of those who are ethically the best.. The ethical progress of society depends not on imitating the cosmic processes, still less in running away from it, but in combating it.'[320]* But what values should we draw upon for this *'combat'* and if our moral nature has evolved over millennia how can we hope to effect rapid improvements? Toward the end, he wrote *Moreover, the cosmic nature born with us and, to a large extent, necessary for our maintenance, is the outcome of millions of years of severe training, and it would be folly to imagine that a few centuries will suffice to subdue its masterfulness to purely ethical ends. Ethical nature may count upon having to reckon with a tenacious and powerful enemy as long as the world lasts. But, on the other hand, I see no limit to the extent to which intelligence and will, guided by sound principles of investigation, and organized in common effort, may modify the conditions of existence, for a period longer than that now covered by history. And much may be done to change the nature of man himself.* In his original lecture, he was forbidden to comment directly on religion or politics. When it came to publishing, he made his politics more clear – his concern was that no regime should turn into a tyranny by misusing science (the individualism of Spenser, social collectivism or eugenics). He himself was caught in a dilemma. If he denounced the ethics of evolution, he was playing into the hands of its religious opponents. He admitted privately, in his letters, that he would rather support the faulty doctrines of predestation and the original sin of human

---

visible being and thence to not being, in which they could discern no sign of a beginning and for which they saw no prospect of an ending. It was no less plain to some of these antique fore-runners of modern philosophy that suffering is the badge of all the tribe of sentient things; that it is no accidental accompaniment, but an essential constituent of the cosmic process.

[320] Idid pp 81, 83

depravity, than the illusion that evolution can morally better us, even through its cruel processes.[321] The issue of whether evolutionary theory has the capacity to produce or be a basis for morality is a hard and much disputed question. Some evolutionary biologists, for example **Theodosius Dobzansky** (1900-1975) rejected the idea that it could produce anything like the ideal of the greatest good. Moral philosophers such as **Henry Sidgwick** (1838-1900) argued that it could not even be used to evaluate or create an ethical system. Many critics rejected the view that *survival* itself is a moral ideal, rather than a *natural* one. If survival was the *summum bonum*, then all kinds of competitive cruelty or criminal behaviour might be morally justified. On the other hand, there is evidence of individuals sacrificing their own wellbeing or lives for the sake of the survival or wellbeing of others. **Michael Ruse**[322] and other ethicists saw the relevance of natural selection as part of the development and adaption of cooperative and altruistic behaviours. Evolutionary psychologists are still debating to what extent humans can harmonise their beliefs and behaviour in relation to their inherited and innate pre-dispositions, as part of the evolutionary process and we will return to this in the chapter on Altruism. Darwin was part of Victorian Society and its values of moral duty. Even some agnostics, and many opposed to religion, were willing to defend its role as a basis for sustaining the values of duty – suspending their own disbelief in order to promote morality! **Huxley** agreed to the baptism of his son and, like **Hooker**, wrote in favour of Bible teaching in schools to inspire respect for knowledge, truth and pure living. Even **Mill,** in his last essay, hinted that belief in a morally perfect divine Being was a more effective guide to morality than any *rational* ideal! There are still those who never go to church, but value its established moral role in society, functioning as if by osmosis. To some extent, there was a continuity of moral assumptions across the nineteenth century which was unaffected by the impact of Darwin on ethics, even though the way people thought of God as Creator could never be the same again. Some who were initially seduced into believing that science had taken over the role of ethics and metaphysics, were still haunted by what Wallace called the *'Unknown Reality'* and Spencer the *'Unknowable'* as the source of ethics. Nature provided no template for morality and Darwin gave up none of his Victorian sense of duty because of his theory of evolution. He had felt himself on safer ground in the *The Origin* and his study of the pigeons and other animals than his work on humans in *The Descent.* The human species was clearly more complicated, not least because, even over long periods of time, they had not evolved physically as other species do. In *The Descent,* he catalogued human attributes related to similar featuers in non-human species – the experience of pleasure and pain, terror, suspicion, curiosity and affection, as well as the powers of imitation, attention,

---

[321] *Life and Letters 111, p 220-221*

[322] *Evolution and ethics; A phoenix arisen.* Zygon Journal of Religion and Science, 21, p. 99.

memory, imagination and reason. He observed his own dog demonstrating evidence of both memory and faithfulness; birds on the Galapagos Islands as behaving 'intelligently' when they avoided contact with human beings; monkeys who were commonly afraid of reptiles; arctic wolves acting rationally when avoiding thin ice; the power of imagination in birds when they 'dreamed'. Darwin saw evidence of logical if not moral reasoning in the case of a retriever dog confronted with one wounded and one dead duck - choosing to kill the wounded so he could retrieve both – a case that could be argued in different ways. Chimpanzees demonstrated their ability to use tools and even primitive architecture – for example a baboon throwing a straw mat over his head for sun protection. Nor did he see language as unique to the human species. One species of monkey had as many as six sounds *'which excited other monkeys' similar emotions,'* and while dogs barked in several distinct tones. His project was to sustain the relationship and connection between human and non-human species, even though he admitted the gap between different kinds of human was never as great as that between humans and non-humans. Himmelfarb is critical here - although he tried to sustain the links within and between species, including the lowest and simplest organisms and the higher apes, in the transition between non-human and human species, his explanation *'petered out.'*[323] He admitted that the difference between human speech and non-human speech could only be explained by the development of a human brain superior to that of the ape. Therefore, there must have been *'some early progenitor of man'* with mental equipment distinguishing him from the ape. In trying to sustain the link between human and non-human species, he used the idea of morals as well as language, but didn't sufficiently distinguish morality from sociability. In a footnote in the *Descent*[324], he argued against **Mill**'s view that the human morality was acquired rather than innate. *'It can hardly be disputed that the social feelings are instinctive or innate in the lower animals; and why should they not be so in man?'* He later re-established a connection between sociability and morality, suggesting that actions were right or wrong depending on how they affected the welfare of the herd, or tribe. Was this sociability and morality instinctive, culturally learnt or a matter of utilitarian reasoning? The problem was that both sociability and morality could be undermined by individual and tribal behaviour that was clearly not utilitarian, either instinctively or by reason. **Himmelfarb** compares his attempt to see language and religion in non-human species. The religious sensitivity was influenced by love, submission, fear, reverence and gratitude - all with their intellectual and moral implications as part of belief in unseen or spiritual agencies. By contrast, she cited a German professor who held 'that a dog looks on his master as on a God'. In the same passage Darwin listed all the

---

[323] page 305.
[324] Part 1 page 71

negatives of human religion, including animal sacrifice and trial by ordeal, celibacy, rules of caste and food as *'absurd rules of conduct'* and *'absurd religious beliefs, operating in complete opposition to the true welfare and happiness of mankind.'* [325] Himmelfarb saw Darwin's practice *'of seeking explanations in the lowest common denominator – morality, in terms of instinct; human motives, in terms of animal impulses, and civilized conduct in terms of primitive customs as a professional failing.'* [326] She viewed his discussion of religion, morality, philosophy and aesthetics as being painfully naïve. **Wallace** reluctantly concluded, despite his respect for Darwin, that natural selection alone could not account for the emergence of man as an intellectual or spiritual creature. Wallace explained the disproportionately larger brain of early humans as only explicable if some special agency, analogous to that which at first produced organic life and then consciousness *'had created a species not once removed from the ape but at least potentially many times removed from him.'* [327] Wallace argued that unless research uncovered the intermediate stages of human evolution, the human brain must have come into existence much later and more rapidly than Darwin's theory allowed. This reminds us of **Huxley's** dispute with **Owen** over brain size and function and we acknowledge the gaps in our own knowledge of the evolution of the human brain. When Himmelfarb was writing, she knew of Piltdown man and paleontological methods to reconstruct skeletons from basic fragments of bone. This was a time when Piltdown was exposed as a fraud because no connection had been established between the ape-like jaw and the man-like brain. Human evolution seems to have developed at a faster rate than some assumptions of the theory of natural selection – where vast eons of time were needed. The Piltdown fraud shouldn't be used, as it clearly was in Himmelfarb's time, to discredit the wider case for natural selection.

Science and culture –difference and division
Sir Edward Burnett Tylor's (1832-1917) important book *Primitive Culture* appeared in the same year as *The Descent* in 1871. Tylor had been influenced by **Charles Lyell's** evolutionary theories and **Mathew Arnold's** (1822-1888) view of high culture in his *Culture and Anarchy* (1869). Tylor defined culture on the first page of his book as *'that complex whole which includes knowledge, belief, art, morals, law, custom, and any other capabilities and habits acquired by man as a member of society.'* [328] His was a non-biological understanding of culture as the factor which distinguished the human from other species. Culture was universal and accumulative. **Franz Boas** (1858-1942), the German-American pioneer of

---

[325] The Descent Part 1 page 99
[326] Himmelfarb page 307.
[327] ibid page 308.
[328] Primitive Culture 1 (3d ed. 1889).

anthropology, applied the scientific method to the study of human cultures. Interestingly, his doctorate was in physics and his post doctoral work in geography. Boas said "*In the course of time I became convinced that a materialistic point of view, for a physicist a very real one, was untenable. This gave me a new point of view and I recognized the importance of studying the interaction between the organic and inorganic, above all the relation between the life of a people and their physical environment.'*-He introduced the idea of separate, cultural development. He saw the danger of multi-culturalism replacing multi-racialism but implying that people are locked into their own separate and monolithic, anthropological cultures, beliefs and values, with little hope of shared connectivity across them. Some definitions of culture, following Tylor, tended to emphasis the difference between humans and other species based on e.g. ability to use tools. The human brain was seen as distinctively different from other primate brains (language etc) and therefore capable of a superior culture. But, we now know that other primates and some other species have their own ability to use tools to some degree, and in that sense share elements of an 'evolutionary' culture.

In Britain, the Royal Society of Arts, founded in 1754, and given a royal charter in 1847, symbolised the multi-disciplinary understanding of culture. Its full title was the Royal Society for the Encouragement of Arts, Manufactures and Commerce.[329] Its notable members included Charles Dickens, Adam Smith, Benjamin Franklin, Karl Marx, Stephen Hawking. The sciences, including theology, had come to mean the reliable sciences. In the 1870s, a sense of distinction between the arts as literary, imaginative production and the sciences as rational, experimental enquiry into nature had developed, and the division became increasingly intense, not least in Utilitarian philosophy and Unitarian circles. The influence of science as a separate and dominant model for education began. This called established assumptions into question and threatened many. In 1880, Thomas Huxley, gave a lecture on science and culture at the opening of University College in Birmingham - which had announced it would not teach classics and the arts, but focus on industrially useful subjects. Huxley claimed the

---

[329] The Albert Memorial in London, commissioned in 1861 and opened in 1872 is perhaps the most impressive physical statement of this inclusive Victorian vision of culture. It shows the historical figures King David and Homer for poetry, Appeles, and Raphael for painting, Solomon and Ictinus for architecture and Phidias and Michelangelo for sculpture. The canopy features eight statues representing both the arts and the sciences– astronomy, geology, chemistry, geometry (pillars) and rhetoric, Medicine, Philosophy and Physiology (niches).. The Freeze of Parnassus round the base depicts 169 individual composers, architects, poets, painters, and sculptors. At each corner there are 2 allegorical sculptural groups – agriculture, commerce, engineering and manufacturing and 4 groups showing the global perspective of Empire – Europe, Asia, Africa and the Americas. At the top of the canopy is a gold cross under which are eight statues representing moral and Christian virtues – Faith, Hope, Charity, Humility, Fortitude, Prudence, Justice and Temperance. The assumption behind the monument's complexity was that its audience had enough cultural education to understand what was depicted and to make the same kind of global and historical connections across the arts, the sciences and manufacturing.

right of sciences over culture. He named **Arnold** in his lecture. They had been close friends, but he saw Arnold as the spokesman for literary culture. Arnold had written in 1882 that he saw a training in the sciences to be of no value to anyone except those becoming a professional scientist. Only the arts could develop a critical social capacity and a cultured personality. Arnold did refer to **Newton**, but showed little appreciation of the significance of the new science. Arnold defended human value based on classical, cultural education. Paradoxically, in the classical period, thinkers and philosophers had understood the importance of the sciences. Euclid's geometry was definitely part of civilization. **John Tyndale** (1820-1893), a friend of Huxley, began to make more of scientific naturalism – as a superior type of knowledge. Tyndall was a prolific scientist and Professor of Physics at the Royal Institution from 1853-1857. He had done much of his work on diamagnetism and thermal radiation. He was elected President of the British Association for the advancement of science in 1874. He produced many lectures and scientific papers on light and energy. Unlike many of his contemporaries (e.g. **James Maxwell**) he wanted to promote a split between science and religion. In his key note speech to the Association's annual meeting in Belfast he supported evolution and mentioned Darwin 20 times. He concluded by saying that religion should not be permitted to *'intrude on the region of knowledge, over which it holds no command.'* Science had rights over all other culture, including religion. The newspapers made this headline news in Britain, North America and in Europe, triggering more debate and disagreement, but in fact increasing support for the evolutionists' position. In 1864, Pope Pius 1X had decreed in his *Syllabus of Errors* that it was an *error* that *'reason is the ultimate standard by which man can and ought to arrive at knowledge'* and that *'divine revelation is imperfect'* and that anyone maintaining those errors was to be anathematised. In 1888 Pope Leo X111 decreed in *Libertas* that *'The divine teaching of the Church brings the sure guidance of shining light. Therefore, there is no reason why true science should feel aggrieved at having to bear the restraint of laws by which, in the judgment of the Church, human teaching has to be controlled'* and *'The fundamental doctrine of rationalism is the supremacy of the human reason, which, refusing due submission to the divine and eternal reason, proclaims its own independence.... A doctrine of such character is most hurtful both to individuals and to the State.... It follows that it is quite unlawful to demand, to defend, or to grant, unconditional [or promiscuous] freedom of thought, speech, writing, or religion.'* Huxley was more willing to compromise than Tyndall, but began a dining club which became known as the X Club. Nine men who supported natural selection and academic freedom and liberalism met for the first time in 1864 - **George Busk**, **Edward Frankland**, **Thomas Hirst**, **Joseph Hooker**, **John Lubbock**, **Herbert Spencer**, **William Spottiswoode**, **John Tyndall** and **Thomas Huxley**. They met once a month in London, united by a *'devotion to science, pure and free, untrammelled by religious dogmas,'* although some of them at least supported liberal Anglicanism, certainly those who supported Darwin. In 1860, a collection of essays by liberal Anglicans on the prehistory and history of Biblical texts appeared under the title *Essays and Reviews to contribute to the debate.* These writers saw the Bible as a subject of rational study, like any other work of literature. In some circles, this created more conflict than the *Origin* itself as the decree from Pope Pius 1X illustrates. The members of the X Club came out in favour of *Essays and Reviews* and Lubbock worked for an alliance between their writers and scientists. Members of the Government became involved in various judgments about the *heresy* of the Essays and Samuel Wilberforce, along with many High Church Anglicans, organised petitions against evolution. They campaigned against the British Association for the Advancement of Science hoping to overthrow Huxley's 'dangerous clique' of Darwin supporters. In 1862, **Bishop John Colenso** of Natal published a scientific study of the Pentateuch showing that the first five books of the Bible were historically unreliable. The X club members supported him in his conflict with the Church of England and dined him at their club. The crux of the two culture debate was its importance for educational policy. **Arnold** had been an H.M. Inspector of schools, Huxley a member of a school board and Tyndall a school teacher, as well as such an influential scientist and lecturer. Tyndall and Huxley were both in awe of Darwin's achievements and obsessed by the challenge of getting science into education, and removing the influence of conservative religion. Arnold had made little, if any mention of science and often confused it with technology. He certainly didn't see science as embracing any kind of creativity or culture. At Oxbridge, a person not knowing the classics or Shakespeare was considered a barbarian, but not for ignorance of the second law of thermodynamics! **H. G. Wells** was a student of **Huxley** and from the 1890s pursued the idea of science, not theology as the organising principle of culture and society. He urged that Britain should become a scientific technocracy. Later, the literary critic, **F.R. Leavis** (1895-1978) saw Wells' work as dangerous and almost evil, because it undermined cultural value, replacing it with modernity and technology. Within a changing understanding of human culture, **C.P. Snow's** 1959 Rede lecture on the differences between science and culture (literary intellectuals) was a significant, intellectual moment. He thought that a society was disabled if science and technology were not integrated into public culture. In the two cultures debate, Leavis criticized Snow for exaggerating the gap, and not allowing for a spectrum of positions in between, but Leavis demonstrated his own ignorance of science in making his points. *The two cultures and the revolution in science* was the full title of Snow's lecture. By now, Snow was addressing a world with far more science students than arts. There was little perceived need to take the theological context seriously, whereas in the time of

Huxley and Arnold the main compass points of culture were theological. Snow had used phrases such as *'Clashing point of two galaxies. ..those in the two cultures can't talk to each other. Very little in 20 century science has been assimilated into culture.'* In some ways, these words were as powerful and controversial as Samuel Huntington's *'Clash of civilizations and the Remaking of World Order'* post 9/11. F.R. Leavis's Cambridge lecture was full of cynical scorn and sarcasm, dismissing Snow's ability as a novelist and intellectual. He called him naïve (a real academic put down) and irresponsible.

Aldous Huxley (1894-1963), author of *Brave New World,* poet, humanist and pacifist was the grandson of Thomas Huxley. Ironically, his mother, Julia, was the niece of the poet and critic **Mathew Arnold**. Aldous strove in his later career to bring the two cultures together. He believed that Snow has oversimplified the distinction. There were many disciplines and cultures in between the **two cultures** of Snow– historians, anthropologists etc. Snow's views could be seen as similar to Wells' in earlier decades and was building on a longer debate. Aldous was concerned, in the 1950s, that science needed a larger place in education. Politicians and policy makers needed to better understand science and technology. Snow may have exaggerated the division, but he effectively raised an important issue - unless the natural sciences and technology were better absorbed into culture, the nation would be disabled, politically and economically. **Simon Shaffer** believes Snow systematically exaggerated the gap between public culture for laudable purposes. Leavis thought Snow was making his judgments based on what appeared in the media. In fact the achievements of science were very visible in the colour supplements of the time and these were the cold war years. It is true that in Westminster there were very few MPs from a scientific background (not so in House of Lords). **Sir Paul Nurse,** President of Royal Society was at school in the 1960s and felt that the arts and culture were seen to be on higher level than technology –certainly the working lives of artisans, engineers, fitters, machine workers. Science, despite being key driver of prosperity and power, seemed to be undervalued. Grammar schools had nourished a culture of history and the arts in ways that sometimes separated the science and arts, as did University departments. Parallel choices were built into the education curriculum at an early stage. Very few people did both. Slowly the new science became more culturally significant (not least through BBC programming). Intellectuals of whatever background took it more seriously. Its worldviews, road maps and metaphors affected other parts of society. A more sophisticated TV audience (and producers) came to appreciate the value of high quality science and arts programmes as the book ends of their daily lives and leisure interests. There is also an increasing clash of civilisations, globally and locally between those with motivation to learn and those who have been discouraged of denied that motivation or opportunity. This is producing new forms of social inequality and division which may prove to be a more profound cultural gulf than any between the arts and science. It undermines the rhetoric and meaning of connectivity. It divides our sense of a shared understanding of the human condition and participation in human society. This matters a great deal to those who believe in Love's energy and its embedded potential.

On the horizon of a future that is already present, we sense the evolutionary emergence of a serious cultural division in our use of cyber-technology. On the one hand, we recognise the value of increased quantities of instant, data access and communications - with all its political and ethical issues about security, secrecy, surveillance and transparency. On the other, we recognise the quality value of knowledge as the wisdom which comes from longer term, cultural and spiritual reflection. No age before ours has had such access to the former, with all its democratising implications. This has changed the meaning of 'life long learning' and it bypasses as well as influences formal, educational institutions and curricula. The learning and resources that were once only available in monastery and then university libraries - all with their own slow procedures and limits – is now mostly available on mobiles, lap tops and desk tops anywhere.[330] These create their own, social cultures and networks of sharing, despite personalised and individualised access. They encourage 'just in time' reactions to news events and personal stories which allow little time for reflection and learning. Knowledge is more than information, let alone entertainment. It still requires hard work in following context and trails, checking data, analysing assumptions and assessing integrity, value and meaning. Scientific knowledge, as we have seen is more than the accumulation of facts. Knowledge can lead to divisive, as well as unifying worldviews. Instant access also means access to the worst of human instincts, stereotypes and propaganda. The judgement, wisdom and discernment needed to identify and reject the dross and the dangerous is a key challenge in the changing tensions of geo-politics. For too many vulnerable people, the images and messages manipulated on their screens is what they come to believe as the truth. These images affect their world maps and diminish their cultural horizons. Even a brief survey of Muslim sites will explain why so many devout young Muslims are recruited to extremism, hatred and violence.[331] Not only is this material so extensive and powerfully presented, but there are very few websites or programmes with the same impact, to counter and challenge its influence. What would be the content of such a challenge? How much depth is there in the heart of true religion, any religion, where all divisive

---

[330] I was recently shown round the library services of the university where I am a governor. Yes, there were a few rooms of books, but intead of the dusty, underground stacks of my university, there were rooms and rooms of computers!
[331] I once did a presentation for the Blair Foudnation at the RSA on this subject so had to do some research.

externals float away and the kernel of love is unveiled? There, deep in the heart of love, seekers will surely find the only power in the world stronger than the violence, fear and hatred such webites encourage and depend on. We now Google global knowledge and information at 'just in time' speeds, from our wide variety of 'on the move' hardware.[332] This is knowledge 'to go' as they might say in the coffee shops. Cinema type, surround screens, and Dolby sound systems in our homes, provide 'all in one' entertainment that can become a totally engaging form of daily distraction. In this clash of cultures, screens have largely replaced bookshelves and paintings on our walls. In some families, they dominate living room and bedroom space and are constantly on. Many spend all day on their screens at work. They walk the streets with groups of friends, heads down and texting. They travel by car with their entertainment and directions controlled by the satellite screen on the dashboard. They walk into their homes, still working, texting or emailing. At home, they turn on their larger screens to access more data, transformed into adverts, soaps, game shows, films and music. Soon, we will tap our watches to turn on a hologram, and look through Heads Up Displays in our glasses or lenses at the data we need to know, about who we are meeting and the latest *just in time* information.[333] Maybe, in the future, it will become even harder to cope without a signal, a technological, interface fix between ourselves and others. Love is not so much an interface between us and other *others* or nature; it is embedded within us and those relationships of participation and potential. It is more intimate than any interface and all intimacy is flawed. Out of the flaws comes a longing to reach out and make something, to make something happen, rather than die. Facing the silence of a direct experience of nature or another person, or the thoughts inside our own heads and hearts, may become an increasing luxury, or one that will feel disruptive and alien in the connected and dependent habits of our social media and communication. Is this electronic interface its own kind of evolutionary change in our species? It is one we have directly created - naturally selected by our brains' ability to make choices, to shape our environment – a process that began many millennia ago and will, no doubt, continue in ways that are unpredictable and unimaginable. As we saw in LE1, the emergence of new technology can be a difficult and divisive transition, as well as bringing a new range of improvements and progress. But curiosity is hot wired into our condition and the evolution of new ideas and technology seems to be irresistible. At any one point, the Luddites might be right or wrong to oppose a particular change, but in the longer term, the technological interface between humans and their environments is increasing. The environmental and other concerns are huge, but our species is evolving in the direction of more, rather than less dependency on devices and

---

[332] Most of which, most of us don't understand enough to make or mend.

[333] Predicted by Michiu Kaku in his book *Visions* in 1999. Many of these 'Startrek' type gadgets are now coming in to use.

machines, especially those that save us energy - even as they consume it - and those that save time – even as many complain of boredom and lost purpose. From a larger perspective, the causes and effects within human, technological, *natural selection* are diverse and connected, indeterminate and random, inconsistent and incomplete, and always subject to a complexifying force with its own proton gradients of transformation.

The Leadership debate - Adaptability and innovation

The rhetoric of innovation is now part of any leadership course or conference, and may even be an evolving part of successful cultures and a test for individual performance. Adverts for senior posts, in all sectors, use the language of creativity and flexibility, even where this is the last thing a particular organisation wants it practice, at least in the public sector. SMART goals and the capacity to be fit and fleet of foot, ever more *fit for purpose*, may not be a direct inheritance from Darwin, but it is certainly embedded in our beliefs and values. The Victorians, too, were great entrepreneurs, innovators and inventors. Somewhat paradoxically those attitudes co-existed alongside the values of solid and reliable, upright citizenship which were often resistant to adaptability. In the struggle for survival, the 'fittest' (Spencer) aren't always the strongest or the largest, but the most adaptable to their environment, to new opportunities and the threat of competitors.[334] Strength and intellect are of course matters of evolution and play their significant part in performance. Darwin did seem to favour adaptability, although as part of a natural selection process of modification over much longer periods of time than the appointment of any individual to a leadership position! To be flexible and adaptable, like all species, humans have to identify and take seriously the nature of different threats and opportunities. Leaders have to make daily, detailed decisions about whether to challenge and overcome a threat or adapt and reposition in relationship to its implications. This requires emotional, intellectual, moral and physical strength, not only to make the decision, but to live with its consequences. Adaptability is a complex process, and we draw on inner resources of learnt experiences and personality to reach out in this way. If Love's energy is somehow embedded in these processes, it is a source of different kinds of reaching out, to respond creatively and sensitively to different situations, in ways that enhance an evolving adaptability. In the natural world, 'intelligence' is not the issue – adaption is about physical and biological change, rather than cognition and rational thinking. A reptile does not *decide* intellectually to grow feathers, but modification does take place over time. Nor is strength the issue, although an animal born weak at birth may well not survive for lack of strength, or protection and support. This need for parental protection,

---

[334] Scholars still dispute the Darwinian provenance of *'It is not through strength that species survive; it is not through intellect that species survive. It is through adaptation that species survive.'*

feeding and nurturing may explain why altruism developed in the higher animals, while the newly born of many non-primate species have to survive more independently, at a much early age. A species that doesn't have the strength to survive/adapt to changing climate, predators or resource limitations may not survive. Humans have had to evolve adaptability skills in response to the short term and immediate threats and opportunities of their situation – often through the use of technology as well as personal (hard and soft) competencies. Leaders in many organisations are now discovering the importance of emotional intelligence in changing organisational culture as the route to improving performance. In the name of adaptability, leadership gurus such as **Tom Peters**[335] used to say that hard times need maverick leaders to succeed and 'flexibility to survive.' We are moving away from such macho-management responses and realising the importance of nourishing human relationships through trust and even love, to create more adaptable and innovative work places where people can draw deep from the well of love's gracefulness, forgiveness and understanding.

Darwinian Economics

It is tempting to find the ubiquitous influence of Darwin wherever we look, if not directly, then through the ideas he inspired in many different people and areas of life. This is true in economics. **Professor Niall Ferguson,** concluded his historical study of the way economies are created and influence politics and civil society with a final chapter entitled *'Afterword.'*[336] In this, he pursued the argument that finance and financial institutions function in a Darwinian way, with institutions and infrastructure evolving around capitalism's 'creative destruction' of species and forms. Without this 'creative destruction,' he argues that new forms of institution could not emerge.[337] *'The fact that a particular firm successfully devours smaller firms along the way is more or less irrelevant. In the evolutionary process, animals eat one another, but that is not the driving force behind evolutionary mutation and the emergence of new species and sub species. The point is that economies of scale and scope are not always the driving force in financial history. More often, the real drivers are the process of speciation – whereby entirely new types of firm are created – and the equally recurrent process of creative destruction, whereby weaker firms die out.'* [338] **Joseph Schumpeter** (1883-1950) in *'Capitalism, socialism and democracy'* had

---

[335] *Thriving on Chaos. A handbook for a management revolution.*1987 Knopf Inc. It is revealingly dedicated to two individuals *'whose **flexibility** of mind and **raging impatience** with inaction have inspired the most dramatic and fruitful revolutions.'* My emphasis in bold.
[336] *'The Ascent of Money: The Financial History of the* World' Penguin 2008
[337] Ibid p. 341-358.
[338] Ibid p. 352

characterised industrial capitalism as 'an evolutionary process.'[339] Quoting *John Maynard Keynes'* (1883-1946)[340] Ferguson argued that the *'general volatility of financial markets is related to the heuristic pieties of individuals and human behaviour.'*[341] In his analysis of how the financial world functions as an evolutionary system, he identifies the following ways in which business practices perform the same role as genes, *'allowing information to be stored in the organisational memory and passed on from individual to individual or from firm to firm when a new firm is created.. spontaneous mutation usually referred to in the economic world as innovation...competition between individuals of any species for resources...natural selection through the market allocation of capital and human resources and the possibility of debt in cases of under-performance...the scope for speciation sustaining biodiversity through the creation of wholly new species of financial institutions...scope for extinction with financial species dying out altogether.'* [342] He argued that evolution, as a model for the development of financial institutions, does not lead to a perfect organism. *'A good enough mutation will achieve dominance if it happens in the right place at the right time, because of the sensitivity of the evolutionary process to initial conditions.'* He then compared this view with the followers of **Herbert Spencer** in biological evolution, who argued that the direction was always progressive. *'Evolved complexity protects neither an organism nor a firm against extinction – the fate of most animal and plant species.'*[343] He said that recent decades, leading up to the financial crash of 2008 onwards, have seen an explosive flourishing of existing financial species, but *'size isn't everything in finance as in nature.... All that matters is that you are good at surviving and reproducing your genes. The financial equivalent is being good at generating returns on equity and generating imitators employing a similar business model.'*[344] The fact that such a notable historian, with global coverage through his television series and books chose to use evolution as a paradigm for understanding change in financial infrastructure and institutions is another demonstration of the lasting power and applicability of Darwin's book. He ends the analogy by pointing out that the creative destruction in the evolutionary process will be thwarted by the (understandable) response of regulators and legislators to times of crisis and crash. In the biological world, natural selection responds in the longer term to eliminate weaker, un-adaptable members who won't survive a time of resource limitation, or environmental change. In the case of finance, crises *'are showing new rules and regulations as legislators and regulators rush to stabilise the*

---

[339] London 1943 p. 80-84.
[340] *'The General Theory of Employment,'* Economic journal 51, 52 1937 p.214
[341] *The Ascent of Money* p 344.
[342] Ibid p.350.
[343] Ibid p. 351
[344] Ibid p 355

system and protect the consumer/voter.'[345] He warned that the possibility of extinction cannot and should not be removed by excessive, precautionary interventions. Again he quoted **Joseph Schumpeter** *'This economic system cannot do without the ultima ratio of the complete destruction of those existences which are irretrievably associated with the hopelessly unadapted.'*[346]

Darwin, religion and liberal democracies

There are many ways of discussing this issue. Here is just one example, not taken from America - where the debate about Darwin and education reached the highest court in the country and even the White House – but from India and Pakistan. After partition with India in 1947, which many see as a cause as well as an expression of many of the problems in the area, **Ali Jinnah's** (1876-1948) vision was to create a liberal democracy for Muslims. But Pakistan became an Islamic state, undermining the whole idea of plurality. In the process, there was a moment when somebody started stapling together or destroying the chapter on Darwin in school, text books. It is said that children who had been taught that if you combine oxygen and hydrogen you get water, were now taught that Allah created the water. Perhaps it was at this point that the liberal democracy project failed. For true science to flourish, religious oppression and domination has to be challenged. We saw in LE1 how the Church, controlled by its powerful clerics, tried to stop the progress of science because they saw it as a threat to religious dogma. In liberal democracies, where the clerics no longer dictate the educational curriculum, the law, and the values of civil society, people ask fundamental questions again. The architecture of a liberal democracy can only be sustained by a belief in a pluralism of options, opportunities and choices. Within that pluralism, competing, and sometimes conflictual values, beliefs and attitudes have to be negotiated by interest groups within a strong civil society. People of faith have to make choices too, as they position themselves within such pluralism, discovering what their faith has to offer wider society, rather than their own internal interests. As we saw in LE2, those interests if they reflect the presence of Love's energy cannot be separated from their emanating ekstasis. None of that is easy and requires its own form of emotional and spiritual evolution – one that cannot be forced on individuals, for Love does not work in that way.

Scientific developments

Just as *The Origin* was going to press in 1859, **David Livingston** reached Lake Malawi in Africa and named it Lake Nyasa.[347] It is allegedly the home of more

---

[345] Ibid p 360

[346] *The Theory of Economic Development,* Cambridge MA 1934 p.253.

[347] Candido José da Costa Cardoso was the first European to 'discover' the lake in 1846

species of fish - including more than six hundred species of cichlids - than any other body of freshwater. The cichlids are very diverse in shape, mouths, colours and breeding habits, but all from one 'family.' Those that scrape algae off the rocks have evolved a different arrangement of teeth. Some bury their heads in the sand to eat small animals under the surface. Others developed larger and sharper teeth to live off other species. All have descended from one ancestor and they prove Darwin to have been right. Of course, the influence of Darwin on the way different biologists and zoologists worked was extensive and not without controversy. Only some figures in this story can be mentioned here. The German evolutionary biologist, **Friedrich Leopold August Weismann** (1834 –1914) was to become Director of the Zoological Institute and the first Professor of Zoology at Freiburg. He work was judged by **Ernst Mayer** to be second only to Darwin's. He introduced the idea of germ plasm theory to show how inheritance took place only in the gametes – the egg and sperm cells – rather than other cells of the body. The material of inheritance was in the nucleus of these cells and so isolated from the influence of and change in other cells. This was his Weismann barrier theory and it challenged **Lamarck's** ideas of inheritance. So, the mutations of natural selection could only work in this way and were random. Weismann revived the case for natural selection against the critics of Darwin. He cited the case of butterflies in Africa. Some species evolved to look like those species which could produce tastes that make them repugnant to birds, and they do this by gradual mutations over generations. His work confirmed **Gregor Mendel's**[348] (1822-1884) studies on peas – that the inheritance of certain traits followed a pattern.[349] In breeding peas, he counted the number in each generation and discovered what Darwin missed – the mathematical laws of inheritance. Meanwhile the Dutch botanist, born a year before the *Origin*, **Hugo de Vries** (1848-1935), Director of the Botanical Gardens in Amsterdam, bred and cross bred thousands of evening primrose plants to show they produce progeny that change and become stable. He called them *mutations*. This happened more quickly than Darwinian evolution, but few other organisms apparently mutate this quickly. The geneticist **William Bateson** (1861-1926) popularised the ideas of **Gregor Mendel** and Vries's study of mutations which Bateson saw as the creative force behind evolution – questioning whether natural selection was necessary. By the beginning of the twentieth century, Mendel's significance was rediscovered, partly by the work of Vries, **Carl Erich Correns** (1864-1933)[350] and

---

[348] He learnt physics in the Olomouc Faculty of Philosophy and entered St Thomas' Abbey but initially failed his exams to become a teacher. He was sent to the University of Vienna in 1851 and studied physics under Christian Doppler. He returned to the Abbey as a qualified teacher of physics. Initially he studied heredity using mice but then switched to plants and bees. He also studied astronomy and allegedly published more on meteorology than on peas!
[349] He had to give up his experiments because of his duties as an Augustinian Abbot in 1863. Many of his papers were burnt when he died but his experiments were rediscovered and he became famous as the 'father of modern genetics.'
[350] Noted for his independent work (separate from Erich Tschermak von Seysenegg and de Vries) on heredity and his rediscovery of Mendel's earlier papers.

Erich Tschermak Edler von Seysenegg (1871-1962). Perhaps, Vries didn't understand the significance of his own work until he read Mendel, but he later lost interest in the latter's work. This debate between the biometricians and the Mendelians became very robust at the beginning of the 20th century. The former claimed more mathematical rigour and the latter a better understanding of biology. The English statistician and geneticist, **Ronald Fisher** (1890-1962) – much praised by Richard Dawkins - played his part in the story by fusing a neo-Darwinian synthesis[351] from each approach, as he calculated agriculture yields. The numbers were complicated and he invented statistical tests. As we shall see, he was also interested in eugenics[352] and was concerned that the 'feeble minded' were out breeding the intelligent. This became an obsession and he calculated the ratios, as he thought about natural selection, producing a theorem equation for it,[353] proved by **Bernard Kettlewell** (1907-1979) not least in the survey of moth adaptions.[354] In this emerging synthesis, some questions remained. Why, in a world of competition, are some animals altruistic? In a termite mound, built by millions of altruistic 'individuals,' the soldiers and workers are devoted to feeding the queen. According to Darwin they are engineered by natural selection to increase the chances of reproduction. Yet most termites are sterile. The queen alone replicates the genes of the colony on its behalf. Natural selection counts the fate of genes, not individuals. So the termite soldiers 'sacrifice' themselves to protect the genes of the colony, showing that they can be *altruistic* as well as *selfish.* These seem strange terms to use of termites, though as we shall see in the chapter on *Sociality and Altruism* the questions about them are important. But it seems that each individual is obeying its gene needs and these shape our behaviour to varying degrees. A force that had seemed weak and uncertain in Darwin is now seen as omnipresent through socio-gene theory, but Darwin had given us a new narrative which showed that the history of life's development was a tale of epic forces which could be used for many different reasons. We must now explore some of those uses and misuses, before we turn again to the subject of genes.

---

[351] Coined by Julian Huxley in *Evolution; the Modern Synthesis* 1942. Other figures included Dobzhansky, Haldane, Mayr. The Population Genetics of 1918- 32 inspired the later synthesis by showing how Mendelian genetics cohered with natural selection. There is evidence of poor communication between different figures and the increasing isolation of the main disciplines - paleontology, biology, genetics, botany, ecology etc..

[352] Along with Horace Darwin, Charles's son and Meynard Keynes when at Cambridge.

[353] He received the Darwin-Wallace medal from the Linnean society in 1958

[354] See earlier reference in the work of James Tutt – the peppered moth experiment.

Chapter 8
Social Darwinism

Sub Headings
Developmental Spencerism; Pattern Paradoxes; 'Natural' intervention or 'natural' chance?; From Race and Slavery to Totalitarianism; Kellog, Bryan and the Scope Trial; Veneer Theory

Developmental Spencerism
Herbert Specer's first book *Social Statistics* of 1851 asserted human adaptability to the pressures and needs of a society in ways that would require less State involvement. Its publisher introduced him to networks of progressive thinkers including John Stuart Mill (*A system of Logic*), George Lewis and George Elliott (Mary Ann Evans, with whom he had some kind of romantic relationship). Herbert intrdoced Thomas Huxley to this same salon and they sustained a close, working friendship. Spencer's 1855 *Principles of Psychology* (influenced by both Lewis and Evans) suggested that the human mind was built on natural laws of biology and so could be part of a general, species development. He used the Lamarckian idea of 'use' inheritance to explain how the association of ideas emboddied in brain tissue could be passed on generationally. He hoped this book would do for the natural laws of the mind what Isaac Newton had done for the forces of matter. Although he disagreed with Auguste Comte's positivism, he followed his commitment to the universality of natural law and believed, unlike Compte, that the principle of progressive development was its single most important driver. So, in 1858, a year before the *Origin*, he produced an outline for his *System of Synthetic Philosophy* which was to take the rest of his life to complete and was to demonstrate that the single principle of evolution could be applied to biology, psychology, sociology (a word used by Compte) and morality. He saw evolution as *'an intgration of matter and concomitant dissipation of motion; during which the matter passes from an indefinite, incoherent homogeneity to a definite coherent heterogeneity; and during which the retained motion undergoes a parralel transformation.'*[355] In Love's energy, we have noted the movement from and pattern paradox relationship between the simple and the more complex, the universal and particular, the singularity and diversity. Now, we find this idea of tranformation in the motion between *incoherence* and *homogeneity* and *coherence* and heterogeneity. Because he believed this law was

---

[355] H. Spencer *First Principles* 6th Edition London 1900, p 367 cited by Paul Elliott in the *Derby Philosphers* p205. He may have picked up the idea of homogeneous to heterogeneous from Karl Ernst von Baer's view of progressive animal development, but Spencer applied it more generally to social organisms in his universal law.

universal, he applied it to the development of galaxies, geology, changes in government and religion, political systems, language, music and the the arts. They were all manifestations of a progressive and universal, developmental law from the earliest shape of the cosmos to the detials of human civilisation and industrial progress. The movement from *homogeneous* singularity to *heterogeneous* complexity and diversity was clear and universal, affecting, for example, the idea of work force specialisation to increase production and a celebration of the diverse benefits from the new science of steam power in mining and transport. Behind these optimistic universality was Spencer the railway engineer, the growth of Derby as an industrial and transport centre and of course the ongoing influence of the Derby Philosophical Society and in particular Erasmus Darwin. He knew that progress of this kind had come at great cost for individuals, if not human society, and nature itseslf was red in tooth and claw (his Principles of Biology) and only the really fit survived, yet he sustained this positive vision of social and scientific development. His phrase *'survival of the fittest'*[356] proved to be influential in the process of adapting natural selection to the field of social change, applying its metaphors to political and cultural contexts. He thought there was a rough equilibrium between animal populations and food supplies with fertility decreasing as civilisation developed. Extinctions and natural deaths were part of nature's way of ensuring progress through *'survival of the fittest.'* However, the *harmony* of society increased in proportion to the growth of *sympathy* to faciliate social congruity.[357] The lack of congruity held social organisms back from its moral perfectability. In his ethics, he was perhaps influenced by Erasmus' evolutionary psycho-physiology and the idea that society was shaped positively or negatively by individual character. Sympathy brought greater altruism and harmony (see chapter on Altruism). As we have seen, he produced his own inclusive and universal theory of evolution, rather than just bolting on some sociological application to the work of Darwin. Some commentators have stereotyped him as the first of the social darwinianists, rather than being, in his own right, a hugely significant thinker and writer, with his own theory of developmental evolution. Like Erasmus Darwin, the range of his Enlightenment interests was much larger than Charles' Darwin's. Unlike Charles, Erasmus and Herbert were fascinated with electricity and were mechanical experimenters and inventors[358]. They came at evolutionary theory

---

[356] Coined by him in his *Principles of Biology* 1864 after reading the *Origin*.

[357] See the very perceptive arguments in Paul Elliott's *The Derby Philosophers* Manchester University Press p 209ff and his general argument in that chapter for the influence of the provinces on the emergence of new theories of development.

[358] He had many times observed and assisted his father in experements on air pumps and electrical machines and so came to see *causation* operating in nature and science rather than supernatural religion. Fro his father he would have heard of the ideas of Erasmus Darwin and the other Derby Philosophers and members of the Lunar society. For him, *causation* was embodied in the way machines and nature functioned internally according to its own universal laws.

from a much wider perspective and arguably this larger context provided a platform for Charles's more specific focus on biological adaption and species development. Herbert integrated cosmological,[359] geological,[360] biological, psychophysiological, political, social and industrial examples of adaption in his universal theory, much more so than Charles. In this sense, his work was closer to that of Erasmus Darwin.[361] His genius was to perceive a universal, natural law across many different kinds of *heterogeneity*. Like Charles, he was convinced that there must be such a single, natural law. Unlike Charles, he wrote of its application more generally and perceived its presence in the functioning of human societies and institutions, as well as in cosmic and natural sciences. Spencer was clearly realistic about the presence of conflict and violence, but, like Darwin, felt a certain optimism about the future. The geological, cosmological, psychological and evolutionary developmentalism of Erasmus Darwin and the Derby philosophers[362] was part of Spencer's inheritance – with their confidence in scientific and industrial progress and political reform.[363] The progressive, Enlightenment confidence of Lamarckianism and others was a driver of the breadth of these natural philosophers' interests, especially Erasmus Darwin and W.G. Spencer (1790-1866).[364] It also opened the curtain on the stage of Victorian public and academic opinion on which Charles Darwin was so hesitant to appear. Academic opinion is divided on the extent and nature of these influences on Herbert Spenser's own thinking.

Pattern Paradoxes

The *pattern paradoxes* of *plentifulness* and *scarcity*, the *oppressors* and *the oppress*ed, the *powerful* and *the powerless, inclusion* and *exclusion, competition* and *cooperation, laissez faire* and *centralisation, individualism and collectivism,*

---

[359] Influenced by Herschel's analogy between the naturalist and the astronomer, Erasmus made connections between what was known of the slow development of nebula and natural development.

[360] Charles Lyell's 3 volume *Principles of Geology* 1830-33 influenced Charles Darwin and Herbert Spencer and popularised James Hutton's uniformitarianism and Lemarck's developmentalism. Its full title *being an attempt to explain the* former *changes of the Earth's surface, by reference to **causes now in operation*** is revealing. My emphasis in bold. Charles was given volume 1 by Robert FitzRoy of the Beagle. While in South America, he received volume 2 and began to see rock formation through Lyell's eyes, though he rejected Lyell's *centre of creation* argument to explain species diversity. Erasmus had seen the formation of rocks (in Derbyshire) by animal decomposition and 'submarine fires' as '*Monuments of the past felicity of organised Nature – and consequently of the benevolence of the Deity!*' cited in The Derby Philosphers p 192. For him this was more evidence of a movement towards progress, if not perfection. See his *Love of the Plants,* 1791

[361] There is scholarly argument about the exact relationship between Herbert and his father's relationship with the ideas of Erasmus Darwin. W.G. Spencer was only 12 when Erasmus died but as secretary to the Derby Philosophers would have known of his work from reputation and his writings.

[362] And their disciples – Forrester, Brookes Johnson and the Strutt brothers as well as W.G. Spencer's deistic evolutionary world views.

[363] In *Zoonomia, Phytologia* and *Temple of Nature* (originally called the *Origin of Society),* Erasmus Darwin was working out how natural waste, destructiveness and productivity could be seen as evidence for an overall theme of development and progress not dependent on a deity.

[364] Inititally teaching mathematics at his father's school, he later became the secretary of the Derby Philosophical society. Both he and Herbert were born in Derby. George was a dissenter moving from Methodism to Quakerism.

totalitarianism and *democracy* abound in the developmental implications and adaptability of social organisms. As we have seen, *pattern paradoxes* are not the same as contradictions or polarities – they are not fixed positions and some move like waves across their connectivity. The basic metaphors of these thinkers, and Charle's adaption of them in his own big idea, became foundational paradigms for the decades which followed. In the social and political sciences, it was assumed that the behaviour of human beings was describable in terms of different drivers of natural selection over much shorter periods of time. Because social and political change took place over weeks, months, years or generations, their causes and outcomes could be documented. The metaphor of longer term, unconscious adaption was replaced by different ideas about social engineering. Politics has been seen as the engineering tool for changing conditions as well as moving human adaption forward. If political interference through public policy, education and cultural messages became the process of adaption, then theories were needed to justify such interference – in whose interests and to whose benefit? What could be said about individual freedom and group, or political responsibility? What most needed to change? What social and individual conditions held people back from beneficial change? Was social mobility a sign of natural adaption out of poverty, ignorance, hopelessness, oppression? Were certain kinds of people, groups, classes, colours, cultures better than others at adapting to power and prosperity? Could we dissect the human brain and find the emotional or rational explanations for behaviours that either helped or hindered us making a better life for ourselves and others? In human terms, what made one finch's beak longer than another?

'Natural' intervention or 'natural' chance?

So, the ideas of natural selection and species adaption were transmitted from biology to society and politics, despite, as we saw, **Huxley's** plea not to confuse evolution with ethics. As far as we can tell, Charles Darwin neither intended nor predicted this development of social darwinism, whatever the influence of **Malthus** and **Marx**[365] on his work. While it was predictable and may be understandable, it seems to constitute its own kind of category mistake. Behaviour in human societies may be comparable to the 'jungle,' but they have their own *freely*[366] chosen agency to intervene *consciously*, for good or ill, in the processes of natural selection. From a longer term perspective, we can observe random processes of social change overlapping with planned interventions, not least through the creativity of artists and architecture, education, scientific, technological and commercial productivity, and political policy making. Society is

---

[365] Marx certainly relied on Darwin more than the other way round and saw his work as providing the natural science for the class struggle in history.
[366] See the arguments for conditionality in LE2

very rarely 'left to itself,' even though there are emerging and evolving *pattern paradoxes* of *continuities* and *discontinuities, proactivity* and *passivity, congruity and incongruity* in the themes and types of social change. Both Marx and **Spencer** believed that *struggle* would, of itself, lead to a dialectic of change and adaption. In his times, these dialectics had a teleological tone. **Engels** commented '*Just as Marx discovered the law of evolution in organic nature, so Marx discovered the law of evolution in human history.*'[367] Both were fixed internal laws and required no external intervention. God was now as powerless as humans to change the self-adjusting, dialectical struggle of nature and society. Darwin later commented '*What a foolish idea seems to prevail in Germany on the connection between Socialism and Evolution through Natural Selection.*'[368]

Other socialists argued for human cooperation rather that struggle. The zoologist and economist, **Prince Pyotr Alexeyevich Kropotkin** (1842-1921) believed in a 'communist' society built around free association, with minimum Government interference. *Kropotkin by F. Nadar c 1900. P-D. Copyright expired.*

In his '*Mutual Aid A factor of Evolution,*'[369] he described evidence of cooperative behaviours in animals even across species, believing that Sociability is as much a law of nature as mutual struggle.[370] He thought that if the struggle for existence was universal, it would lead to the destruction, rather than the survival of species. It was partly written to counter **Thomas Huxley's** *The Struggle for Existence,* 1888. Others challenged any link between the 'fittest' and the 'richest,' arguing that the 'fittest' weren't biologically, but socially determined. **William James** (1842-1910), (see LE2) commented, '*The entire modern deification of survival per se, survival returning to itself..naked and abstract with the denial of any substantive excellence in what survives except the capacity for more survival still, is surely the strangest intellectual stopping place ever proposed by one man to another.*'[371] At the heart of this debate about Darwinism turned into socialism and economics is the continuing question about our capacity to influence the broader sweep of social change. The current debate between **Max Hastings**[372] and **Niall Ferguson**[373] over whether it was right for Britain to go to war in 1914 is a good case study. Would it 'just' have been a continental, not a global war if Britain hadn't intervened? Was it caused by

---

[367] Marx. Selected works 1 p 16

[368] Darwin to Dr. Scherzer *Life and Letters* 111, p 237. 1879

[369] Written in exile in England 1902 based on earlier drafts 1890-1896.

[370] See my later chapter on Altruism

[371] Hofstadter *Social darwinism in American Thought p 201*. See also James' essay *Environment and Mental Evolution in International University Reading Course Text Matter* Section 7 p 172 where he contrasts least '*evolved functions of the mind ..and their higher orders.*'

[372] *The Necessary War*, broadcast on 25.02.14

[373] *The Pity of War* broadcast on 28.02.14

design or historical chance, or predetermined by all that had preceded it, or a result of specific *butterfly effects*, not least in the Balkans and Belgium? Was it part of a longer term, cyclical or systemic process of shifting alliances, imperial, industrial, technological and commercial change that no one could stop? Was it nature's way of pruning itself, using the human propensity for violence? Once it happened was the law of unintended, negative consequences unpredictable or inevitable - the Totalitarianism of Soviet Socialism and Hitler's National Socialism, the horrors of the Cold War and its surrogate wars. What is the relationship between systemic forces and contingent decisions? The fact that the questions remain, even after a Hundred years, illustrates the ambiguities of such *pattern paradoxes* and how they weave their way into human reflection and memory.

From Race and Slavery to Totalitarianism
Spencer and the 'Social Darwinianists' had begun to lay the foundations for the social determinism which was to influence evolutionary psychology and the social sciences in different directions. While Darwin emphasised the slow pace of evolutionary mutation, based on natural selection across '*an interminable number of intermediate forms*,' some early, social Darwinists believed society could evolve or be engineered quickly for the better, if led by those with leadership gifts and self interest in survival often at the expense of those less able to compete. This kind of thinking led to the eugenics of **Francis Galton** (1822-1911), Darwin's half-cousin, based on racial theories.[374] This was an age when white colonialists were still justifying their sense of superiority over indigenous populations. **Anthony Trollope** (1815-1882) took up the notion of weaker nations withering when hardier plants arrive, and Darwin probably believed more civilised nations would eventually replace the less civilised. The Augustinian Friar, **Gregor Mendel's** (1822-1884) genetic discoveries had added a new vision of biological 'engineering.' **Galton** became obsessed with applying the selection idea to humans. In the face of poverty and poor education, he suggested the Government should encourage the best families to have more children. He thought intelligence and culture could be inherited and that clever people had bigger heads. He circulated a questionnaire to judges and painters testing his theory. Darwin replied giving his hat size! With the data, Galton published his breeding programme for humans, but of course any kind of compulsory scheme was impracticable, as well as unacceptable. Some of **H. G. Wells's** writing reflected this concern about a master race avoiding the degeneration of the human species. When recruiting for the Bohr wars at the end of the century, the British army discovered it was time to toughen up and improve the levels of fitness in the general population. In 1912, a bill came before Parliament to control the 'feeble minded' and segregate the 'half witted.' Churchill called for

---

[374] See description in LE2 of emergence of Afrikaner and Apartheid racist theories in people like Paul Kruger.

sterilisation programmes.[375] Eugenics was taking hold in Scandinavia, Germany and America. In the latter, it may have been affected by alarm at the number of immigrants. In 1910, the Carnegie Foundation was looking at the preservation of human quality programmes. The Darwinian influence in **Charles Davenport's** (1866-1944) eugenics creed is clear,[376] and he drew up family trees to identify criminality in the past. These ideas were appealing economically. Why should Americans go on supporting the costs of 'defective' humans? The courts were beginning to consider cases of compulsory sterilisation to prevent the birth of more 'defective' children. It was thought that the death penalty or prison were inhumane treatment for such people. Prevention was better than these bestial cures or punishments. Thousands were forcibly sterilised on this argument. The *survival of the fittest* was taking hold in social policy in America, well before National Socialism developed its theories of Aryan superiority, although the German Society for Racial Hygiene had been set up by **Alfred Ploetz** in 1905 and its work enlarged by **Bauer, Lenz and Fischer**. In 1933, Fischer's inaugural as Rector of Berlin University argued that what Darwin had failed to do, eugenics would accomplish. In that same year, the Nazi party adopted much of the American code for sterilisation, and thousands more were sterilised for various social 'impurities' or mental disabilities.[377] The 'science' of the 'defective' was expanded to include homosexuals, gypsies, religious dissidents, mentally disabled and communists, as well as Jews. Some Nazis thought that the Jews who survived the camps would prove 'natural selection' and to prevent them 'seeding' a new generation they too must be eradicated! What a contrast the UN founders' idea of universal, human rights must have seemed to that generation. Now, that other Darwinian theme could be reasserted – that all humans belong to the same species, regardless of their 'variations.' Slowly the emphasis shifted away from Nature to Nurture, as the best way to improve and 'modify' human behaviour. Evolutionary psychology had to battle with these and other aberrations. Given Darwin's concern about slavery, the immediate context of the American Civil War (1861-1865) and ambiguities in the *Origin* about *connection* and *separation* between varieties and species, it is not surprising that conflicting views about

---

[375] 'The improvement of the British breed is my aim in life,' letter to his cousin Ivor Guest January 1899, reinforced by his experiences abroad in the army.

[376] 'I believe in striving to raise the human race to the highest plane of social organization, of cooperative work and of effective endeavour. I believe that I am the trustee of the germ plasm that I carry; that this has been passed on to me through thousands of generations before me; and that I betray the trust if (that germ plasm being good) I so act as to jeopardize it, with its excellent possibilities, or, from motives of personal convenience, to unduly limit offspring. I believe that, having made our choice in marriage carefully, we, the married pair, should seek to have 4 to 6 children in order that our carefully selected germ plasm shall be reproduced in adequate degree and that this preferred stock shall not be swamped by that less carefully selected. I believe in such a selection of immigrants as shall not tend to adulterate our national germ plasm with socially unfit traits.'

[377] The operational code for the programme was T4 (Tiergartenstrasse 4, Berlin) and it became the model for an expanded extermination in the camps following the Wannsee conference in 1941

racial differences[378] became a stepping stone to Social Darwinism. While Darwin saw all species as evolving from a common ancestor, he didn't deny differences between them including different races. The subtitle of the *Origin* was open to interpretation *'The Preservation of **Favoured** Races[379] in the Struggle for Life.'* In a letter to W. Graham in 1881,[380] he wrote *'I could show fight on natural selection having done and doing more for the progress of civilisation than you seem inclined to admit. Remember what risk the nations of Europe ran, not so many centuries ago of being overwhelmed by the Turks and how ridiculous such an idea now is!..Looking to the world at not very distant date, what an endless number of the lower races will have been eliminated by the higher civilised races throughout the world.'* It is, therefore, easy to see how Darwin's *natural selection* led to a plethora of social theories on race, Empires, nationalism, competition, race and power. Some have argued that an extreme version of social darwinism led to, or was used by, racial theories of superiority in the Totalitarianism of National Socialism and Stalin's Socialism. From such social movements flowed the idea of a super leader or a super race as the instrument of social providence leading to a higher state of 'civilisation.' Both Marxist teleology and Hitler's Aryanism could draw on these assumptions, but so too could laissez-faire individualism to a different end. If nature found its own way of evolving through the internal processes of struggle and adaption, so we should keep political interference in human affairs to a minimum, went the argument. Let us trust those (social) processes to lead to the good, by themselves. **William Graham Summer,**[381] (1840-1910), one of Spencer's strongest supporters in the United States, thought that humanitarian interventions would only dilute the vitality of civilisation and its natural capacity to survive and improve.[382] He translated Darwin's *'struggle for existence'* in the animal world into *'competition for life'* in the human world.[383] We can sense the resonance of this with many current American challenges to the role of Government – for example in welfare and health policy. **Spencer** himself wrote *'The wellbeing of existing humanity and the unfolding of it into this ultimate perfection are both secured by that same beneficent though severe*

---

[378] A secretary to the American Legation in London reading a *Times* warning that abolitionists would turn the population of the South into a mixed race which 'tends not to the elevation fo the black but to the degradation of the white man,' apparently said *'This is bold doctrine for an English journal and is one of the results of reflection on mixed races aided by light from Mr Darwin's book and his theory of Natural Selection.'* See Hammelfarb p 343

[379] Presumably referring to varieties or species.

[380] *Life and Letters* 1 p 316

[381] The first Professor of Sociology at Yale

[382] In his *The Forgotten Man and other essays*, he wrote 'As soon as A observes something which seems to him wrong, from which X is suffering, A talks it over with B, and A and B then propose to get a law passed to remedy the evil and help X. Their law always proposes to determine what C shall do for X, or, in better case, what A, B, and C shall do for X... What I want to do is to look up C... I call him the forgotten man... He is the victim of the reformer, the social speculator, and philanthropist, and I hope to show you before I get through that he deserves your notice both for his character and for the many burdens which are laid upon him.' P 466.

[383] Particularly evident in war. It was probably Richard Hofstadter's *Social Darwinism in American Thought*, of 1944 that most influenced this Social Darwinian view of Sumner.

disciple to which the animate creation at large is subject; a discipline which is pitiless in the working out of good....The poverty of the incapable, the distresses that come upon the imprudent, the starvation of the idle and those shouldering aside of the weak by the strong which leave so many 'in shallows and in misery' are the decrees of a large, far seeing benevolence.' [384] After all, hadn't **Adam Smith** advocated the virtues of the invisible hand working through the natural processes of individual self-interest, and wasn't this being proved by the industrial progress of the eighteenth and nineteenth centuries? Common to these interpretations was the notion of struggle, but the dilemmas went deeper than the rhetoric. Given the human capacity and motivation to struggle for improvements, why was that not seen as part of human, or social, natural selection? In which case, a radical laissez-faire approach could be defended as much as a radical, interventionist argument. The latter could be seen as serving the former and as part of the internal processes of human, natural selection. The struggle to achieve could be understood as a natural part of human behaviour requiring all kinds of 'intervention' in other, natural behaviours! **Spencer** and others saw a teleology in this struggle for a better future, hoping a better kind of human nature would evolve, naturally, as a result. Nietzsche, of course, pushed this to its logical conclusion in the 'will to power.' The *will* and its interventions to change things in the Apollonian – Dionysian, *pattern paradox* was what mattered, not natural or unconscious species adaption. If Marx had translated Darwin's idea of species struggle into the history of class struggle, **Freidrich Nietzsche** (1844-1900) went one step further. He believed that Darwin had torn down the basis of Western Christianity, culture and morality, leaving its architecture bereft and thread bare - by breaking faith in God we were breaking the whole, so that nothing essential is left. Western civilisation was doomed and God is dead. Nietzsche's ideas are, of course, more complicated than this in their somewhat torturous detail and much has been written about his relationship with Darwin.

Kellogg, Bryan and the Scope Trial

In WWI, the causalities on all sides were on an unimaginable scale of almost Industrial slaughter. How could any advanced civilised nation commit such atrocities and barbarism, including the gassing of soldiers and the shooting of women and children? In 1915, **Professor Vernon Kellogg** (1867-1937), the American pacifist and entomologist, and his wife Charlotte, came to organise humanitarian aid to victims in Kaiser Willhem's HQ in northern France. By night he dined with the German high command and held extensive conversations about evolutionary theory. He produced a sobering account,[385] published in

---

[384] Spencer, Social Statics. pp 322-323
[385] Relevant to historians evaluating the culture behind the Kaiser and the German army's mentality and motives. My emphases in bold

1917, entitled *'Headquarter Nights'* in which he writes *'The creed of the Allmacht of a natural selection based on violent and fatal **competitive struggle** is the gospel of the **German intellectuals**; all else is illusion and anathema. The mutual-aid principle is recognized only as restricted to its application within limited groups. For instance, it may and does exist, and to positive biological benefit, within single ant communities, but the different ant kinds fight desperately with each other, the stronger destroying or enslaving the weaker. Similarly, it may exist to advantage within the limits of organized human groups as those which are ethnographically, nationally, or otherwise variously delimited. But as with the different ant species, .. ruthless struggle is the rule among the different human groups. This struggle not only must go on, for that is the natural law, but it should go on. So that this natural law may work out in its cruel, inevitable way the salvation of the human species. By its salvation is meant its **desirable natural evolution.** That human group which is in the most advanced evolutionary stage as regards internal organization and form of social relationship is best, and should, for the sake of the species, be preserved at the expense of the less advanced, the less effective. It should win in the struggle for existence, and this struggle should occur precisely that the various types may be tested, and the best not only preserved, but put in position to impose its kind of social organization its Kultur on the others, or, alternatively, to destroy and replace them. This is the disheartening kind of argument that I faced at Headquarters; argument logically constructed on premises chosen by the other fellow. Add to these assumed premises of the Allmacht of **struggle** and **selection** based on it, and the contemplation of mankind as a congress of different, mutually irreconcilable kinds, like the different ant species, the additional assumption that the Germans are the chosen race, and German social and political organization the chosen type of human community life, and you have a wall of logic and conviction that you can break your head against but can never shatter — by head work. .. The danger from Germany is, I have said, that the Germans believe what they say. And they act on this belief. Professor von Flussen says that this war is necessary as a test of the German position and claim, if Germany is beaten, it will prove that she has moved along the **wrong evolutionary line**, and should be beaten. If she wins, it will prove that she is on the right way, and that the rest of the world, at least that part which we and the Allies represent, is on the wrong way and should, for the **sake of the right evolution of the human race,** be stopped and put on the right way **or else be destroyed as unfit.** Professor von Flussen is sure that Germany's way is the right way, and that the **biologic evolutionary factors are so all-controlling in determining human destiny,** that this being biologically right is certain to insure German victory. If the wrong and unnatural alternative of an Allied victory should obtain, then he would prefer to die in the catastrophe and not have to live in a world perversely resistant to **natural law**. He means it all. He* *will act on this belief. He does act on it, indeed. He opposes all mercy, all compromise with human soft-heartedness. Apart from his horrible academic casuistry and his conviction that the individual is nothing, the State all, he is a reasoning and a warm- hearted man. So are some other Germans. But for him and them the test of right in this struggle is success in it. So let every means to victory be used.. There is no reasoning with this sort of thing, no finding of any heart or soul in it. There is only one kind of answer: resistance by brutal force; war to a decision. It is the only argument in rebuttal comprehensible to these men at Headquarters into whose hands the German people have put their destiny.'* This is not a quotation from Mein Kampfm or a speech from Nazi ideologues about the superiority of the Aryan race. It is evolutionary theory about German high Kultur, as understood by and discussed in the German High Command much earlier. **Kellogg** returned home to campaign for American intervention in the war despite his pacifism. He was well connected with the American elite and on April 6th 1919, the US declared war on Germany. For many Americans, Darwin's ideas were tainted with this context of militarism and war. **William Jennings Bryan** (1860-1925), an opponent of Darwinism, who stood three times to be President of the US and was Secretary of State at the beginning of WWI said that the *'same science that manufactured poisonous gas is teaching Darwin.'* Bryan was a great believer in moral progress, but opposed the theory of evolution because it undermined the Bible and morality, and promoted conflict! Bryan was particularly influenced by Kellogg's book. In 1918, he also read *'The Science of Power'* – an interesting title in itself – by the British social theorist, **Benjamin Kidd**. This work saw Nietzsche's philosophy as encouraging German nationalism and militarism based on the outworking of social darwinism. At a meeting of the World Brotherhood Congress in 1920, Bryan described evolution as *'the most paralyzing influence with which civilization has had to deal in the last century'* and that Nietzsche, in carrying the theory of evolution to its logical conclusion, *'promulgated a philosophy that condemned democracy,... denounced Christianity,... denied the existence of God, overturned all concepts of morality,... and endeavored to substitute the worship of the superhuman for the worship of Jehovah.'* [386] More, he saw Darwinism as a major threat to the US, particularly in its education system. He took very seriously the argument made by **James Leuba** in his *The Belief in God and Immortality, a Psychological, Anthropological and Statistical Study* (1916). This argued that a high proportion of college students were losing their faith, apparently because of the latest science teaching. He also looked on in horror at the presence of liberal theology in the churches. **Bryan's** response was to launch a campaign against 'evolution' in 1921 with his James Sprunt Lectures at Union Theological College which were later published and included an attack on *'The Menace of the theory of evolution.'* As a Presbyterian

---

[386] Bradley J. Longfield, *The Presbyterian Controversy: Fundamentalists, Modernists, and Moderates* (1991) p 68

elder, as well as a famous politician, he then stood for election as the Moderator of the General Assembly and only marginally lost to the **Rev Charles Wishart** who supported the teaching of evolution in theological colleges. It would probably be a mistake to see Bryan as a creationist in the modern sense. He took the 'days' in Genesis to be geological ages and allowed for the possibility or some organic evolution. But he insisted that Adam and Eve were the original and divinely created humans and he continued to campaign against the teaching of evolution in schools. In Tennessee, the Butler Act was passed in 1925 *'prohibiting the teaching of the Charles Darwin's theory of Evolution, Theory in all the Universities, and all other public schools of Tennessee, which are supported in whole or in part by the public school funds of the State, and to provide penalties for the violations thereof...it being unlawful to teach any theory that denies the Story of the Divine Creation of man as taught in the Bible, and to teach instead that man has descended from a lower order of animals.'* **Butler**, a Tennessee farmer had been influenced by reading Bryan's lecture *'Is the Bible true?'* comparing it with what he understood of Darwin's *Origin*. The High School teacher, **John Scopes** (1900-1970) broke this law and it was then challenged by the American Civil Liberties Union in the Scopes Trial of 1925. Scopes was found guilty and the Tennessee Supreme Court found the law to be constitutional because it did not break the separation of church and state. **Judge Grafton Green** wrote *'We are not able to see how the prohibition of teaching the theory that man has descended from a lower order of animals gives preference to any religious establishment or mode of worship. So far as we know, there is no religious establishment or organized body that has in its creed or confession of faith any article denying or affirming such a theory.'* It then reversed its ruling on a technicality related to the amount of the fine. The Butler Law lasted until 1967 when **Gary Scott** of Tennessee was dismissed from teaching for violating the Act. He then sued for reinstatement based on the First Amendment right to free speech and eventually the Butler Act was repealed. It is interesting to see the way free speech in American culture of the time was pitted against this prohibition on teaching Darwin. A state that opposed the ideas of evolution being taught in its schools was defeated by appeal to individual freedom of speech, rather than acceptance of Darwin. In the Scopes Trial, **Bryan** had represented the *World Christian Fundamentals Association* and was asked to take the stand by defense lawyer **Clarence Darrow** to be questioned about his views on the Bible. These covered a range of issues raised by Genesis – was Eve made out of Adam's rib; from where did Cain find a wife etc. When asked when the Flood occurred, Bryan consulted his Ussher's Bible Concordance and gave the date as 2348 B.C. He was then asked if he realized that Chinese civilization went back 7000 years. Bryan replied that he did not. When asked if other religions had recorded a great Flood, Bryan replied, *'the Christian religion has always been* *good enough for me - I never found it necessary to study any competing religion.'* All of Bryan's answers to Darrow's questions were expunged from the trial record on the instruction of Judge Raulston. The trial figured prominently in the national media and the phrase 'monkey' was sometimes used to describe Bryan. Some ridiculed Bryan as a symbol of Southern ignorance, but in fact his views, with all their ignorance of the bible and other religions, let alone history, were shared by millions of Americans. There are still many who would answer questions about the Bible as he did and the trial still represents divided opinion in that country. In his statement to the press at the end of the trial, Bryan didn't focus on Darwin at all, but made a passionate speech about the need to defend public morality in the face of the horrors of wars. By implication, he was associating Darwin and evolution with the worst excesses of science. *'Man used to be content to slaughter his fellowmen on a single plane, the earth's surface. Science has taught him to go down into the water and shoot up from below and to go up into the clouds and shoot down from above, thus making the battlefield three times as bloody as it was before; but science does not teach brotherly love. Science has made war so hellish that civilization was about to commit suicide; and now we are told that newly discovered instruments of destruction will make the cruelties of the late war seem trivial in comparison with the cruelties of wars that may come in the future. If civilization is to be saved from the wreckage threatened by intelligence not consecrated by love, it must be saved by the moral code of the meek and lowly Nazarene. His teachings, and His teachings alone, can solve the problems that vex the heart and perplex the world.'* I, for one, understand and share this last point of view, but not his rejection of evolution, nor his type of creationism.

As a 'revolutionary' idea, Darwinism has been used to justify war, competition and conflict as if it has a life of its own, separate from anything Darwin thought at the time. When the Twin Towers came down on 9/11 **Adnan Oktar** (pen name Harun Yahya), a very influential creationist based in Istanbul, wrote this caption under a picture of the event *'No matter what ideology they may espouse, those who perpetuate terror all over the world, are in reality Darwinists. Darwin's is the only philosophy which places a value on and thus encourages conflict.'* He called on all scientific supporters of Darwin to recant and apologise for *'deceiving humanity and hijacking the world of science.'* God has created all living things as they are. The evil in the worlds are the responsibility of Darwinists. When Darwin visited the Wollaston islands in 1832, he came across the Fuegan people whom he described as the *'most abject and miserable creatures I have ever beheld – their gestures are violent, they are a primitive state of humans.'* After a month, a Christian mission was established by the **Rev Richard Mathews.** Ten days later, Darwin returned to find all the mission's property had been stolen or destroyed.

Rev Mathews was in terror of a group of Fuegans. Darwin observed that three of them who had been previously 'Christianised' were slipping back into their previous state. Darwin wondered whether our ancestors were such as these and whether our civilisation was as thin and fragile as this. This Fuegan experience haunted him. It illustrated how changeable humanity can be, and how much part of the rest of nature. He recorded these impressions in *The Descent.* Darwin had visited London zoo many times and compared orang-utan behaviour with the Fuegans. He remembered observing a particular orang-utan, called Jenny, in 1838, as she threw a tantrum after being teased with an apple by her keeper. Jennie cried like a naughty child. When she pulled herself together and the keeper gave her the apple Darwin observed her 'contented countenance,' and said *'Let man see the intelligence of the orang-utan.'* **Queen Victoria** also visited the zoo and allegedly said the *'orang-utan is painfully, frightfully and disagreeably human.'* Darwin wondered what Jenny was thinking and feeling. He realised that her behaviour seemed almost human-like in its reaction. He saw her looking into a mirror and wondered if she recognised herself in the reflection. He later looked into a mirror himself and stared at his own face, comparing its expressions with Jennie's. The unthinkable became thinkable. It was in 1838 that Darwin first began to sketch out a new family tree, looking for common ancestors, speculating bravely that all life might have a single common source. In 1970, the psychologist **Gordon Gallup** developed his mirror test to discover an animal's ability to recognise itself with some level of self-awareness. It is said that Gallup had heard the anecdote about Darwin's experience. In Europe, **Sigmund Freud** (1856-1939) took Darwin seriously and integrated much of evolution into his own thinking, playing his own part in developing, challenging and propagating it. He had grown up in the Darwinian revolutionary years. Evolution was seen by many intellectuals as a helpful stage in our understanding of the world. Freud took the motivation to survive, adapt, reproduce and manage our primary and conflictual relationships from our common evolutionary past, and turned this into a theory of universal sexual urges, operating just under the surface of our conscious minds and the *veneer* of our civilised lives.

From Freud/ Chagall's love, war, and exile, Jewish Museum. P-D. Copyright expired.

In accepting the relationship between evolution and survival, he thought the latter alone weren't enough to explain the human condition. He added the search for pleasure and the frustrations that arise in its un-fulfilment. Freud's work has itself been criticised for developing self-referential, one sided assumptions and explanations for more complicated varied human needs. He is a good example of those who made anthropological or other types of reduction when applying Darwin to their own special interests. However, the motivating force of a survival instinct remains as an important dimension of the human condition. We are also a species consumed by curiosity, the desire for new learning, understanding, relating, ways of cooperating and the search for different kinds of pleasure and fulfilment, even, or especially, in times of un-fulfilment, frustration, dissatisfaction and disease. It is all too easy to see one crucial aspect of our condition as a description of the whole or to use it as a metaphor for the whole. For example, we are learning more about the electrical basis of the chemistry of cells and understand the body as an electrical system, composed of different kinds of charges, circuits and connections based on ion channels inhabiting the pores of our cells. So, at the beginning of life, as the sperm and the egg draw closer, an ion channel in the sperm causes its tail to beat faster in order to enter the egg. It is no denial of this truth, or the electrical functioning of the brain, or other parts of the body, if we also say that human beings are more than dynamic, electrical 'machines.' Further discussions between between philosophers, literary scholars and scientists may help here. However, it is true that by the time of Freud, science had inflicted three significant wounds on humanity's pride - three questions which challenged our sense of distinctive place and agency in space and time.

1. Copernicus - the earth is not at the centre of the universe.
2. Darwin - humans have evolved from other animals
3. Freud - individuals cannot be in complete control of their behaviour because human nature is based on and began with animal nature.

In the face of such significant and emotive turning points, however simplistically described, it is understandable that many found it difficult to adjust their self-image and beliefs, resorting to what they saw as the safe territory of religious fundamentalism, with its certainties and promise that human beings were created in a special way by God, separately from other species on this, or any other planet. It would take time for theology to catch up, and demonstrate that this defensive view of human distinctiveness was a misplaced way of opposing evolution; that there were other ways of constructing our human sense of place and purpose in relation to God, within creation and to each other. This must include our freedom to make decisions and shape our own environment in ways that are not totally predetermined by our evolutionary past. Echoes of social determinism still occur in the social sciences and political policy making. In one extreme form, it is argued that the behaviour and outcome of an individual's life is causally determined by his/her economic environment. This kind of view leaves little space for human freedom, for those caught in different kinds of 'predictive,' social matrix, as I suppose we all are, in one way or another. Yet, humans seem genetically motivated to do more than pass on their genes. Cultural and political 'natural selection' take us way beyond the gradations of biological changes into the complex organisation of ideas and societies. Ever new forms of technology extend individual reach in relation to our wider social and natural environment and its use and misuse. Cultural selection is deliberately manipulated for political and other ends, as well as being a natural or subconscious process. Cultural constructs become overlays on biological or genetic selfishness or altruism. The latter both seem to be present within certain animal species.

Certainly indices of deprivation correlate with higher mortality rates, family breakdown and negative, social behaviours. As patterns, these are predictable and measurable in ghettoed, cultural areas, though not at an individual level where we like to think that adaption and change are still possible. Marx had argued for the superstructure view that behaviours and ideas are the products of economic context, not the other way round. Darwin's view of natural selection allowed for different kinds of sideways and middle way, intermediate adaptation over long periods of time, with no pre-ordained or predetermined outcome. Ironically, social determinism seems closer to the pre-Darwinian, theological and Newtonian view of the universe that everything was ordered and immutably in place, whatever the response of individuals to their environment.[387] It is true that there was a certain teleological utopianism in Darwin's work. *'And as natural selection works solely by and for the good of each being, all corporeal and mental endowments will tend to progress towards perfection.'* [388] He did see this as part of a natural law and he was speaking out of the assumptions of his times - a complex mixture of rapid progress, social problems and perplexing questions about the human condition. Industrial and imperial expansion reflected a confidence in the possibilities of change and innovation, building on the great science of the previous century, and recognition of the need to reform the worst aspects of human nature through religion, education, harsh corrective penal interventions and public policy. **Huxley** wrote[389] *'That which lies before the human race is a constant struggle to maintain and improve, in opposition to the State of Nature, the State of Art of an organized polity; in which, and by which, man may develop a worthy civilization, capable of maintaining and constantly improving itself.'* Yet, he believed this assumption about constant improvement was a central fallacy in the 'the ethics of evolution.' If the struggle for existence and **Spencer's** 'survival of the fittest' led to improvement in animal species, there was no guarantee in the case of humans. Huxley saw this fallacy as emerging from a misreading of the 'fittest' to mean the morally 'best.' In fact, the State of Society as opposed to the State of Nature required self-restraint rather than competitive self-assertion. He repudiated a gladitorial theory of existence. Society needed not so much the survival of the fittest, as the fitting of as many as possible into an ethic of altruism, respect and reaching out to enhance the

[387] http://www.walrusmagazine.com/articles/2008.09-the-other-darwin-mark-czarnecki-creationism-origin-of-the-species-evolution/
[388] The *Origin*. Chapter 14
[389] *Evolution and Ethics* (1893)

wellbeing of 'other *others.' The gardener, on the other hand, restricts multiplication; provides that each plant shall have sufficient space and nourishment; protects from frost and drought; and, in every other way, attempts to modify the conditions, in such a manner as to bring about the survival of those forms which most nearly approach the standard of the useful, or the beautiful, which he has in his mind.*[390] The moral gardening had to not only chop down the weeds, but create the conditions of growth for ethical behaviour to counteract what he called the *cosmic process of nature,*[391] as the outcome of millions of years of growth. It would be folly, he said, to imagine that a few centuries would suffice to subdue the masterfulness of the cosmic process to purely ethical ends. Yet, his optimism shone through in that rather pithy, but eloquent summary sentence in *Evolution and Ethics - The intelligence which has converted the brother of the wolf into the faithful guardian of the flock ought to be able to do something towards curbing the instincts of savagery in civilized men.*

Thomas Hobbes (1588 –1679) had insisted that societies needed to protect themselves through law and institutions from the worst aspects of the human condition, as if these were indeed pre-determined.
Internet Encyclopaedia of Philosophy. Source unknown

The ordering of society implied the possibility of rational as well as emotional behaviours, while much in human behaviour is prompted by spontaneous, instinctive reactions. The role of morality, as a servant of the law, was articulated and sustained in rational systems as cultural constructs, created to control or modify emotional behaviour. One interpretation was that such constructs evolved as societies became more sophisticated and self-reflective, which in turn depended on measures of economic security; the other that morality was directly embedded in the social instincts we share with other animals.[392] Morality (religious or not) can therefore be seen as both the product of social evolution and a cultural controlling of behaviour within that evolution. Hobbes, might have said that morality and culture are imposed to organise and control our natural instincts, because we are not *naturally* moral creatures. Civilisation is a thin *veneer* over our selfish and brutish nature. It would be wrong to see **Hobbes** and **Huxley** as the originators of such an approach. As we have seen, Huxley had an ambiguous relationship with evolution, after, if not before, the Oxford debate. He struggled to reconcile his negative view of nature with the morality and kindness found in human society. *The garden is in the same position*

---

[390] *Prolegomena* (1894) Section 1V – the garden image is extensively developed in this work.

[391] In *Evolution and Ethics* he saw *'the word evolution as now generally applied to the cosmic process.'* In *Prolegomena* Section 1 Section 11 he continued *'Thus, it is not only true that the cosmic energy, working through man upon a portion of the plant world, opposes the same energy as it works through the state of nature, but a similar antagonism is everywhere manifest between the artificial and the natural. Even in the state of nature itself, what is the struggle for existence but the antagonism of the results of the cosmic process in the region of life, one to another?'*

[392] See the chapter on Altruism.

as every other work of man's art; it is a result of the cosmic process working through and by human energy and intelligence; and, as is the case with every other artificial thing set up in the state of nature, the influences of the latter, are constantly tending to break it down and destroy it... Since the cosmic process of evolution is amoral, there is no guarantee that good will win out over evil, that human society will thrive rather than fail. There is no guarantee — there is only the constant struggle of the gardener.[393] In this metaphor, he seemed to be following Hobbes. Humanity was moral because of its ability to control and oppose its own evolutionary behaviours from natural selection. Again, the historical and contemporary evidence is clear.[394] The reaching out of Love's Energy in the direction of the other can so easily turn, in the human, as well as other species, to a reaching out to dominate, destroy or harm. In Prolegomena he wrote, *Since law and morals are restraints upon the struggle for existence between men in society, the ethical process is in opposition to the principle of the cosmic process, and tends to the suppression of the qualities best fitted for success in that struggle.* Here, we find ourselves at the heart of a *pattern paradox.* Where did this ability to struggle against human nature come from, if humans were the products of natural selection? About twenty years before Huxley's *Evolution and Ethics,* Darwin's *Descent* had embedded morality within human nature and its natural evolution. The *pattern paradox* of Social Darwinism winds itself through the implicit relationship between these two works and these two great thinkers. Twenty years after the *Descent,* Huxley wrote, *For his successful progress, throughout the **savage state**, man has been largely indebted to those qualities which he shares with the ape and the tiger; his exceptional physical organization; his cunning, his **sociability**, his curiosity, and his imitativeness; his ruthless and ferocious destructiveness when his anger is roused by opposition. But, in proportion as men have passed from anarchy to social organization, and in proportion as civilization has grown in worth, these deeply ingrained serviceable qualities have become defects.*[395]

Veneer Theory

Like **Darwin**, **Huxley**, had experienced the cruel hand of nature in personal loss and suffering. In championing Darwin, he had to find his own way of compensating for the unpalatable dimension of nature's ruthless and random processes. **Freud** used the contrast between the conscious and the sub conscious, the ego and the super ego, love and death, as ways of understanding and reconciling the *pattern paradoxes* of positive and negative human

---

[393] *Prolegomena* 1894

[394] There have been too many victims of feral 'wild pack' group behaviour in history – from bullying to murder and genocide – to ignore the need for social controls of different kinds to complement our socialising interventions.

[395] *Evolution and Ethics*

behaviours. The super ego was Freud's explanation of how we gained control over the forces of nature. In *'The Selfish Gene'* (1976), **Dawkins** waited until the very end of the book before giving some compensatory reassurance – *'we alone on earth can rebel against the tyranny of the selfish replicators.'* In 1996, he said *'what I am saying, along with many other people – among then T H Huxley, - is that in our political and social life we are entitled to thwart Darwinism, to say we don't want to live in a Darwinian world'.* The Huxley form of dualism has had many supporters ever since, although with different variants on what **Francis De Waal**[396] called *veneer theory*. De Waal quotes **Michael Ghiselin's** (1974) model for concentric circles with bad behaviour at the centre and morality at the edges and is known for asserting *'stretch an altruist and watch a hypocrite bleed.'* He also quotes **Robert Wright's (**1994) and **Badcock's** (1986)[397] views that *'Those who think they are at times unselfish are accused of self deception, hiding their true motives from themselves.'* There are echoes of Hobbes and Huxley in these arguments. **Petr Kropotkin** (1902) and other colleagues, working in much harsher climates, argued for cooperation and solidarity within a Darwinian system. In this, they were influenced by Huxley's dilemmas. Kropotkin, as we saw, was finding in these cooperative, evolutionary behaviours the possibility of a moral society - one not dependent on veneers and dualisms, but embedded within the nature of different species. We will pursue this in the chapter on Altruism. Some distinctions need to made when using the same language about natural processes and human behaviour. For example, *selfishness* implies a certain moral choice or intent, not least to serve oneself at the expense of others. *Self interest* is often used to imply legitimate interest that may be served either by helping others, or acting selfishly. Plants have no moral intentions, but grow in ways that serve their own interests, either at the expense of other species, or to benefit their own or indirectly other species, including humans. They do this through selected self interest. Sometimes this involves cooperation for example with birds and bees to spread their seeds and pollinating nectar. If we cannot apply ethical choices to plants, can we talk of moral intent in certain species of animals when they act in ways that promote their own self-interest, but indirectly benefit others? The evolution of altruistic and empathetic behaviours is not inconsistent with self-interest. **Edward Wilson** and **De Waal** [398] remind us that many animals, not just the higher primates, evolve well developed, social instincts, based on their parenting or extended family roles. As Darwin saw in The Descent, this evolves into a moral consciousness as rational capacity evolved in humans and their immediate predecessors. While intellectual growth leads to moral capacity, immoral, amoral and conflicting behaviours continue in human groups and

---

[396] See chapter on Altruism
[397] *Primates and Philosophers.* P 11
[398] see chapter on Sociability and Altruism

individuals. One group's self-interest can exclude altruistic behaviours towards other *others*, while in another group, it can lead to cooperation. Darwin noted that many animals sympathise with each others' distress or danger. This is an evolved and protective behaviour within a group, which sometimes extends to other groups. Without the imagination to see that 'others *others*' suffer in the same way as oneself, there can be no morality in our cultural constructs. If this imagination evolved over time, why is it eroded in certain human groups over time? In violent computer games, young people seem to be honing their aiming skills and killing instincts. In these 'games,' unlike their gladiatorial predecessors in the Colesseum, it doesn't help that those killed get up again to be killed again without evidence of the suffering involved. These 'games' seem to be undermining the imagination necessary to see the negative results of our actions, therefore decreasing our moral sensitivity towards the results of our choices on other people. At least, in the Colosseum, people watched the vivid and bloody reality of this suffering. Perhaps some felt empathy towards the victims, but they were being programmed by crowd behaviours to overcome it. After all, the gladiators had been chosen and trained for their killing skills which were highly prized in those times. But this barbarism continues, as we turn from the horrors on the news to watch 'entertaining' war films and thrillers with guns and explosives. Those who are shot appear to bleed, suffer and die. We know they are only acting, but the more *real* this suffering and killing appears to be, the more we like it! Can we invent a computer game which evokes higher levels of empathy when others suffer or are killed? If we did, would people bother to buy it? Certainly, particular species of animals are affected by the emotions of other animals – again probably as an extension of in-group, parenting experience which socialises males or females into understanding the vulnerability of those they feed and protect. As this experience of responsible care is extended beyond primary relationships to unrelated, or out-group members of the same or other species, we see the development of altruism. In this limited sense, we can see an evolved altruism in non-human species, even in the absence of any cognitive morality.[399] Religion has often depicted moral goodness as a polar opposite to moral badness (e.g. Augustinian notion of sin) rather than as part of a changing continuum in the human condition, in which, at different points, we can see different combinations of motives and behaviours. To argue, as **De Waal** does, that morality is explicable in evolutionary terms is not to ignore the way humans have added levels of cultural and cognitive reflection. These have taken instinctive and *natural* morality to new stages of social organisation, not least in the law. Cognitively, we have developed a holistic connectivity between beliefs, values and behaviours, but there is evidence for *discontinuity* as well as *continuity* in this *pattern paradox*. We are capable of reflection on our

---

[399] See fuller discussion in Sociability and Altruism

behaviours to discern their *discontinuity* with our values, beliefs or experience of consequences. Evolutionary biologists use cognition as if it were a simple opposite to emotions such as empathy. Cognition is certainly more than the process of decision making, but that is a central part of its processes in any discussion of morality. As we saw in the work of **Daniel Kahneman**,[400] and his System 1 (quick) and System 2 (slow) thinking, the relationship between emotional and rational choices is more complicated. People can be reasonable, but not in the strict sense, rational. System 1 is mostly dominant in our decision making because of his 'law of least effort,' but they overlap. Many prejudices spill over into System 2. System 2 can rationalise, as well as override system 1 decisions. True cognitive expertise can be found in System 1, when professional knowledge e.g. in brain surgeons, comes effortlessly. System 1 specialises in providing quick or approximate answers where there is no System 2 solution. When faced with difficult cognitive problems, System 2 resists System 1 reactions, operating almost as in veneer theory. This raises many questions, not least what does this mean in the courts? In crimes of passion (System 1) are we less morally and legally responsible than for our planned crimes (System 2)? The processes in the Law are biased towards the rational[401] when often the decisions made around the crime are not rational. Is the system 1 decision experienced as being rational in some cases? In states of delusion there can be a rational decision. As we get older, we make decisions on the basis of more experience – so system 1 reactions are informed by changes in the feedback to system 2. In a culture dominated by a dangerous and harmful ideology e.g. National Socialism, did it become *rational* to conform with emotionally irresponsible choices? What did that culture do to our notion of free will? If the latter is a fiction, it is a necessary fiction. Can system 1 or 2 be located within the physical chemistry of the brain? Clearly not, as the electrical chemical processes are so dynamically interconnected and we are so ignorant of how consciousness or cognition arises from this physicality.[402] Rationality implies that we make decisions with the maximum calculation of consequences, but we are not good at imagining, let alone calculating them, and conflictual choices put huge emotional strains on cognition. Politicians often have to choose between conflictual *losses* rather than simple *gains* or *losses*. Deciding on risk in surgery is complicated enough, but when presenting risk to a patient, it becomes more complicated. Optimistic personalities can overestimate the positive and pessimists can overestimate the difficulties. If we had accurate, predicative cognition would we ever take any risks? **Kahneman** says that when we take risks, we are loss averse which is

---

[400] Prospect theory in LE2. See his *Thinking fast and Thinking slow.* Penguin 2012

[401] It is assumed that the jury and judge act will rationally on the basis of 'factual' evidence, when in practise our emotions are shifting the basis of that cognition.

[402] The Ancients thought the consciousness of the soul's instincts or decision making was located physically in other organs – the heart, stomach, liver and we still say that 'we feel it in our guts.'

compensated by optimism. Some surgeons and doctors say that to make a **rational** decision you have to close down **empathy**. But this creates tensions - it conflicts with the golden rule of only treating people as you would like to be treated yourself. If you like a patient too much, you can make the wrong decision. Sustaining an emotional distance from the feelings of the patient is in conflict with a good bedside manner of relational care, which communicates empathy and helps positive patient attitude. Some professionals choose to display a decisiveness beyond their actual, rational knowledge; others, often more senior and more experienced, have the confidence and humility to admit they don't know all the answers. Either can be the right tactic, dependent on the patient or client need. We also know that gender makes a difference – for centuries the prejudice was that women act more emotionally, while men act more rationally. Therefore women's voices should not be heard in public as opposed to private or domestic matters.[403] I am only touching the surface here, but hopefully enough to show that, in any discussion of morality, the relationship between emotions such as empathy and the rationality of cognition is a hugely complicated subject - one I can only see as a *pattern paradox* rather than a simple polarity. We can adjust and adapt our moral compass, but still we lose our way, doing the things we *shouldn't* and not doing the things we *should*. We can attempt to develop our moral and spiritual intelligence, but the pendulum keeps swinging back and forth. Understanding more about our evolutionary inheritance may rescue us from some of its negative traits, and justify our need for imposed laws and controls, but controls don't, of themselves, help us understand who we are, and why we behave as we do. Despite the controls, we seem to repeat our offending, and there is conflicting evidence about the possibility of linear improvement over time. We need the type of controls that educate and enhance our moral compass not make us more dependent upon their regulatory practices. Despite what has been said about a parenting experience increasing evolutionary altruism, human families can be fractured, tense places of learnt violence and dysfunctional behaviours which are passed on through generations. Is this also part of our evolutionary inheritance? As we saw in LE2, this reflects the inherited *pattern paradoxes* of human freedom. We are free to harm each other as well as to help. **De Waal** uses *Helping* and *Not Hurting* as categories of moral behaviour.[404] He views anything which falls outside of this pair of 'H's, as social conventions,[405] not morality. Surely this is too rigid a distinction, as many social conventions evolve out of and influence moral beliefs and behaviours.

---

[403] See for example the work of Professor Mary Beard at Cambridge University on examples from classical literature and myths (muthos – authoritative narratives of meaning).

[404] Ibid p 162

[405] He sees social conventions as cultural constructs and quotes same sex relations and nudity, pornography and burping after eating as examples

Controls need institutions and hierarchies of order and legitimacy – their quantity and type very much a continuing, political debate. Hierarchies can be found in many non-human species, as can specialised roles and responsibilities, many of which can be conflictual. Do non-human species manage better than us in their 'controls,' by apparently having fewer, if any institutions? Humans seem to need ever more sophisticated and complicated infrastructure to protect us from our worse selves as well as to enhance what is best in our evolutionary inheritance and potential. Reference to the presence of that inheritance, within the human species, undermines some of our higher claims for human freedom. In the thesis of Love's Energy, the same amount of radical freedom is present in the development of all species. There is no evolutionary boundary point in this ontological chain of freedom and love's energy which makes it possible. However, of course huge distinctions need to be made between different non-human species, let alone between them and the human species. Our overall, long term inheritance is shared in common, but, at certain times in history, significant changes have emerged. Nor does the fact that the Love's energy thesis also believes that Love is present within the working out of this evolutionary freedom protect it from its own innate randomness, ambiguities, risks and threats. If that were the case then the nature of Love – let alone freedom – would itself be compromised. For, as we have seen, Love does not, in its longing for the other to exist, in all fullness of abundant flourishing, intervene in ways that restrict or control the things which prevent this. This must be the case in the ontology of our evolutionary story, into which the energy of Love is embedded, Love *lets be* the factors which use the struggle for survival as dimensions of natural selection. But it does more. Love also inspires individuals and species to relate through the same ontological energy which helps them adapt in the processes of natural selection. We adapt, not just to survive, or even to compete in the struggle for survival, but to do something more with our surviving. This *more* includes Love's longing for the other to exist in ways that enhance its existence and through that the existence of others. In our struggle for surviving, we act in ways that can harm or benefit the same struggle in others. Alongside the sociality of group behaviour, triggered by survival needs,[406] we learn new purpose and possibilities. We develop concerns for the wellbeing of the other, because we've discovered non-utilitarian benefits to cooperation and altruism in the higher levels of self-actualisation and self-expression through the arts and the sciences. This experience of actualising more of ourselves is also part of how we enact the energy of Love's longing for others, within the shaping and adapting of our larger and inter-relational environment. Sociality, therefore, differs in difference species, both in its causes, processes and results. As coral reefs form out of the

---

[406] Making the first primitive toolsand weapons in the Rift Valley may have been a cooperative effort, helping motivation and language and developing a capacity to visualise their use and patience and determination in the process.

bodies of trillions of small animals they contribute[407] to a food chain that is invisible to their earlier stages of development. The many animals, including herring, which feed of the plankton around the reefs, do so without the reefs longing for these others to exist. However, there is a connectivity in the way these species link. It happens by their nature and its processes, not by any other artifice or design. In Darwin's search for connections, he too studied coral reefs, though he was not a good swimmer. He saw they were built by an infinite number of minute 'architects' and spent two years writing a detailed paper on their different kinds. He saw that changes to the composition of the sea may destroy them and indeed reefs all over the world now face a more uncertain future. They are a global warning system of the effects of climate change. Choral reefs absorb carbon from the ocean and lock it away in their structures, helping to regulate surface temperature. As Darwin studied choral and rainforest[408] he was in awe at their different levels of complex functions and structures. Darwin showed the balance between extinction and the birth of new species. If the rate of extinction occurs disproportionately, this has a huge impact, as when catastrophic events caused mass extinctions. If forests are suddenly destroyed, the species dependent on them are poorly adapted to their new environments.[409] As we saw elsewhere, Darwin's work on connectivity has influenced many environmentalists. Humans are capable of observing and consciously making, developing and celebrating the connections. We choose to improve our individual lives by creating social constructs that benefit others as well as ourselves – for example the internet. **Adam Smith's** theory of self-interested sociability is still with us, and he certainly inspired Darwin and the Victorians. A century before the *Origin*, Smith had written in 1759, '*however selfish so ever man may be supposed, there are evidently some principles in his nature, which interest him in the fortune of others, and render their happiness necessary to him, though he derives nothing from it, except the pleasure of seeing it.*' Selfishness is contrasted with a pleasurable empathy in contributing to the happiness of others. He wrote in an age when the American Constitution (1787) would produce grandiose statements about the pursuit of happiness, without necessarily implying any selfishness in the self-interest of that pursuit. The *pursuit* required its own freedoms and responsibilities, including those paradoxically provided by the State. **Rousseau's** '*Born free, but everywhere in chains*' was clearly wrong. We are born with an evolutionary inheritance, but with the potential of the evolutionary process to adapt and change, not least by removing those chains. The constitution makers in America had longed to throw off their chains, but also to build a society where responsibility for mutual

---

[407] They also compete and fight for space

[408] Covering about 60 % of the planet's land surface and home to about 50% of its species.

[409] There has been massive forest clearance in the past few decades

interest could include the free pursuit of individual happiness and prosperity, with minimum interference by the State. If altruism begins with cooperation for the sake of survival (prompted by the parental, caring role for the vulnerable and extended through closely knit groups) - then it is a very mature State that learns to create the kind of freedom where this can takes place and encourage moral responsibility to underpin it. Once family group, or tribe based altruism is stretched to include other species or strangers within the same species, the self-interested moral benefit may appear diluted or threatened. In Christianity, this stretching is taken to its extreme. Love reaches out to benefit the other, even the stranger who one might never see again, or the enemy who works against one's own self-interest and survival needs. Can such altruism draw on anything in our evolutionary inheritance, or does it transcend it? Does Jesus's example extend an evolutionary process or break with it? Is this a new kind of veneer theory – not imposing moral order on chaos, but unveiling the Kingdom of God's Love and Love's energy embedded within us? Love's energy longs not just for the radical freedom of the other, but models ways by which each *other* can benefit every other *other* by *longing for* and working for their benefit. This requires enormous self-sacrificing energy in order to make the morally innate real or the invisible visible. Do we have the reational will and emotional capacity for higher levels of empathetic understanding, forebearance, forgiveness and sensitivity to make possible these behaviours, even when inspired to do so? Darwin integrated the idea of morality into mainstream, evolutionary adaption far more than Huxley and his followers. Without the veneer emphasis of Huxley and Hobbes, our societies might be facing even more risk and threat. Without the basic Darwinian inspiration that the capacity for moral behaviour can be found within evolutionary processes, the state would have to trespass even more on individual freedoms, and become more draconian and dictatorial in its controls. Many have misused Darwin to establish the totalitarian power of one group's superiority over another. There is a story that when Jesus and his disciples were walking through a field they asked whether they should separate the wheat from the tares. Jesus' reply was to let these things grow naturally together. Can we read back into this parable the value of avoiding veneer type dualisms? Perhaps the very image of a veneer[410] with its structural layers of control is necessarily too simplistic, given the organic and interconnected nature of species behaviour.

As we have seen, Adam Smith's *'invisible hand'* functioned as its own kind of veneer theory, connecting self-interested, individual behaviour with wider, social benefit. The baker bakes bread to earn a living and in the process provides others with bread. In his 1759 *'Theory of moral sentiments,'* written exactly a century before the *Origin*, this process had wider implications within the moral and social

---

[410] We will return to this theme in the chapter on *Sociality and Altruism*

infrastructure of society. Well before his *Wealth of Nations*, he was commending these structures to protect human beings from their worst behaviours,[411] but also, more positively, provide the facilities and resources for humans to enjoy their sociability as a framework for their own self-development. Sport and culture provide their own opportunities for such development,[412] and show that we desire to do more, with our sociability, than simply prevent aggression and harm. These evolutionary traits unite us with non-human species as well as distinguish us. The discontinuities can be subtle, compared with the genetic similarities, but they are significant. All life may come from a common ancestor, but its diversity is extraordinary. Although humans seem to share certain global models of commercial organisation and technology, there is still great diversity physiologically (no human face is the same, yet all are recognisable as human faces), psychologically and culturally. This diversity is a crucial expression of the radical freedom of the Other, as created by Love's longing for the Other to exist in all its varied, self-making and often fractured discontinuities. Love's energy spontaneously bursts into a longing of life for the other, which then adapts as it shapes the diversity of its interconnected forms, environments and forces. As we saw in LE1, this operates at the smallest levels of sub-particle behaviours and in the cosmological forces that shape all things. In humans, we see the longing of love's energy to express itself ever more creatively, not just in response to survival needs, but through the diverse organisation of work based and artistic productivity, relationships and cultures, to share new learning, fulfilment and pleasure. These are expressions, in all their entangled competition and cooperation, of the radical freedom which emanates from Love's energy.

---

[411] Not least through the criminal justice system. The figures for looked after children alone remind us of the extent of negative family behaviours in domestic violence, neglect and abuse and their tragic consequences.

[412] Though examples of the former e.g. Rugby provide a ritual framework for aggressive and skilful behaviours. It was the novelist Sir Kingsley Amis (1922-1995) who controversially said (escuse the gender language) football was a gentleman's game played by hooligans; rugby a hooligan's game played by gentlemen and cricket a gentlemen's game played by gentlemen!

CHAPTER 9
Cells, DNA and Genes

Sub Headings

Evolutionary biology; Wilkins, Franklin, Crick and Watson; Brenner, Sanger, Sulston and Venter; Membranes and Mitochondria; The Quantum Leap of Life; RNA to DNA;

Evolutionary biology

Himmelfarb described Darwin as a Victorian conservative[413] who didn't so much invent something revolutionary, but recognised it in the work of others. While all evolutionary biologists see natural selection as a line in the sand, some say it cannot be observed to function as an explanation for everything. What it has done is to set a cultural and scientific template within which much later work has developed, in its own diverse ways. As we seek to understand more about the story of significant, scientific achievements, we are not just studying the past. We are in the middle of a continuing revolution in human understanding. As this continues through the present into our future, we are invited to remain open to its possibilities and implications, many of which are hard to grasp, or controversial as we saw in the previous chapter. Without Darwin, many later discoveries about cells and genes would not have been possible. If our science was still boundaried by fixed theological categories, limiting our search for truth, we would not have made such progress. Our children and grand children will take for granted medical and technological capabilities which Darwin's contemporaries, and those that preceded him, would not have thought possible or even imagined. It is now commonplace to hear of gene and stem cell coding, sequencing and therapy that has affected many areas of our lives. It is literally saving lives in medicine and helping our criminal justice system to correct its past mistakes and become more forensically accurate. However, our knowledge is never complete and we may be on the cusp of answers to questions we haven't as yet clearly articulated. Genes are part of our makeup and the processes which explain our evolutionary inheritance. We are now at the frontier of Darwin's dangerous idea - new knowledge about our DNA explains more about our evolutionary history and predicts much of what will happen to us in the future. This may become a controversial issue for future policy makers, insurers and employers. How will this affect ideas of equality, if evolution links different kinds of intelligence, brain types and sizes with different gene pools in different parts of world? As more genes are better understood, more predictions will be made for different types of behaviour, personality and intelligence.

---

[413] He was generally apolitical except in the issue of slavery, but seemed to have supported Mill and Gladstone. *Darwin and Darwinian Revolution* p 357ff.

Darwin would surely have been thrilled by the latest genetics as the missing connection in the puzzle of how naturally selected modifications are passed on and all life is related. The discovery of DNA also confirmed his theory of the common descent of all life. His contribution to the question we have been asking in this trilogy is very significant – *when we say we believe in a God who created everything, what do we think we mean.* If a creationist answer to that question is untenable, how will we talk about the processes of nature and creation in a world that 'makes itself?' In the spirit of Darwin, the new field of evolutionary biology is reminding us that human behaviour results from historical imprints. Crudely put, it is commonly argued that many, male behaviours come from our hunter gatherer ancestors. The work of evolutionary biologists seems relevant to many branches of psychology, with its different disciplines and approaches. As we saw earlier, in the work of **Oparin** and **Haldane**, organic life emerged from inorganic life, and crucial to that story is the appearance of the first living cells. Jack Haldane was a committed atheist. He is reported as saying that when he set up an experiment he assumed no God, Devil or Angel would interfere! A friend asked him if he would sacrifice a life to save a brother from drowning. He replied no, but I would for two brothers or eight cousins! Now, we see the mathematics of goodness being based on genetics. Meanwhile, even in the mid 20th century, the Catholic Church was still struggling to resist evolution. In *Humanis Generis* issued by **Pius X11** in 1950, we find '*Some however rashly transgress this liberty of discussion, when they act as if the origin of the human body from pre-existing and living matter were already completely certain and proved by the facts which have been discovered up to now and by reasoning on those facts, and as if there were nothing in the sources of divine revelation which demands the greatest moderation and caution in this question.*' This was preceded by '*if the human body takes its origin from persisting matter, the human soul takes its origin directly from God.*' They were solving the problem of Darwin by introducing a dualistic, territorial claim, as if warning science that while they might have the body, the church still had the soul! It is time to remind ourselves of the progress made genetics since Darwin, and because of Darwin. We have begun to understand the physical processes of species adaption through the study of genetics, but how was this discovered; how is adaption triggered; how are genes transmitted and of what material are they made? As I try to research these questions, I do so only in a generalist sense and without any formal training or specialist knowledge. I trust those with that knowledge will bear with my obvious inadequacies, as I weave in and out of these exciting subjects.

While the great British chemist and physicist **Ernest Rutherford** (1871-1937) said '*there is only physics or stamp collecting,*'[414] **Schrödinger** asked the question '*What*

---

[414]. Quoted by B. Birks *Rutherford at Manchester* (1962)

is life?' in his little book with this title, written in 1943.[415] This remains a crucial and disputable question, particularly for those looking for evidence of 'life' on other planets. The diversity of life from bacteria to whales on this planet illustrates the difficulty. Diversity itself seems to be an expression of 'life.' Life carries information in DNA, but that may not be essential for all forms of life. The chemical capacity for energy extraction is crucial for all metabolic processes. Without that, there is no energy to create, reproduce or sustain life forms. Schrödinger covereds vast territory in physics, chemistry and biology and his answer seems to have been that 'life' is order in a sea of chaos, based on the process of entropy. Cells will die but pass on a code. **Mendel** asked himself how species characteristics are passed from one generation to the next and he laid down a template for our later understanding of the role of 'genes' in the process. How do inherited genes trigger physical changes which then get passed on, in turn, via the genetic coding? Because this takes place over many generations, if not thousands of years, observation of real time adaption is practically impossible in any single case, as far as I understand it. What are the physical processes that carry the message *'you need to adapt in order to survive?'* How does a finch's beak grow longer to reach certain foods? Darwin asked himself this question many times, and produced his own descriptions if not explanations, based on observations. The finches who did not extend their beaks died, and only the ones who adapted could live in a particular (changing or new) environment. This we know for certain, but for the layperson, it is very hard to understand how the finch who 'first' developed a longer beak transmitted this need into the smallest levels of genetic change that made the longer beak possible for others. This is the detailed challenge still being explored by evolutionary scientists. We know that the answer goes right down to the quantum level of cells, where change is unobservable and a function of the interactions of the waves and particles where physics and micro-biology meet. The ground breaking discovery of the physicist Crick and the biologist Watson was the crucial link in our understanding of this puzzle, but there were others working on the same questions at the same time. They could not have made progress without leaning on the neo-Darwinian synthesis discussed earlier.

Wilkins, Franklin, Crick and Watson

In the years just before WWII, progress in the biological sciences was mostly slow because of the pressures of developing more practical benefit from physics and chemistry in the war effort. While **Schrödinger** was talking about the physics of biological life, many of his physicist colleagues were working on how to kill great numbers of people more effectively. So, after the war, and partly for moral and philosophical reasons, scientists returned to biology. The New Zealand born

---

[415] My copy is a combined *What is life* and *Mind and Matter*, combined Cambridge University Press 1967 with a foreword by Roger Penrose

Maurice Wilkins (1916-2004), a Nobel Laureate, had been an active communist in his youth, and yet was later 'cleared' to work at Los Alamos and Berkley on sensitive material in more ways than one! He had a wide interest in physics, including isotopes. At the end of the war, he left the Los Alamos project to work on what he called 'life' again, and, in particular, crystallography and how to shoot X rays at salt structures. Then he transferred the idea to biology and molecules, especially DNA, which was then thought of as the *stupid molec*ule. It was everywhere, but didn't seem to do anything much! So began, in the most unlikely of areas, the work that led to the next, huge jump in our story – our understanding of how *connectedness* works in the relationship between generations. In terms of Love's Energy—how can we understand the relationship between other *others* (between A and B)? How is that relationship carried and triggered? If Darwin had changed our understanding of connectedness and diversity across species, how could we explain the process of transfer, transformation, modification and mutation of different characteristics at a deeper biological level? At at the end of the 1940s, in King's College London, Wilkins began to explore the structure of DNA using X rays. He finally produced a comparatively clear image, showing that if DNA was the gene material, then genes could be crystallised. He then presented this result to a conference in Naples, in 1951, which was also happened to be attended by **James Watson**.

It is claimed that this excited Watson's interest in chemistry and there was talk of a possible collaboration with Wilkins, who also introduced **Francis Crick** (1916-2004) to the importance and potential of DNA. Crick believed that Wilkins should find a 'good protein' to work on. Wilkins realised better equipment (new x-ray tube and a micro-camera) was needed and, then, by 1952, managed to produce a 'B form' X shaped image using the sperm from squid which he sent to Watson who simply commented '*Wilkins... has obtained extremely excellent X-ray diffraction photographs*' (of DNA).[416] At this time, there were two types of photographs – 'A' and 'B.' B showed a perfect helix pattern, albeit in 2D format, of course. Wilkins

was convinced that DNA was helical, even though **Rosalind Franklin** (1920-1958) disagreed. Ironically, she had already and unexpectedly produced a quality picture of 'B' form DNA which she largely ignored to focus on other work.
Franklin date and author unknown P-D

The original, crucial, Xray diffraction image is known, rather unromantically, as Photo 51, and was taken by **Raymond**

---

[416] James D. Watson, *The Annotated and Illustrated Double Helix* p180

Gosling, in May 1952, in the lab of Rosalind Franklin. Gosling was her PhD student. It has become one of the most famous photographs in all of science, and well used in many scholarly and popular articles.

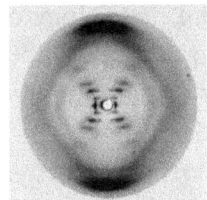

Photo 51. there are many versions of this photo on the web. The original by Gossling is still copyright.

She instructed Raymond Gosling to give this image to Wilkins who used it to confirm his own results and showed it to Watson. This seems to have been an important moment in a complicated, and not always happy story. It galvanised Crick and Watson to 'work together' on the model building of DNA. Wilkins continued to test their model and to work on the structure of RNA. All three were awarded the Nobel prize in 1962 *'for their discoveries concerning the molecular structure of nucleic acids and its significance for information transfer in living material.'*

Collaboration, Confusion and Competition

This is only one way of telling the story. It is, of course, far more complicated at all levels and not just what happened in their various labs! These were four very different people and the word *'collaboration'* is not the first that springs to mind, though the contribution of each was significant for the others. For example, Wilkins was away when Rosalyn Franklyn finally arrived from Paris in 1951. She was led to believe that DNA would be her project. Wilkins thought it would be his, with her collaboration. The tension between them was partly a result of the confusion in her appointment and the terms of reference negotiated for her by **Sir John Randall** (1905-1984). He was famous for his work on Radar and its role in the British war effort, but it seems his own radar missed the effect of the confusion he created organisationally. He hadn't informed Wilkins of her brief, and the tension only built over time. Both were competing for samples of DNA which was a real problem and a block to progress. Meanwhile Crick and Watson were certainly collaborating, but neither describes the nature of this working relationship as being particularly warm or close. They were competing with the work of the American bio-chemist, **Linus Pauling** (1901-1994),[417] afraid that he would soon correct his incorrect model of the structure of DNA and beat them to it. I can almost hear Darwin chuckling at this case study of academic species competition for survival! In the early days after Franklin's arrival, no one really shared their work. Franklin's personality was said, by some who knew her at the time, to be somewhat brittle and she saw the others as competitors and less serious than herself. She admired their brilliance to some degree, but despised their lack of solidity. Sadly, she died of ovarian cancer in 1958, aged only 37. Of the 'group' of four, her story has been the least heard, and in recent years estimates of her contribution and character have been revised. After all, she was working when the contribution of women was

---

[417] Winner of Noble Peace Prize for his peace work as well as a Noble Prize in chemistry.

very much undervalued.[418] There have been many stories about and books written on the relationships between these incredibly important scientists. Many who knew them well, including **Lisa Jardin** and **Stephen Jones** now openly discuss the extent of the tensions between them. This has become part of the mythology of the double Helix DNA story. Perhaps we will never really know, in any full detail, what they felt about each other, and to what extent their discoveries were helped or hindered by this. If Wilkins had worked at Los Alamos, **Francis Crick** had been a bio-physicist at Cambridge, and his work was hardly motivated by the horrors of war, even though he also worked in the Admiralty as a senior, science, civil servant, working on different types of mines.

Crick Photo: Marc Lieberman. Creative Commons Attribution 2.5 license

He became increasingly interested in the border line between living and non-living things, but had a reputation for being an over confident, and wild personality. He was still working on his PhD thesis at the time.

Jim Watson was a brilliant, American, Harvard based, molecular biologist and zoologist who gained his PhD when only twenty two. When he came to Britain to work in the Cavendish Laboratory, he had a bad

experience of the poor quality of accommodation and life in this country.

Source National Human Genome Research Institute. P-D

Watson never really managed to clinch a proper collaboration with Wilkins. His style was to sit back in lengthy discussion about how the parts of DNA might fit together. Meanwhile, Wilkin's and his group at King's were doing something more practical. The early results were interesting, but disappointing. Without **Rosalind Franklin**, progress might not have been made. Not only did she, or rather Gossling, take the crucial photograph of the 'B' form of the helix of DNA, but she had 'independently' worked out the location of the phosphate groups on the outside of the structure. When Watson started making new models of the DNA helix, he was putting the phosphates on the inside - it was Crick who suggested trying them on the outside. It is claimed that they worked with some excitement for four days, without a break, before discovering that it fitted. The structure they had modelled showed how DNA could make copies of itself and this was the crucial issue – it revealed its method of reproduction. If the helix pulled apart, each *base pair* could only pair up with themselves. Crick *said 'we just sat and looked at the model because it was so beautiful'* and then, apparently, he just went home to bed. They published their double helical structure in *Nature* in April 1953, when public attention was all on the new technology of televisions, the climbing of Everest and the Coronation. The press coverage was fairly minimal in both Britain and the US. They both

---

[418] I n some University 'common' rooms, women were still banned.

acknowledged they had been *'stimulated by.... the unpublished results and ideas'* of Wilkins and Franklin, which is, of course, only one way of putting it. Although the discovery which changed our understanding of how coded information is passed through generations will always be associated with them, historians of science are only too aware of the contribution of Wilkins and Franklin and indeed others in their teams and labs.

Brenner, Sanger, Sulston and Venter

We also need to remember the contribution of a **Sydney Brenner**, a medical doctor, born in 1927 to Jewish immigrants who came from Lithuania to South Africa.

Courtesy of Salk Institute of Biological Studies University of California

He gained a scholarship to do his PhD at Exeter College Oxford and then spent twenty years working at Cambridge. He was one of the first, in 1953, to see the model of Crick and Watson, and he ended up working with Crick at the Cavendish Laboratory. Watson had already returned to America. Brenner helped explore the question - how do mindless chemicals turn the code of DNA into personality? Brenner had a humorous, larger than life personality and this added its own chemistry to the next stage in the puzzle. DNA seemed to be made up of 3 letter codes. In proteins, it had 20 letters. What was the right number? Crick introduced, in 1961, the possibility of three mutations which opened the door to the idea that the whole gene could be read in its entirety. Brenner, working with **George Pieczenik,** created the first computer, matrix analysis of nucleic acids. Then working again with Crick, as well as **Aaron Klug** (b 1926) and **George Pieczenik,** he produced a paper on the origin of protein synthesis and its *'five base'* interaction, creating a triplet code translating system. This is the only published paper in scientific history with three independent Nobel laureates collaborating as authors. Brenner worked in a patient way on the details of the DNA codes. He showed that all overlapping genetic coding sequences were impossible, and so the coding function could be separated from structural constraints. Crick was then able to propose the idea of the *adaptor* or the transfer RNA or tRNA. The physical separation between the amino acid on a tRNA and the anticodon enables the principle that information flows in one way only - from nucleic acid to protein, and never the other way round. Then Brenner produced the idea of a messenger RNA, based on the work of **Elliot Volin** and **Larry Astrachan** in 1956. In 2002, the Nobel Prize in Physiology or Medicine was awarded jointly to Sydney Brenner, **H. Robert Horvitz** and **John E. Sulston** *'for their discoveries concerning genetic regulation of organ development and programmed cell death.'* Brenner's choice of the tiny worm, Caenorhabditis elegans, was a crucial element in the success of this work. **John Sulston** was to be a key player in the next stage of the story, but first we turn to the look at **Frederick Sanger**, (1918-2013) working in his Cambridge lab on DNA. He was crucial to the story of the reading of the genome code, turning genetics into the modern day equivalent of anatomy.

Sanger. 2006 Source unknown. P-D image

He was a British biochemist who won the Nobel Prize for Chemistry twice, the only person ever to have done so - in 1958 *'for his work on the structure of proteins, especially that of insulin'* and in 1980, with **Walter Gilbert** *'for their contributions concerning the determination of base sequences in nucleic acids.'* At the same time, **Paul Berg** received a shared prize *'for his fundamental studies of the biochemistry of nucleic acids, with particular regard to recombinant-DNA.'* In 1943, Sanger joined the group of **Charles Chibnall,** a protein chemist who had recently taken up the chair in the Department of Biochemistry in Cambridge. Sanger's first triumph was to determine the complete amino acid sequence of bovine insulin in 1952. This showed that proteins have a defined chemical composition and that every protein has a unique sequence. This was to influence Crick's sequence hypothesis in relation to the DNA of proteins. Sanger, a member of the Medical Research Council, moved to the new Laboratory of Molecular Biology in 1962 and became head of the Protein Chemistry division. He began work on the possibility of sequencing RNA molecules and methods of separation. The challenge was to find a pure piece of RNA. While searching with **Kjeld Marcher** in 1964, he came across formylmethionine tRNA which initiates protein synthesis in bacteria. He missed being the first to sequence a tRNA molecule, as the **Robert Holly** group at Cornell University published in 1965. By 1967, Sanger's group had worked out the nucleotide sequence of the %S ribosomal RNA from Escherichia coli. Sequencing DNA was a different challenge, however. In 1975, working with **Alan Coulson** he published a sequencing procedure he called the *'Plus and Minus'* technique. This could sequence up to 80 nucleotides at once and, although a definite step forward, required very patient, detailed work. Despite this, they managed to complete a sequencing of most of the 5,386 nucleotides of a single-stranded bacteriophage which was the first fully sequenced DNA based genome. The main surprise was that the coding regions of some of the genes overlapped. In 1977, they produced the 'dideoxy' chain-termination method for sequencing DNA molecules, known as the Sanger method. Now, longer stretches of DNA could be

more rapidly sequenced. For this, he received his second Nobel Prize in 1980, shared with **Walter Gilbert** and **Paul Berg.** The new method was used to sequence human mitochondrial DNA which is made up of 16,569 base pairs. **Sir John Sulston,** FRS is chair of the Institute for Science, Ethics and Innovation at Manchester University.

Sulston, from his personal details page at Manchester University

After some time at the Salk Institute, he returned to Cambridge and the Medical Research Council Laboratory of Molecular Biology and then moved to be the first Director of the new Sanger Institute in Cambridge. He retired from there in 2000 and has held a variety of roles ever since. As we have seen, he shared a Nobel Prize for cataloguing cells in the very small C. Elegans worm. For about twenty years, this involved very intensive, daily observations of the order in which those cells divided. To understand their genes, he had to first sequence them. This work then became part of a global effort to read the human genome for the benefit of science everywhere.[419] At a 1996 summit in Bermuda, he joined other leading scientists and given the huge amount of data and research on the genome, they agreed three ground breaking principles.

> - Automatic release of sequence assemblies larger than 1 kb (preferably within 24 hours).
> - Immediate publication of finished annotated sequences.
> - Aim to make the entire sequence freely available in the public domain for both research and development, in order to maximise benefits to society.

But then **Craig Venter**, the American entrepreneur and biologist appears in our story. In 2010, the *New Statesman* listed him as one of the world's top 50, influential figures.

Venter 2007. Creative Commons Attribution license. Photo appeared in Article by Liza Gross, but no photo credit given.

He founded the Celera Genomics, the Institute for Genomic Research and the J. Craig Venter Institute. He remains passionate about using genomics to transform health care, and became frustrated about the slow speed of progress and the lack of funding – his motivation for turning to the private sector. At Celera Genomics, the plan was to sequence the whole human genome as quickly as possible, and to release it into the public domain for non-commercial use, while subsidising this via subscription access to a data base. As team leader of a US genome group, he decided that money could and should be made out of patenting the human genome. This went against everything Sulston believed in - that knowledge should be publicly owned. The tension was obvious, to all in the networks and beyond. The pace of competition between them increased, and John Sulston became even more anti-Corporate. There were two programmes running in parallel; one in the private and the other in the public sector, but the (financial) relationship was more complicated than this simplistic division implies. In 2000, Venter and **Francis Collins**, Director of the Human Genome Project and the National Institute of Health

---

[419] He has strongly held humanist and anti consumerist views and believed everyone should release their data under what became known as the Bermuda Accord.

in Maryland[420] announced they had mapped the human genome, well ahead of other 'publically' funded teams, albeit using techniques which were described as 'shotgun' by some. Then the politicians became involved and the story hit the press. Ultimately both **President Clinton** and **Prime Minister Blair** came down on the side of Sulston. They supported public access to the genome project, because of its obvious medical benefits. Some commentators, including **Lisa Jardin**, saw Sulston as the more traditional, public spirited scientist and others viewed this as a cultural division between American and British values. Again, this is probably too simple a division. There is evidence that Venter only turned to the private sector because of the difficulty of getting public funding, and that he always intended to make the knowledge publically accessible. He certainly had his own difficulties with the running of Celera Genomics and one of its main investors, and was fired in 2002. In May 2010, Venter led a team to create 'synthetic' life, by synthesizing a very long DNA molecule containing an entire bacterium genome which was then introduced into another cell. Increasingly, he popularised the science in many media appearances. By 2013, he went even further, claiming that scientists would soon be able to use 3D printers to create synthetic life forms and that, over the next 20 years, synthetic genomics would become the standard for making anything. He may well be right. In an interview with The New Scientist in 2007, he claimed that *'The chemical industry will depend on it. Hopefully, a large part of the energy industry will depend on it. We really need to find an alternative to taking carbon out of the ground, burning it, and putting it into the atmosphere. That is the single biggest contribution I could make.'* Meanwhile the Hungarian born Professor **Veronica van Heyningen,**[421] said, in April 2014, that the tracing of a genetic defect that would, in the 1980s, have taken a few years, now only takes thirty six hours.[422]

I try to imagine what it would have been like for Darwin to witness the sequencing of the human genome. This provides a complete set of genetic information, encoded as DNA, within the 23 chromosome pairs in cell nuclei, and in a small DNA molecule found within individual mitochondria. What would Darwin and Sanger, Culston and Venter have said to each other and how long would it have taken for Darwin to grasp the physics and chemistry involved? The human genome includes both protein-coding DNA genes and non coding DNA. The genomes contained in human egg and sperm cells (haploids) consist of three billion DNA base pairs, while somatic genomes (diploids) have about twice that number. If the significant differences between the genomes of human individuals is in the order of about

---

[420] The first official funding for the Project came from the US Department of Energy's Office for Health and Environmental Research in the Reagan Administration budget submission to congress. Collins had followed Watson as Director of the Project.

[421] Before retirement she was head of Medical Genetics at the Edinburgh MRC Human Genetics Unit and President of the Genetics Society. An expert on the eye and the disease aniridia.

[422] Interview with Jim Al Kalili BBC Radio 4 *The life scientific* 1.04.14.

0.1%, the differences between humans and their closest living relatives, the chimpanzees, is greater, but still only about 1%. On hearing this, Darwin would probably have thought of **Wilberforce** and **Sedgwick** and all those who were so scandalised by his thesis. By 2012, thousands of human genomes had been completely sequenced and more have been mapped at lower levels of resolution. This is vital information for our understanding of how evolution works, but more work is needed to understand the biological functions of protein and RNA products and the biochemical activities of non coding DNA. What effect do they have in regulating gene expression and the organisation of chromosome architecture and inheritance? It seems that protein-coding sequences account for only about 1.5% of the genome, and the rest is associated with non-coding RNA molecules, regulatory DNA and sequences for which, as yet, no function has been identified. As the English geneticist, **Paul Nurse,** President of the Royal Society and CE and Director of the Francis Crick Institute[423] reminds us - many other challenges remain. He also believes, along with others, that many issues of inheritance depend on how we understand the physics. There is always more to explore in the wonder of *life longing for life*, working continuously and invisibly in our cells.

Membranes and Mitochondria – their part in the story of Love's Energy

All cells are bounded by a membrane. This immediately raises the question, in our story, of how A relates to B across a membrane. Life had to emerge before the boundary appeared, but it is difficult to know how, or when cells without a boundary emerged, if they didn't do so at the same time. Inside the membrane, a series of metabolisms take place, assembling catalysts (inorganic initially and then proteins). Within the process, as we have seen, there has to be some kind of a code or 'book of instructions' - however crude that language appears - stored in the deoxyribonucleic acid or DNA, or in the ribonucleic acid or RNA. The code must act as a catalyst as well as a template. Catalysts produce energy and energy is central to our story in Love's Energy. Across the membrane, there is a chemical/electrical mechanism for producing energy for inflation and growth. I compare this to what we saw in LE1 about the process of inflation in the micro seconds after the Big Bang, and the oscillating waves that were imprinted in the background radiation. The physics of life influences its chemistry. In Schrödinger's *'What is Life?,'* he defined life as a patch of order in a sea of chaos. The movement of sub-particle balancing across a membrane creates life - building tiny bits of energy to self-organise and cause movement of the *'proton gradient'* kind. The story of life's development is the story of the movement of energy to organise the probability of increased energy efficiency, distribution and connectedness. The story of life's stability, within the oscillations of change and development, is that of

---

[423] Awarded the 2001 Nobel Prize in Physiology or Medicine with Leland Hartwell and T. Timothy Hunt for their discoveries of protein molecules that control the division (duplication) of cells in the cell cycle.

homeostasis. There is, it seems, a self-regulating function within the external and internal environments of living things, but it comes at a price. Temperature regulation in warm blooded or endothermic animals enables them to still function in varied conditions, but this requires extra energy. If the external temperature rises, the body loses heat by sweating or panting; if it falls by shivering. Living systems (and mechanical ones too) have evolved sensors, central controls and effectors to react to bring stability in change.[424] Negative feedback e.g. lack of food (energy) causes the body to lower its own parameters so that it can still function, albeit at a slower rate. This self-adjusting is automatic – its chemical processes function without our conscious controls. This parallels the way evolution functions at the sub-particle and genetic levels. The relationship between stability and change is is own dynamic *pattern paradox* in the story of energy and its *autonomous* processes.

The Quantum leap of life - Eukaryotes

Bacteria don't seem to have done much for the first billion years. They were *energetically inefficient*. Then, it seems that one cell type took over other cell types, producing increased *energy efficiency* in its own kind of *pattern paradox*. Once a nucleus appeared in a cell, it could protect the code of RNA or DNA and facilitated growth, but how did this quantum leap happen? We don't know. All larger organisms share a common ancestor, so the quantum leap presumably only happened once. Some kind of event/process caused the appearance of **Eukaryotes** cells with the energy generators – mitochondria. **Eukaryotes** get larger as bacteria become specialised and more energy is created. Mitochondria (mitos = thread; chondria = granule) develop increased *energy capacity*, compared with bacteria and so develop the energy to expand and increase. Cells inside another cell had to evolve a process to become more efficient. The separation of genetic material into a nucleus was clearly important, but then, at some point, cells separated into smaller separate membranes, or sub-compartments. The process of symbiosis enabled photosynthesis, which in turn enabled the use of oxygen, which in turn enabled more efficient and larger growth with the capacity for mutation. Larger capacity allows for further development of a larger DNA 'library' to contain more RNA 'books.' This, at least, is one way of looking at it, and new research is constantly refining our understanding of this highly sophisticated, physical, chemical and biological process. Meanwhile, membrane 'bags' became more complicated and self-organised into different functions and specialisations. All of this opened *energy pathways* to increased capacity for evolution. As we have seen, there is still much to discover about the selectable advantage at each stage of the

---

[424] If the oxygen content of the blood falls, or carbon-dioxide increases, the heart works harder to increase blood flow and lungs work harder; kidneys remove excess water and ions from the blood, PH measures of alkalinity and acidity; blood glucose with insulin and glucagon.

appearance of the early cell. Questions continue about the stages of early development of life and energy and different *'inflation points.'* Clearly mitochondria are crucial for any work on energy. They are the *energy power house* of the cell which generates most of the cell's supply of adenosine triphosphate (ATP) which, as we have seen, is an essential source of chemical energy. A mitochondria is a membrane enclosed organelle, or subunit, with a specialised function found in most eukaryotic cells. It seems that in addition to being an energy power house, they play a significant role in signalling, in differentiation, in controlling cell cycles of growth and death (and so play a role in the ageing process of all living things including humans). Symbiotic events of connectedness across the relationship between A and B lead to the 'take over' of one cell by another. Mitochondria behave as if they have a yearning for an independent existence, fighting with the nucleus. They contain the tricarboxylic acid cycle, or Krebs cycle - a series of chemical reactions used by all aerobic organisms to generate energy, which may have been one of the earliest, established components of cellular metabolism. Mitochondria probably began as form of bacteria and its relationship with its host cell was crucial. They appear to fight with each other for reasons that are not understood, but this clearly has a profound influence on cell development. Eukaryotes cells seem to reproduce by dividing in the cell cycle. Chromosomes appear at points in this cycle, and move to opposite *electrical poles*. Then they disappear and cell division continues. Sperm and egg cells are made in different ways through a process of meiosis in eukaryotes cells. Whilst the process of meiosis is similar to mitosis, the latter separates the chromosomes in its cell nucleus into two identical sets, in two separate nuclei. In meiosis, the chromosomes recombine, which makes possible a different genetic combination in each gamete. Meiosis begins with one diploid cell containing two copies of each chromosome (one maternal and one paternal)—and produces four haploid cells containing one copy of each chromosome. Each of the resulting chromosomes, in the gamete cells, is a unique mixture of maternal and paternal DNA, resulting in the genetic diversity which makes *natural selection* possible. Again, I imagine Darwin sitting at his desk wondering how natural selection really worked, hearing a knock at his door at Down House and finding there a stranger, come to explain the processes we are beginning to understand, nearly two hundred years later.

RNA to DNA

But how do cells specialise? Higher organisms certainly require greater, cellular specialisation. For example, without the ability to transport water up their structures, plants can't grow on land. There must be a kind of metabolic *specialisation of energy* and the way this is done must have begun at a very early stage, perhaps even in the original, common single cell. Is the RNA code read from the DNA library to put things in linear sequence? Bacteria only have a small number of genes, while eukaryotes have many more – some as many as 40,000. This increased possibility of switching on and off different genes produces different organisms e.g. the kidneys or liver. But how does the switch on process work? The theory is that before the appearance of DNA, there were self-replicating RNA molecules which stored genetic information in a similar way to DNA.[425] The transition from RNA to DNA needed intermediate stages, using enzymes such as ribosome and ribozymes. It seems that DNA *evolved* greater stability[426] and capacity for data storage, and that proteins replaced RNA because they contained a greater variety of monomers or amino acids. Supporters of this theory see RNA in existing cells as an *evolutionary remnant* of an early RNA era. RNA is made up of nucleotides - made up of a sugar phosphate backbone. This combined in series of long strands, so that their sequence could carry information. The nucleotides would have been free floating in the *primordial soup* and would have bonded with each other, depending on how low the energy levels were. Again, it appears that *energy* is a crucial element in the bonding process. As increased bonding took place, so the chains grew longer, more nucleotides were attracted and attached, reducing the instability and break up. Were these chains the first form of primitive life? Did a form of natural selection operate here, with the more efficient molecules surviving and producing their own reproductive capacity? Competition may have led to some levels of cooperation between different RNA chains leading to the first proto – cell, and to random combinations which helped amino acids bind together. The reaction that bound amino acids together into proteins is known as peptide bond formation, and it is thought this reaction was triggered by an adenine residue in the ribosomal ribonucleic acid, or rRNA which is essential for protein synthesis in *all living things*.

Since the work of Crick, we know that the genome in all animals reveals some surprising correlates. Studies done on animals as small as fruit flies, or as large as whales, show the significance of *common factors* within the early embryo. As we have to keep reminding ourselves, Darwin's understanding of natural selection came before modern genetics. He described *what* without always knowing *how*. No doubt our present DNA driven understanding of *similarities* and *differences* will be helped by subsequent discoveries. The study of embryos and their development demonstrates the wonder of naturally selected *differentiation* from what appears to be a very similar and *connected*, early cellular formation. Small

---

[425] Walter Gilbert first used the phrase *RNA World* in 1986 when commenting on the catalytic properties of various forms of RNA. The concept of RNA as a primordial molecule can be found in the earlier work of several people including Francis Crick, Leslie Orgel and in Woese's 1967 book *The Genetic Code*. The nucleic acids essential to understanding RNA were discovered as early as 1868 by Friedrick Miescher who called the material 'nuclein' since it was found in the nucleus. The presence of RNA in protein synthesis was suspected as early as 1939. Severo Ochoa and Archur Kornberg received a Nobel Prize in medicine in 1959 for discovering an enzyme that can synthesize RNA in the laboratory.

[426] RNA molecules are by comparison more fragile and can be broken down by hydrolysis. This may not have mattered in early forms of life.

changes in the DNA coding and sequencing, account for apparently huge differences in bodily form, adaptation and ability. So, evolutionary biologists have to trace the historical imprints back much further than the early primates or the lower animals to the most amazing appearance of single cell animals, where life was present without stomachs or brains.[427] It is from animals like these that we see the big birth evolving into the lower animals, the primates and eventually homo erectus. Homo sapiens cannot be isolated from this process of development in the way some creationists propose.

The question remains as to what extent, and how, humans developed their nature and behaviour beyond, as well as through, physical, evolutionary processes at the cellular level. Some Darwinians, and their successors, have too easily been viewed as implying only deterministic, linear development. We need to be cautious in making too simple a connection between early and later forms of life, even though the echoes and imprints of the past are clearly present in the latter. Whatever the influence of early RNA, DNA and genetic coding, we evolve in ways that transcend those imprints, even though they always remain. This allows for modifications and creativity not present in earlier species. The earlier imprints do not always predict later developments, not least because of indeterminacy and random effects and environmental changes. So, there must be space for potentia (LE2) and propensity in the processes of natural selection and genetic inheritance. Religion argues for the possibility of this potentia in the evolutionary constructs of the human being (and I would add all life), claiming the possibility of real moral formation and choices over predictable, determined ones in human beings. It also claims that humans have evolved a sense of the transcendent or the 'divine,' as well as the 'good'. An extreme view of Determinism substantially challenges any connection between love's energetic potentia and freedom, as we saw in LE2. I believe that Love has to be discussed as much in relationship to DNA coding and the physical processes of energy transformation as in romantic poetry. In this Trilogy, Love's freedom is not posited as some kind of abstract edifice above DNA and evolutionary biology. Many of the ways of love are imprinted, not least in different views of the 'natural' sociability of many animals, including human beings. The evolutionary biologists know the limits of this sociability, both in terms of the numbers (about 150 'relationships') and their ambiguous motivation and diverse nature. From the Stoics and Epicureans onwards, people have argued that sociability gives pleasure, and this must have some kind of evolutionary explanation. Reaching out to the other or *oikeiosis*[428] may be motivated by genetic, as well as

---

[427] Many of which have been found amongst thousands of fossilised species from the Cambrian – 400 million years ago in the Burgess Shale – see earlier references
[428] identification with someone or something

subconscious factors. Christianity has talked much about duty and responsibility, certainly in those forms of Christianity influenced by **Augustine,**[429] as **Adam Phillips** and **Barbara Taylor** argued in their much discussed work '*On Kindness.*'[430] To understand the social reasons for cooperation is not to reduce them to merely functional phenomena. Pleasure, fulfilment and aesthetics, as well as utility, can be observable and sometimes even measurable, qualitatively if not quantatively, certainly in human beings. Nor should we easily dismiss the idea of duty where it takes us beyond the rewards of pleasure. A motivation to help others, even where this demands significant personal cost (time, energy, worldview, role or status), may have its own evolutionary antecedence or unconscious motivations. It seems that such behaviour, despite all the examples of its opposites in human history, is a crucial part of the way human beings evolved, albeit mostly within their own extended family and tribal groups. Christianity, like some other religions, and certain pre-Christian philosophies, argues for the same behaviour being extended beyond the tribe to strangers and even enemies. The latter is part of the radical and transforming claim of Christianity based on the reported words and actions of Jesus of Nazareth. Given that these words and actions were and are so costly, it is hard to see how they can be totally explained by an evolutionary genetic code, although this too may be possible as science becomes more sophisticated. It is time now to explore the evolutionary inheritance of sociability and altruism in more detail.

---

[429] And its affect on philosophers like Hobbes
[430] Penguin 2009

Chaper 10
Sociality and altruism

Sub Headings

Moral Goodness; Social Evolution; Morality and Human superiority; From Empathy to sympathy; The Russian Doll of Continuity and Discontinuity; Descriptive and Normative morality; Embedded altruism.

On the tomb of Byron's dog *Botswain* is written a longer epitaph than on his own tomb. *Epitaph to a Dog* 1808 includes the line *'all the Virtues of Man without his Vices.'* It was the Victorians who invented the phrase *'a man's best friend.'* Not only was the social reformer, anti-vivisectionist and women's suffrage campaigner, **Frances Power Cobbe** (1822 –1904) self educated and descended from Archbishop Charles Cobbe (1743 to 1765), Primate of Ireland, but she lived in a kind of marriage with the sculptor **Mary Charlotte Lloyd** whom she met in Rome. She disliked religion, but she liked dogs. Darwin too liked his dogs,[431] though perhaps not so much his cats. His relationship with Polly[432] and his other dogs may have influenced his belief that animals could express emotions. He saw them as straddling nature and the domestic and human worlds; emotional and intelligent as well as being savage predators. Polly was the model for the illustrations of dogs in his last work *Expression of Emotions in Man and Animals,* 1872. Emma Darwin said Frances Cobbe was very agreeable when they met in 1868. Frances persuaded Charles to read **Emmanuel Kant** and **John Stuart Mill** and her writings about the latter. But then she published a letter Darwin had sent her. She did it without his permission, and he wasn't pleased. *Darwinism in Morals* was her critique of his *Descent*, published in T*he Theological Review* in 1871. [433] She found it difficult to accept that morality was inherited in evolution and tried to persuade him that it was intuitive.[434] A significant question about human nature, certainly from Aristotle onwards, was to what extent humans are social animals, or just different combinations of individuals, living in formal or informal contracts to keep the peace? The *'peace,'* of course, contains many issues of social justice internationally and locally, as well as the prevention of violence and war. The phrase *'struggle for survival,'* made its own contribution to our assumptions about the human condition. As we saw, **Hobbes,** and to some extent **Rawls,** argued that different kinds of social contract are essential to

---

[431] No least because of his obsession with hunting in his youth as well as it being a popular Victorian habit. See *The Hunter's Gaze: Charles Darwin and the Role of Dogs and Sport in Nineteenth Century Natural History*, by David Allan Feller.

[432] Perhaps named after Polly Howard the first wife of his Grandfather. This was his favourite dog.

[433] See also her *Scientific Spirit of the Age* 1888 and other books on social themes.

[434] See *The Theological and Ethical Writings of Frances Power Cobbe, 1822-1904*, by Sandra J. Peacock Lewiston, NY: Edwin Mellen Press, 2002,

prevent us from tearing each other apart. Hobbes did not know much about our evolutionary inheritance from our non-human ancestors. It was only after, and because of Darwin, that we could frame the question in that way. If Aristotle saw humans as essentially cooperative beings, Darwin and the evolutionary biologists point us to back to that longer history of inherited connections.

Moral Goodness

On the 6th January 1975, in a squat near Euston station, the body of the Harvard Chemist, **George Price** (1922-75) was found and then identified by **Bill Hamilton** (1936-2000), a famous, Oxford based biologist who was to influence Richard Dawkins. Price had been investigating the genetic basis of good and evil. He had contributed to the Manhattan project; developed a plan to free Hungary from Russian domination, based on buying well made shoes, and started an unpublished book on Soviet American relations called *No Easy Way* - but events were changing faster than he could write. He turned to the haunting question of human morality and came to London in 1967 to work at the Galton laboratory, influenced by Hamilton's work and equations on kin selection, and **Maynard Keynes** on evolutionarily, stable strategy or game theory.[435] He studied empathetic reactions in and across animal species and eventually invented an equation for the evolution of altruism. He showed that animals are more likely to show altruism toward each other as they become more genetically similar; that the basis of altruism was related to genes pursuing their own self-interest and how self-sacrificing behaviours are passed on, even though they harm some carriers of the genes. Altruistic actions benefit relatives of the altruist - who are themselves likely to share the gene in question. But, he also realised that the calculations were descriptive, not prescriptive as in religion's 'love your neighbour.' In early 1970, he attended his first Christian service at All Souls Church in London and decided to become his own experiment in altruism, giving his money randomly to the homeless. He ended up giving everything he had. His life disintegrated into chaos, depression and schizophrenia. Those he was helping stole from him. He became homeless and finally committed suicide. Until recently, his equation was largely ignored, but now is regarded as a helpful model for showing why altruistic behaviour thrives. Darwin had implied that human sympathy is the noblest part of our nature, and Price's story is but one chapter in the development of Darwin's influence on our understanding of the human condition. As we saw in the chapter on Social darwinism, **Dr. Frans De Waal** and his colleagues[436] are notable for the questions they ask about altruism in primates and humans. **R.W.Wrangham** 1980 and **Van Sehaik** 1983 argued that human companionship and sociality offers '*immense advantages in locating food*

---

[435] Both were at his sparsely attended funeral.

[436] '*Primates and Philosophers - How Morality Evolved*' Princeton Science Library 2006

and avoiding predators.[437'] **Silk** (et al), in 2003, pointed out that group orientated individuals leave more offspring, so that sociality is even more deeply ingrained in primate biology and physiology. Whereas **Hobbes** had seen sociality achieved *'by covenant only which is artificial',* social contract theory linked us with the behaviour of other species. Hobbes popularised the Roman proverb by **Titus Machius Plautus** (254-184 BC) *'homo homini lupus.'*[438] and **Lucius Annaeus Sennecca the Younger's** (4 BC-65 AD) response *Homo, sacra res homini'* in his own *De Cive* of 1651 saying *'To speak impartially, both sayings are very true; That Man to Man is a kind of God; and that Man to Man is an errant Wolfe. The first is true, if we compare Citizens amongst themselves; and the second, if we compare Cities.'* This assumption about our selfish and violent, wolf like nature,[439] fails to do justice to the gregarious and cooperative behaviour found amongst wolves themselves and many other animals,[440] as well as denying the social instincts of the human species. De Waal cites society's use of solitary confinement in prison, not just for its deprivation of liberty, but for removing a person from social connectivity. It seems that our physiology and emotional intelligence has adapted for life in the presence, not the absence of others; our health and wellbeing deteriorates without social support and depression soon follows. This needs qualifying, not least in terms of significant gender and personality variants. Particular emotional and social dimensions of connectivity seem to be more naturally understood by and seen as important to women, although men have their own forms of relationship bonding. Perhaps all of this is an evolutionary trait. It seems that mammalian females with caring instincts have out produced those who lack them, over the past 180 million years. De Waal argues that, for men, the best chance of living longer is to stay married. Statistics show that this increases their chance of living beyond the age of 65 from 65% to 90%.[441]

Social Evolution

The lifelong work of **Professor Edward Wilson** on Ants [442] shows the way what he called *sociobiology* and social evolution develops in insects - with ants, bees and termites being the most altruistic invertebrates. Significantly for the nineteenth century Darwin debate, these are millions of years older than the primates. He traced the pre-human history of eusociality,[443] demonstrated the labyrinth of

---

[437] ibid page 5.

[438] First used as *lupus est homo homini* by Plautus in *Asinaria* 195 BC,

[439] See earlier quote by Huxley p 220

[440] Schleidt and Shalte 2003

[441] See Taylor 2002.

[442] *The Social Conquest of the Earth.* 2012. Liveright Publishing Corporation New York

[443] Used in 1966 by Suzanne Batra to describe cooperative behaviour of Halictine bees which enabled specialised responsibilities. In 1969, Charles Michener developed her work by observing different kinds of bees. Characteristically this specialisation includes division of labour; overlap of generations; and cooperative working on common tasks. Wilson included other social insects.

evolutionary, mazes which led to the unlikely event, six million years ago, of hominids splitting off from chimpanzees and evolving through homo erectus to homo sapiens. He described reasons for the extinction of the Neanderthals, traced the outbreak of hominids from the African Savannahs and listed the probable, evolutionary causes for the appearance of homo sapiens – a combination of food, climate, territory, use of fire and tool manipulation, triggering the next stages of brain development, communication, language and culture. He described the latest anthropology of group and tribal, social organisation and the archaeology and palaeontology of social culture, with its sense of honour, the creative arts, morality and religious rituals. His writing is not, of course, without its critics, but his work on group selection mechanisms is supported by **Martin Nowak** and **Corina Tarnita** of Harvard, who argue that the gene for high order sociality is not linked to kinship, but to social organisation per se. Wilson believes group selection shapes the instincts which tend to make individuals altruistic toward one another. This explains the development of moral virtue to balance the negative behaviours retained in individual selection. This debate is of relevance to how Christians rethink their understanding of the causes and functioning of the doctrine of 'sin,' at least in its heavy, Western Church usage. The American evolutionary biologist, **Robert Trivers** uses the concept of reciprocity rather than kin or group selection, as the key ingredient in human sociality, showing how cooperative behaviour is selected. According to his work and his followers, including Richard Dawkins, it is reciprocity, rather than group selection which leads to moral emotions, such as guilt. The genius of Edward Wilson's work is that he allows for a multidimensional approach to the processes of evolution and gene selection in individual, kin and group selection. He believes that genes are selected to benefit participation in a group. Sometimes the drivers produced by the different sets of genes conflict with each other. This produces negative and positive moral behaviours in individuals and groups, but the direction is towards the altruism of eusociality. Sociality has its clear evolutionary purpose (and history) and brings with it the behaviours of altruism beyond tribal and species loyalties, as well as within them, in both cooperative and conflictual groups.[444] Altruism can be expressed as loyalty within a kinship group in ways that lead to conflict with a competitive group. Altruistic acts can reinforce participation in wider groups, and so reinforce and express different kinds of identity. In asking why advanced social life exists, Wilson believes that creation myths are Darwinian devises for survival. They hold the tribe together and distinguish its identity over against other identities. The myth gives individuals a reason for their existence, both as individuals and in

---

[444] Karen Armstrong argues that we each possess evolutionary traits of a reptilian (fight flight and survive with no altruism), limbic (protection, nurture and alliances for survival) and neo cortex brains (enabling self awareness and thought beyond instinctive passions). See her *Fields of Blood, Religion and the History of Violence*, Bodley Head 2014.p 4ff

relationships. This is crucial for their wellbeing and sense of purpose. It is also a vehicle for forming and sustaining social morality and encouraging social and altruistic behaviours. We are asked to take seriously the biological prehistory of human nature and our social evolution, with group selection being at least as important as kinship or individual, natural selection of the Dawkins type. Wilson argues that human survival and evolution very much depended on the functioning of group selection and so broadens the debate away from the 'selfish gene' reductionism.

Morality and Human superiority

In the debate that followed Darwin's work, one way in which theologians rescued the idea of a God who *directly creates* was through the notion of the soul and human morality. 'Let's accept Darwin is right for the biological and physical development of species through natural selection' went the argument, *'but the divine spark or presence of God in the soul was added directly by God – infusing as it were the physical with the spiritual'*. As we saw earlier, in the 19th century British confidence in the superiority of its civilization was increasing with scientific and industrial progress. This was self-nourishing because it justified itself by developing a perception of the inferiority of other societies and cultures, not least across its Empire. Such confidence also depended on its value systems, undergirded by culture and morality. Most academics, politicians and industrialists correlated their sense of civilisation with the values of Christian morality. If confidence in the creator had been undermined by Darwin's work, then they increasingly resorted to morality alone as the defining gift of the creator. If the soul is invisible, then moral behaviour is tangible. These moral instincts and behaviours distinguish human beings from all the other species from which we have evolved. After the Oxford debate, some church leaders re-grouped around this idea. What distinguished us from the primates, from which, they grudgingly admitted, we may have evolved, was our superior moral nature. Of course this was evidentially true. The human sense of culture and morality was far more sophisticated and could be articulated and discussed by humans in ways that are not possible in other animals. The very idea of being related to primates was laughable for this very reason, but Darwin had sowed the seeds of doubt. What if the discontinuity between humans and other species wasn't as complete as they thought? What if the cultural values we have developed evolved from much earlier forms in non-human species? De Waal demonstrates that our human capacity to act morally has its evolutionary origins in behaviours and feelings that we share with the higher animals. He insists on the essential or natural goodness of human beings and explains how this is inherited, along with much else from non-human ancestors, in the processes of natural selection. His studies were mostly based on chimpanzees, but also on some other primates and non primate social animals. He takes those aspects of *natural* behaviour which contribute something good to species experience, and argues that this morality has a common evolutionary source. Of course, human behaviour is more complex and has 'evolved' in different directions within diverse and complex cultures and societies. Along with his collaborators, he does not believe that humans uniquely possess a transcendent, soul based morality which distinguishes them metaphysically from other species. He defines goodness as *'taking proper account of others'*. Badness includes the sort of selfishness that ignores the interests of others. His approach is markedly different from those who begin with the assumption that humans are *essentially* selfish and self-interested.[445] Theology has contributed enormously to this assumption by a dominant paradigm of original sin 'infecting,' all all humans and other species like a universal disease. This implied that, even before they act immorally and even as newly born babies, they are, in their very nature, ontologically sinful. The empirical evidence for negative behaviours has often been used to justify this Christian doctrine. But, Christianity is divided over this, as John Hick describes in his *'The God of Evil and the God of Love'*[446] where he contrasts St Irenaeus and St Augustine. For the former the human being is like a wounded child needing to grow into maturity, making mistakes along the way. For the latter, we start with the huge disadvantage of inherited, original sin as a physical, ontological fact from which we can only be rescued through the action of Christ. For St Irenaeus, humans are vulnerable in their development and immature in their potential, but in Christ can grow into fuller maturity. Put crudely, we are mostly, if not totally bad and can only become good once saved from our predetermined condition. There is an inevitable exclusion and dualism hidden within this latter approach. The world can be divided between those who are saved and those who are not and of course this division required a church authority to define its boundaries. If the Catholic Church did this through its institutional power, many Protestants did it with the Bible in their hands. The crude simplicity of this position has been part of its appeal for those who respond to the sin and immorality of what they called *'the world the flesh and the devil'* by building an ark of salvation and escape in church going, or personal acceptance of Jesus as Lord. This simplistic dualism played into the hands of other category separations – the sacred and the profane, religious and secular, church and state. There may be various New Testament references that can be used to support such a dualism, but they don't sit comfortably with the example of Jesus' own actions. The horrors of the 20th century are evidence enough to justify a radical view of sin and to take evil very seriously. For centuries, Christians blamed such evil on an external, metaphysical force, who became, in some theologies, an alternative, ontological power to God.

---

[445] See the arguments against this by Dr. Nayef R.F. Al-Rodhan in *'Emotional amoral egoism'* LIT 1998
[446] MacMillan and Co 1966

If we did such terrible things to each other, it was evidence of Satan's or the Devil's influence over us. After all, Jesus had cast out *demons*. We can of course translate this first century demonology into different diagnoses of mental illness and we can trace the way the story of the snake'[447] temptation in Genesis has been wrongly interpreted. There are other biblical references to 'Satan' and personified Evil, not least in the book of Revelation, with its complicated political symbols and codes where the dualistic battle between Good and Evil was seen to have metaphysical implications. This was enough for the medieval period to accumulate its own mythologies around this dualistic theme. Belief in the Devil and all his works removed any sense of human responsibility and introduced a problematic dualism into theodicy. As we saw in LE1, parts of the first century world saw sin as the cause of suffering and both results of demonic forces. We can adopt such pre scientific views or spiritualise and reinterpret them, or in the light of what we now understand in medical sciences, challenge them. Jesus was after all a person of his times and acted from within its assumptions. Any view of morality based on our scientific knowledge of the human condition has to manage without metaphysical excuses which have dehumanized moral responsibility and facilitated certain kinds of power that church and state have wielded to propagate their own positions. In the history of Christian art, the Apocalypse of Mathew 25 was a defining icon of this metaphysical dualism. I once spent time looking at the beautifully painted, external walls of many Romanian monasteries and churches, all of which depicted the last judgement scene. This must have functioned like a horror story or film to the people of those times. It was designed to inculcate moral behaviour and belief by inducing fear. On one side of the painting, all the enemies of the (Orthodox) Church, including famous Protestant and Catholic figures, are being dragged down by the torturing devils into the fires of hell. On the other side, all the heroes of church and state are being taken up to heaven by the Angels. Any credible theology in the 21st century will have to manage without such metaphysical dualisms and explanations for human behaviour. It will still take seriously the radical extent of our capacity for hurt and harm as part of the story of evolution, as well as the cultural and moral values we have learnt to construct out of our better behaviours.

Humans differ from other species in their ability to create cultures based on intellectual constructs. We can reflect and order the meaning and interpretation of our experience through every kind of technology, since language and the written word made possible the more efficient and creative communication of ideas as an essential part of human identity, awareness and learning. In the

---

[447] A common story in the ancient Near East, but not in Genesis associated with 'Satan.' The snake is part of God's good creation but acts as a religious tempter by asking a religious question.

earliest emergence of cities, for example in Mesopotamia about 4000 BC, we see for the first time the creation of surplus value as combined, agricultural output exceeded that of the individual, family or tribal group.[448] Its organisation required new bureaucracies and social infrastructure. This in turn required its own myths to support authority and decision making. Surplus value also provided the luxury of extra time for reflection and cultural interaction. This is when many religious myths, including those of creation, first appeared. Through the trade of surplus value, these stores were shared beyond the geographical confines of the immediate area. Trade transcended traditional, tribal groupings and became the source of expanded ideas and world views. From the experience of comparing different human behaviours and beliefs, we came to reflect on the choices and causes of human behaviour. Poets, politicians and philosophers could produce their differing interpretations. Shared oral stories about the good or bad behaviour of famous people could be spread and developed. The great epics of heroic virtue and wise leadership could eventually be written. So a whole industry of human morality developed through economic growth and the expansion of ideas across tribal groups and their evolved behaviours. Slowly humans carved out their different assumptions about the human condition, sustained by cultures and myth making which sought to explain the origin of the contradictions in our moral nature and choices.

From empathy to sympathy

De Waal, quoting Edward Westermarck (1862-1939) argues that moral choice involves strong convictions, rather than a cool rationality alone. Learning about how we act and react emotionally to others is the basis for the development of any morality. Westermarck classified some of the different, delayed and spontaneous 'retributive' emotions involved. There is a growing literature on different kinds of reconciling, or revenge based behaviours, delayed or spontaneous, in primate behaviour. This shows how they have evolved complex forms of interrelationships with retributive, kindly emotions (Westermarck's *'desire to give pleasure in return for pleasure'*) as well as retributive resentment emotions. He sees retributive, kindly emotion as being close to reciprocal altruism. De Waal sees Westermarck's work as pivotal for a whole new trend in evolutionary ethics. Westermarck lays down the basis for distinguishing human morality from the non-moral emotions expressed by other primates. Morality is always at some distance from the immediate situation, even though it provides resources to deal with the particular. It is constructed of general judgements of approval and disapproval in the abstract, as a framework within which to judge the particular. Empathy leads to sympathetic reciprocity in non-human species and morality could not be constructed or understood without that experience.

---

[448] *A History of the Ancient Near East* . 3000-323 B.C. Marc Van De Mieroop. Blackwell 2004

Darwin's view of evolution was that as species adapt, they mutate existing forms. The fins of a fish become the limbs of an animal which may turn into hoofs, birds' wings or human hands. Of course, some structures lose all function and become unnecessary, but may still leave rudimentary traits rather than disappear altogether. He spoke of descent with *modifications,* but these took place gradually over thousands of years. So, by analogy, empathy and reciprocity in primates is the basis of what evolved in humans as morality. Advanced forms of empathy grew out of more rudimentary ones. **Peter Singer,** Professor of Bioethics at Princeton agrees with **Kant** that morality is based on reason not emotion alone.[449] There are clearly many different states of emotion in primates and humans and there are different emotions motivating our rationality. There is a *pattern paradox* between *reason* and *emotion* in the process of moral development and sometimes De Waal makes too simple a distinction in his concern to soften veneer theory and view altruistic behaviour, if not morality itself, as an inherited trait or 'direct outgrowth' of evolution. Singer accepts that even as we use reason to check or control our emotional responses, imprinted in our biological nature, we may be using certain other emotions responses to do this. How much then of human morality is a veneer over those emotions and how much is it part of an underlying structure? The question affects, for example, the debate about animal rights. Singer sees our use of this as selective – we don't hear much about *'the rights of rodents to take over our homes'*[450] and agrees with De Waal's emphasis on our obligations to them, rather than their *rights,* although clearly both recognise a legitimate concern to reduce the pain inflicted on animals by human behaviour and need.

Of course, once language and culture evolves, it can, in turn, shape and re-shape our evolving experience of emotions. The experience of empathy is pre linguistic, but through culture and language humans can discuss empathy in new ways and shape its development and adaption. Culture and language have increased our awareness of the danger of anthropomorphism – the projection of human characteristics backwards into pre human species. De Waal argues that when that was paradoxically combined with the assumption that all pre-human life is about struggle and combat, rather than social connectedness, it is not surprising that we've only just begun to understand the evolutionary basis of human morality. Again, it is important not to over compensate and introduce false correctives. The natural freedom of non-human primates to respond to each other for reciprocal revenge and altruism is different from the freedom of humans to reflect on their own and non-human species' abilities and needs.

---

[449] *Primates and Philosophers* p 150
[450] Ibid 154 see his book on *Animal Liberation* 1975

Within the freedom that humans experience when they seek to understand and then manipulate their genetic and evolutionary inheritance, there is a new kind of freedom, without which the idea of morality would be greatly reduced. The evolved human brain allows us to reflect on choices before, during and after they've been made to a greater degree than seems to possible in any non-human species. Language based reason and communication produces more sophisticated forms of connectedness and morality. While animals can spend just as long in each other's company and display just as strong social bonding as humans, language gives humans the possibility of greater levels of understanding the moral implications of their choices and behaviours. Communication between non-human primates has developed in different physical ways - sight, hearing, smell and touch. Many of these are far more acute and sophisticated than in humans. They seem just as capable of expressing emotions through different facial expressions as humans. These forms of communication have evolved as natural behaviours, but are deliberately chosen to have particular emotional affects on other animals – fear, threat, attraction etc. **Isenberg and Strayer,** in 1987, argued that humans add cognitive layers to '*the emotional contagion*' which develops into empathy.

Many studies of animal behaviour have focused on the relationship between mental and physical ability, such as numerical competence and tool use. Natural selection has favoured mechanisms by which the emotional state of others can be evaluated and responded to. **De Waal's** work shows that some primates are capable of high levels of sympathy[451] at the distress of another, and express this in ways that communicate more than just empathy. It takes human children at least a year before they can show such empathy or sympathy for others and it seems that this is also true of social animals. De Waal shows that **Wechkin and Masserman** (1964) found that rhesus monkeys refused to pull a chain that delivers food to themselves if the action also delivered an electric shock to another monkey. One monkey stopped pulling the food chain for twelve days after seeing the results of electric shock to another monkey. They were,

---

[451] The all-important emotion of sympathy as a physical instinct is connected to, but morally distinct from that of love. Darwin argued that '*the basis of sympathy lies in our strong retentiveness of former states of pain or pleasure. … We are thus impelled to relieve the sufferings of another, in order that our own painful feelings may be at the same time relieved. In like manner we are led to participate in the pleasures of others. But I cannot see how this view explains the fact that sympathy is excited in an immeasurably stronger degree by a beloved than by an indifferent person. The mere sight of suffering, independently of love, would suffice to call up in us vivid recollections and associations. Sympathy may at first have originated in the manner above suggested; but it seems now to have become an instinct, which is especially directed towards beloved objects, in the same manner as fear with animals is especially directed against certain enemies. .. With strictly social animals the feeling will be more or less extended to all the associated members, as we know to be the case. however complex a manner this feeling may have originated … it will have been increased, through natural selection; for those communities, which included the greatest number of the most sympathetic members, would flourish best and rear the greatest number of offspring.*' The Descent p 82-3

therefore, starving themselves to prevent pain being inflicted upon another. [452] This implied emotional connectedness with a high degree of empathy. De Waal argued that the difference between monkeys and apes is not in empathy per se, but in its cognitive development into active sympathy, including consolation behaviours.[453] **Robert Wright** takes issue with De Waal's use of cognitive language to describe chimpanzee behaviours,[454] though it is hard to see how humans can escape completely from the anthropomorphic trap, and this may apply to what Wright calls emotional as well as cognitive language. **Christine Korsgaard** questions the whole basis on which the morality of 'self-interest' is used in these discussions of both veneer theory and De Waal's belief that morality has its roots in our evolutionary past.[455] She claims that *'the principle of pursuing your own best interests as a principle of practical reason has never been established.'*[456] She disagrees that self-interests can be separated from the interests of others and follows Kant that other people are ends in themselves not just means to our own ends. She believes it is *'absurd to think that nonhuman animals are motivated by self interest'* because this would require the capacity to calculate the future and the abstract concept of the longer term good.[457] She makes the telling point that if we thought animals were like us, because of common evolutionary traits, it is as unlikely that we would eat them, wear them, perform painful experiments on them, hold them captive as we would do these things to ourselves. Denying continuity with animals, therefore, does nothing to show that there are either similarities or discontinuities. She accepts that De Waal's work shows that *'animals can be intelligent, curious, loving, playful, bossy, belligerent creatures in many ways like ourselves,'* but insists that humans are *'set apart by our elaborate cultures, historical memory...literature, science, philosophy and of course telling jokes.'* She agrees with Freud and Nietzsche that *'human behaving seem psychologically damaged in ways that suggest some deep break with nature.'*[458] This is a strange distinction, given what we now about similar damage in animals. Her other concern is to distinguish different kinds of intention or agency purpose, based on assessment and choice of behaviour emanating from the 'affective state' of functional organisation, and resulting in action of different kinds. Again she relies heavily on Kant in this discussion which warns us against too simple an interpretation of moral intent. She believes that animals cannot be judged for following their 'strongest impulses' in ways that humans can be. She quotes both **Adam Smith** and **Darwin** as believing that

---

[452] Ibid p 29
[453] Ibid p 33
[454] Ibid p 85
[455] Ibid p 98ff
[456] Ibid p 100
[457] Ibid p 102
[458] Ibid p 104

*'giving an account of the capacity for normative self government is essential to explaining the development of morality because it is essential to explaining what Darwin describes as "that short but imperious word **ought** so full of high significance."'*[459] Both Smith and Darwin explained *ought* in terms of human social nature, but this doesn't explain its absence (according to her) in social animals. This takes her into a discussion of Smith's understanding of sympathy[460] as how we are seen through the eyes of others. *'The internal spectator transforms our natural desire to be thought well of and praised into something deeper, a desire to be worthy of praise.'*[461] So, she argues, normative self government and intentional control must be the basis of morality and unique to humans. These things, rather than altruism and the pursuit of the common good are the essence of morality. A form of life governed by 'ought' is very different from one governed by instinct and emotion, even in intelligent and social animals. So, she disagrees with De Waal and tends towards her own version of veneer theory - that humans struggle to overcome their instincts not because morality is a thin veneer on our animal nature. *'It is the exact contrary; the distinctive character of human action gives us a whole different way of being in the world.'* Our distinctiveness is as much our capacity for evil as for good, but unlike other animals we know something is wrong. Social animals may well have altruistic instincts as De Waal describes, but she says *'Even if apes are sometimes courteous, responsible and brave it is not because they think they should be.'*[462]

Phillip Kitcher's disagreement with De Waal is based on the latter's heavy handed way of dismissing veneer theory and his vagueness about what he means by morality *stemming* from *traits* present in the higher primates.[463] In dealing with the former, he takes up **Huxley**'s use of the gardening metaphor in his 1893 lecture - we are capable in a way that animals are not of pruning out our worst behaviours from the *entangled bank* of our inheritance. He claims Huxley was going too far away from Darwin at this point. The latter knew very well that we have inherited weeds, as well as flowers, but might have seen all as part of the *diversity* of the same *entangled* bank within our own makeup, as well as in nature's. Morality is part of our inherited capacity to do the pruning, rather than coming from outside of our nature. Kitcher agrees with De Waal's starting point that much of our nature is inherited from the dispositions of our non-human ancestors, but suggests he hasn't thought enough about where it has taken humans in their moral development. He criticises the early supporters of *'Solid to the Core Theory'* who saw morality as essentially present in social animals, but

---

[459] Ibid p 114 citing Darwin in The Descent p 70. She also refers to Freud and Nietschze in the same context.
[460] In *The Theory of Moral Sentiments,* 1759
[461] Ibid p 115 , my emphasis
[462] Ibid p 117
[463] Ibid p 121 ff

only in a reduced version of the sophisticated, moral sensitivity of humans. The end point is qualitatively different from the starting point of morality. **Kitcher** says we must accept **Darwin's** and **De Waal's** dictum that humans are descended from the apes, so long as we allow for the discontinuities, as well as the continuities. This, he argues, we must do, if we are to be true to Darwin's own *'descent through modification.'* He then goes on to accuse De Waal of vagueness in relation to what has been *modified*. Kitcher is equally critical of the socio-biologists at this point. He describes this as the **Hume-Smith** lure into a sentimental view of the role of *sympathy*. Given that, according to De Waal, sympathy can be found in animals, the *lure* enables the conclusion that animals have morality. He then investigates the types of psychological altruism[464] found by the evolutionary theorists. He rightly reminds us that there are many different kinds of altruism in humans, let alone in animals - not least the difference between the psychological and the biological. I am not qualified to comment on these distinctions, nor their overlaps, but agree with Kitcher that they need deeper exploration than De Waal provides. Kitcher distinguishes paternalistic altruism (which responds to the *needs* of others) from non paternalistic altruism which responds to their *wishes*,[465] but I suspect in this *pattern paradox*, there is even more to be considered than Kitcher manages. Desires, needs, wishes and hopes merge into different kinds of *separate* and *overlapping* interpretations according to the perception of different individuals, in different situations and diverse cultures. Such is the flexibility of human experience, as filtered through our different attempts to use language to define and express these emotional and physical experiences.

He goes on to identify four dimensions of altruism – *intensity, range, extent and skill*, though again others could be discussed. He believes altruistic profiles can be mapped across this kind of matrix, and then asks what we would consider as the best combination in humans, let alone in non-humans. Until we are clearer about this, he maintains De Waal's criticism of veneer theory only leaves us with questions about the relationship between psychological altruism and morality. The former is clearly present in the animals De Waal studies, but this alone doesn't justify the use of the latter term. He believes both **David Hum**e (benevolence) and **Adam Smith** (sympathy) would recognise the value of De Waal's work, but agree with him (Kitcher) that it is limited in intensity, range, extent and skill. Kitcher interprets Hume's *'Enquiry concerning the Principles of Morals'* as saying humans have the capacity to refine inherited emotional dispositions, and to apply them beyond our in-groups. Without Smith's *impartial*

---

[464] He uses this as a kind of catch all phrase for any non human behaviour which is not 'egotistical.' The use of this language of course raises its own problems.
[465] Ibid p 128

spectator, or Kant's *inner reasoner*, human sympathy is altruistic, but falls short of morality, and rarely has the capacity to adjudicate in situations of conflict. Kitcher sees **De Waal's** conclusions about non-human altruism[466] as limited in *intensity*, *range* and *skill*, even within the coalitions and alliances of different in-groups where there is clear evidence of cooperative behaviours. The problem is that the peacemaking behaviours, e.g. animal grooming, often break down, fracturing the social fabric in the face of conflictual situations – a process that is hardly discontinuous from human peace-making! In Adam Smith's terms, they lack what Kitcher calls a 'little chimp in the breast' to provide them with a more sustainable use of their energy. Chimpanzees may have a richer, social organisation than other animals because of their altruism, but the latter is so limited that it cannot develop into larger societies of more extensive cooperation. Again we might comment that for all our achievements in the latter, humans demonstrate ample replication of the former limits. Humans often behave as wantonly in their desires and appear to be as vulnerable to their impulses. The fact that we have evolved culture, law based social infrastructure, moral philosophy and story sharing to overcome what **Kitcher** (after **Harry Frankfurt**) calls our *wantonness* is a huge differentiator, but if that is a form of veneer theory, the jury is out on its success. He challenges De Waal's high doctrine of non-human altruism as a basis for morality, but perhaps underplays how much human behaviour is affected by aggressive and other traits inherited from non-humans, despite the 'veneer' protections of cultural advances and normative infrastructure. There seems to be a *pattern paradox* here between non-human and inherited human behaviours which make the idea of a veneer appear sometimes too *solid* and sometimes too *porous*.

De Waal asks the important question – given that genetic selfishness is a primary mechanism of natural selection, how did humans come to be attached to the value and benefits of goodness? Why isn't it good to be bad? De Waal argues against *'veneer theory'* and addresses **Thomas Huxley's** defence of Darwin against his opponents. De Waal recognised that veneer theory didn't identify the source of the 'veneer' of goodness which protects us from our evolutionary inheritance. Does it come from a causation outside of nature? If so, it must be rejected as 'myth' by anyone committed to a scientific explanation of natural phenomena. If moral goodness is based on a myth, then how do we explain the examples of goodness that do exist? It is at this point that De Waal reverses the model. He sees human beings as essentially good, because this *goodness* has been inherited from our non-human ancestors through processes of natural selection. If goodness can be found in the behaviour of non-human animals and if, occasionally, humans act as if we, too, were good, then goodness is not only

---

[466] In his *Chimpanzee Politics 1998,* and *Peacemaking among Primates 1989*

real but natural. Therefore, the moral causation of this natural goodness must come from a common source within natural selection. Of course, human goodness takes on many different forms and good behaviour often has mixed motives. De Waal observed emotions we shared with other animals which are physiologically expressed as responses to the circumstances of others. Because these are physiological reactions they are assumed, in most animals, to be pre-rational or an involuntary form of 'emotional contagion.' By identifying with the circumstances of another, an animal feels his or her pain. Empathy at this level can still be selfish. We seek to comfort the other, through our empathetic reaction, as a way of dealing with the pain we ourselves have felt or feel. At a more advanced level, *empathy* can lead to *sympathy,* which he defines as the recognition that the other has specific needs or wants that are different from mine and to which I can respond. His famous example is the compelling story of a chimpanzee trying to help an injured bird to fly. Because flying is something the chimpanzee can never do, it is responding *sympathetically* to the other's distinctive needs, not its own. The incident took place at Twycross Zoo in 1997. *'A Bonobo female called Kuni captured a starling. Out of fear that Kuni might molest the stunned bird which appeared undamaged the keeper urged the ape to let it go....Kuni picked up the starling with one hand and climbed to the highest point of the highest tree where she wrapped her legs round the trunk so that she had both hands free to hold the bird. She then carefully unfolded its wings and spread them wide open, one wing in each hand, before throwing the bird as hard as she could towards the barrier of the enclosure. Unfortunately, it fell short and landed onto the bank of the moat where Kuni guarded it for a long time against a curious juvenile.'*[467]

This story has now become part of the narrative of evolutionary ethics. Many questions arise. What, in natural selection, has produced this behaviour? It's impossible to interrogate the animal and ask about motive, but significantly its action crosses a species boundary. Higher primates helping, as well as fighting with other primates is fairly common. This is different. From the perspective of Love's Energy, this is a good example of a non-human species *reaching across* to another non-human species, longing for it to exist within the freedom of its own ontological nature. Kuni is aware that birds fly in the air and wants this bird to be able to do that again. Perhaps this is a learnt behaviour resulting from watching humans interact. In animal terms, this is the story of the Good Samaritan who crosses the road to help another. In this case, the racial and religious enemy is replaced by another species, who is radically alien because it can fly. There seems to be no short, or longer term, evolutionary gain to the action. Something else is happening. The fact that Kuni acts from within her own feelings makes the

---

[467] *Primates and Philosophers*. P 30

story very compelling. Do other Bonobos do this? If not, what made this one different and what effect might this have on others or even other birds? These are perhaps pointless speculations, for they cannot easily be tested. Granted, Kuni doesn't give up her own security or survival for the sake of the bird, it is still an extraordinary act, reminding us that struggle for the survival of the self or the species may not be the only motive in animal behaviour. De Waal speculates that *'having seen birds in flight many times, Kuni seemed to have a notion of what would be good for the bird, thus offering us an anthropoid version of the empathetic capacity so enduringly described by Adam Smith (1937) as 'changing places in fancy with the sufferer.'* Of course, every dog owner will describe some level of emotional bonding, including examples where the dog seems empathetic towards human needs.[468] De Waal also quotes **O'Connell's** 1995 analysis of thousands of qualitative studies of empathy, and found that apes had more complicated reaction to the stress of another than monkeys seem to demonstrate. **Ladygina-Kohts** (2002) wrote about her young chimpanzee Joni. She discovered that evoking his *sympathy* was a better way of getting him off the roof of her house than either reward or punishment. *'If I pretend to be crying...Joni immediately stops his plays or other activity, quickly runs over to me all excited and shagged, from the most remote places in the house, such as the roof or the ceiling of his cage from where I could not drive him down despite my persistent calls and entreaties. He hastily runs around me as if looking for the offender; looking at my face, he tenderly takes my chin in his palm, lightly touches my face with his fingers as though trying to understand what is happening and turns around, clenching his toes into firm fists.'[469]* De Waal believes the main difference between monkeys and apes is *'not in empathy per se but in the cognitative overlays which allow apes to adopt the other's viewpoint.'[470]* From the viewpoint of Joni, she is picking up signals of distress and translates them into assumptions about threats which might cause the distress. She can empathise with this, and then act sympathetically towards the human concerned just as she might with her own species. This episode shows that Joni reacted much more quickly to this signal than to reward or threat signals. This is perhaps surprising and significant.

The literature discussed by De Waal mainly sees empathy as a cognitive matter requiring certain levels of self awareness. As with human children, the brain in other species has to develop to a certain stage, before it can differentiate its own self-awareness from an awareness of others. Animals have to recognise another

---

[468] My sister in law is sure her dog knows when she is about to have a diabetic incident and comes to warn her with the same, repeatable behaviour.
[469] Ibid p 2-30
[470] Ibid p 30

animal as differentiated from themselves and therefore as having different, as well as similar needs. It is only at that point that empathy can become sympathy. At this point, A recognises the difference of B and has the imagination to see B within its own differentiated autonomy, context, behaviours and need. A can project imaginatively from its own experience about what B might be feeling. This establishes bonding and empathy and may lead to sympathetic action. The literature De Waal is quoting implies that when the subject other A can attend to the object other B's needs, it happens because the subject's neuro representations of similar states are automatically activated. The closer the relationship between subject other A and object other B, the more likely this is to happen. Physiological changes can be observed at this point in heart rate, skin conductants, facial expression, body posture, etc. Communication of these needs from the object *other* to the subject *other* is clearly a sophisticated process. The brain has to be able to pick up and process these 'signals' quickly, empathetically translating them as implying a need and desire to help. The signals have been described as *'the somatic marker hypothesis of emotions.'*[471] The work on empathetic response goes back to the beginning of the 20th century, particularly in Germany. **Preston** and **De Waal's** 2002 idea of *'Perception-Action-Mechanism'* (PAM) is a helpful tool for understanding what is happening at an involuntary level. PAM implies that empathy is a natural process, demonstrated by electromyographic studies of invisible, muscle contractions in human faces, in response to pictures of other human, facial expressions. It seems that these reactions are instantaneous and unconscious. Seeing another's pain or anger is physiologically very similar to being in pain or experiencing anger oneself, at this spontaneous level. Effective, language communication creates, or increases, the transmission of similar physiological states between subject other and object other. Between humans and some primates (without language advantage), there seems to be an instantaneous or involuntary PAM which does not require higher levels of empathy cognition in order for the subject *other* to react to the object *other's* emotional state. How this has embedded itself in evolution is a more complicated story.

The Russian Doll of Continuity and Discontinuity

If Love's energy is embedded in matter and the connectedness of things in creation, then De Waal's science helps us understand more about how that connectedness evolved. The *energy* which transmits empathy across subject *other* and object *other* happens involuntarily because of physiological stimuli and, at a conscious level, through cognitive empathy. We are only just beginning to understand the neural basis of PAM. It is possible that the quantum physicists,

---

[471] See Damasio's 1994 research on the link at the cellular level between perception and action in Pelligrino's 1992 work on mirror neurons

working on the connected existence of waves and particles, will have something more to contribute. If empathy is a form of connectedness between other *others*, within and across species, it will have a quantum dimension below its cellular and physiological interactions. De Waal's Russian doll metaphor is a simple picture of how emotional contagion is linked to cognitive empathy through the PAM and what he calls *attribution* – the full adoption of another's perspective. He argues that autism is an example of a dysfunctioning Russian doll where different levels do not connect. We need to understand why many other examples of dysfunction prevent one subject *other* from sensing enough of what the object *other* is experiencing to want to respond sympathetically, particularly in humans. The sympathetic response is a vivid example of how Love's longing for the other to exist is embedded in the higher levels of organic life. Sympathy – or *suffering with*, is an expression of Love's energy and certainly uses the kind of emanating, ekstatic energy we discussed in LE2. It implies that pain can be shared by physiologically and emotionally experiencing a measure of it oneself. This causes its own *reaching out* towards the other. I understand *sym-patheia*, even before it expresses itself in acts of *reaching out* concern, as an expression of Love's connected presence in the exchange between the subject *other* and the object *other*. Sympathy has to deal with the radical freedom of the otherness of the other as a differentiated being and yet relate across this difference as love relates across boundaries, because of its longing for the wellbeing of the other. Sympathy then is an expression of the Perichoresis of Love that we saw in LE2

The connectedness of all life, at a DNA level, is one of the most significant recent additions to our understanding of how Love works, not just as a human emotion, but in all physical reality. When we look at the biodiversity of our *'entangled bank,'* we know that all life has a common ancestor and, under a powerful microscope, there are similarities under the surface of differentiated forms. Knowledge of a sub particle connection doesn't prevent the differences turning into conflictual differentiation, or futher prejudice, ignorance, indifference, separation or isolation. Despite a common ancestor, animal species still kill and feed off each other, in the complicated but connected food chain that holds life together. Mutual inter relationship and dependence, paradoxically and tragically, includes a subject *other* threatening and harming an object *other* for many different reasons, not least a struggle for survival. This seems to be part of the nature of our cosmology and evolution. Whether or not any one particular animal feels any kind of cognitive or instinctive empathy for the animal it feeds on, is a larger question. Some indigineous peoples developed rituals of gratitude for the life of the animal they killed and ate, showing empathy for their connectedness. Despite predatory and competitive behaviour, some primates and other animals display the capacity to feel empathy in different ways. This reminds us that Love is embedded not just in the human species but its energy reaches out across all species, in ways we don't as yet understand, or maybe cannot fit into our anthropomorphic perspectives. It is not surprising that many species exhibit behaviours which are the opposite of 'Love's longing for the other to exist', so absorbed are they in their own survival or other needs, which sometimes result in the destruction of the object others' chances. We find this same behaviour in humans, albeit veiled in different cultural forms. Both threat and indifference block empathetic cognition and even instinctive behaviours in some cases. De Waal's Russian doll is cracked in all directions, and some sections don't fit neatly with others, but as we have seen, neither cracks, nor poor fit disprove the presence of the doll nor the internal relationships of its different parts.

De Waal has also studied the detailed evidence for what humans call *gratitude* in the behaviour of chimpanzees. Nearly 7,000 interactions over food events were carefully recorded by observers, and entered into a computer according to strict definitions established by him. Chimpanzees were more likely to share food with individuals who groomed them anything up to 2 hours earlier. This emotional exchange was only with the individual who'd groomed them, not part of a more general goodwill to all chimpanzees. This act depended not only upon an emotional response, but the ability to *remember* what had happened earlier. In the case of chimpanzees this implies a certain cognitive, emotional sophistication. If humans have a common evolutionary ancestry in non-human species, then it would be surprising if we found no trace at all of such reaching out in other species. It is De Waal's work which has proved this to be the case. In the New Testament, Jesus is portrayed as crossing the boundaries of human diversity and division - race, ethnicity, gender, age and background, to stand alongside the needs of the other, modelling the nature of Love reaching away from self towards the other, and the other's needs. While De Waal's *'emotional contagion'* is commonly observed in many species, sympathy is only found in the higher primates. This raises more questions about evolutionary continuity and discontinuity in natural selection. There is ample evidence of empathetic reaching out which transcends selfish self interest. Is there a genetic explanation for this reaching out away from self to other *others*? Is this possible only in species with a larger brain? Certainly elephants, dolphins and capuchins demonstrate emotionally motivated forms of empathy if not altruism. It's possible that our definition of altruism is too narrow, and we haven't as yet recognised its physiological expression in a wider selection of species – as **Edward Wilson's** work on social insects indicates. Certainly, empathetic reactions in some (higher) animals, is more elaborate than the *'emotional contagion'* found in others. For De Waal, humans did not cover their animal instincts with a veneer of goodness, but are more like those Russian dolls - whose moral selves are ontologically continuous with a nested series of inner, pre-human selves. But why, if goodness is naturally inherited through natural selection, has it not evolved in humans to a much higher degree? There is evidence of this in many individuals, but not universally, and not at all times in all situations. The caring capacity of humans extends beyond the immediate needs of child rearing and parenting, the needs of extended families, and even beyond the needs of the tribe. The *Not For Profit* or Charity sector is ample evidence of this, as are the personal examples we can all quote. The exceptional cases of self sacrifice in war, or social crisis, seem to occur across cultural and other conditionalities. They point to something foundational in our nature, which has not been selected out in our evolutionary inheritance. The motives may vary and the feeling of wellbeing that seems programmed into altruistic acts is rarely achieved in any other way, at least to the same quality and degree.

As we saw in LE2, Love seems to replenish and fulfil itself by being given away. Is this learnt and inherited behaviour? In the future, we may be able to trace its presence in certain combinations of genes. This *reaching out* is fragile and indeterminate in most of us – it is part of a moral struggle, as well as being, in some cases, purely instinctive and therefore apparently 'natural.' Whatever our genetic coding, there is no simple, causative relationship between inherited instincts, personality and behaviour in all cases. Much is predetermined, but we can learn to compensate for those behaviours, not least through observing other role models and through exposure to education. We can be made to behave differently through stigma and political control, legislation and sanctions. The equivalents of this are also present in the higher animals. Certain behaviours can clearly be taught and learnt as well as being innate. There is growing evidence of their ability to reflect and self evaluate, as they learn new skills and react to the responses of others. They may have their own way of passing on learnt behaviours, even without the constructs of cultures and institutions that carry beliefs and values in humans. The latter add layers of complexity to what in non-human species appears to be more of a direct, physiological and un-reflected 'natural' response. This complexity in human relationships illustrates the ambiguities, dilemmas, overlaps and contradictions we experience in conflictual, moral choices. These may follow un-reflective and spontaneous reactions to threats, needs and opportunities. In evolutionary terms, humans seem caught between their ability to adapt to different cultural and contextual situations and their instinct to conserve and defend traditional positions. Adaption has to happen comparatively quickly in many situations of moral learning, and as with non-human species, we seem programmed to react instantaneously to a diverse range of situations. Our brains have retained automatic emotional responses as well as the capacity to consider and reflect. In the New Testament, St Paul described the conflictual nature of human morality as *'for the good that I would I do not: but the evil which I would not, that I do.'*[472] It is hard to align intellectual, emotional and physiological reactions when it comes to making moral choices. Too often we are at war within ourselves, let alone between ourselves. The idea of Love as the energy which transforms our wellbeing, morally and physically, by reaching out to benefit the other, leads, of course, to its own conflicts. This reaching out risks rejection, indifference and opposition. We know that doing good to, or for, the other can be patronising and undermine the essential liberty of the other. As Love reaches out towards the other, the other may react against it for many reasons. We know that goodness, like success, is often a threat to some others. In this sense, Love's longing for the other to exist, goes beyond our normal categories and morality. Because it insists on the radical freedom of the other, it has to take seriously the other's capacity to behave badly and do harm, including its rejection of this kind of love. We see a vivid expression of this in the suffering and crucifixion of Jesus, and in some of his stories and parables[473]. In many examples of human behaviour, it is hard to find altruism reaching beyond a circle of narrow self interest. For larger, complex societies to function at all, this necessitates the need for legislation to enforce a 'veneer' of moral standards over the predicted, bad behaviour of citizens. For example, without the 'veneer' of a thirty m.p.h. speed limit, it can be assumed people will drive to fast. With it, people's worst behaviours will be controlled, and hopefully they will learn to reflect on and internalise the reasons for the control.

De Waal considers how far the 'circle' of altruistic morality can be expanded without becoming porous to bad behaviour. Theoretically, this circle should be inclusive of all human behaviour to support his thesis, but the evidence is very mixed. In some non-human, social animals, we see the best traits of empathy extending within and beyond their own species, as well as a partiality for 'insiders.' According to De Waal this latter is an evolutionary threat to human morality. If moral behaviour is only limited within the narrow circles of extended kinship and cultural groups, to what extent is it really 'altruistic?' Certainly humans demonstrate a capacity to enlarge these narrow circles through policy making, international institutions and individual and national acts of humanitarian assistance, aid and development investment. With evolved language and communications, in a networked world, humans can see and empathise with the outer circles of need, across different cultural boundaries, but this doesn't guarantee effective *reaching out* across them. *Continuity* and *discontinuity* between human and non-human evolutionary behaviours is part of

---

[472] Romans 7.19

[473] See Luke 20- the owner of vineyard and the wicked tenants

the *pattern paradoxes* in our human condition. Animals are not humans, but humans are animals, and both share a common evolutionary inheritance. The argument for continuity can inspire the wrong kind of anthropomorphism. This is an important discussion for any ethics of animal behaviour or questions of legal status. *Continuity* should increase our sensitivity towards and treatment of non-human species. They are more than passive recipients of human need or domination. On the other hand, popular culture and particularly children's films may go too far in giving animals human personalities and voices to endear them, and evoke more empathy. The reality of animal behaviour, in natural habitats, is different from the romanticised Walt Disney version, while domesticated animals may genuinely feel and act like human companions in a more continuous sense. Humans have evolved a close, empathetic relationship with such species with high levels of communication in a *reaching out* between them, if not deeper psycho-emotional bonds, well beyond any utilitarian function. The *owner* – a revealing term in itself - may perceive signs of his/her animal pet being confronted with moral dilemmas which go well beyond natural, need based, reactions. A dog presented with choices may choose its 'owner's' needs, or commands, rather than its immediate physiological needs. Does this imply some kind of moral cognition in the animal brain and, therefore, some continuity with humans? Moral choice implies the freedom to choose, and a self awareness about alternatives and their consequences. If non-human animals never experience any cognitive awareness or perception of alternatives, their choice can hardly be described as moral. De Waal's work certainly implies there is choice, if not cognition as humans experience it. Domesticated animals are capable of learning new patterns of response to different human behaviours, whether by observation, command or empathetic bonding. They do this without any shared verbal language. This implies that animal brains have evolved a sensitivity to other signals we may not be capable of as humans. If animals can modify and adapt their behaviour in response to a human relationship, they must have evolved the capacity for choice between alternative actions and this aids the mutual bonding (connectivity). This motivates the human to invest more in their side of the connectivity, and in turn experience more empathy in its developmental stages.

Normative or descriptive morality

We can choose to raise the bar of normative, moral definitions so high that it becomes beyond the capacity of any non-human animal. There must be a connection between the *normative* and *descriptive reality* and only humans, though shared cognition and discourse can make that kind of connection. Humans are often described as being able to construct the normative, while animals can only experience the descriptive, including responsive choices between alternatives. **Peter Singer**[474] alludes to the studies of brain scans of humans faced with difficult moral choices – e.g. *'killing one to save five'* in the 'Trolley Problem,' which has been tested in different versions. Those who responded that the highest value is *'not to kill'* showed brain activity associated with emotion; those who chose *'kill the one,'* showed increased activity in the parts associated with rational cognition. Which is the more 'moral' response and how does the human brain and personality manage the tension between the two – the *normative* and the *formative*? From a rational perspective, it is surely right to save five lives, even if this means killing one? But most people hesitate to do that because of an evolved, emotional response. This tension between the descriptive or formative, emotional reaction and the normative, cognitive reaction is surely another *pattern paradox* for our list. Humans seem to move in and out of this pattern in ways that animals cannot. De Waal's work has been primarily with the descriptive, emotional responses of altruism rather than cognitive questions about the highest moral good, either in the abstract, or for the highest number of people in a more utilitarian sense. Human brains have developed their normative capacity in ways that other animals, as far as we know, haven't, or perhaps can't. However, Singer doesn't deal with higher levels of moral complexity in his runaway trolley car example. He doesn't, for example, imagine a situation where the one who needs to be pushed, in order to save five others, is one's own wife or child. He doesn't compare the differences between pushing a stranger rather than a friend, or an enemy rather than an ally. He doesn't consider whether any of the five are strangers, friends or enemies. If they were all enemies it is unlikely that anyone would push a friend over a bridge to save them. This would add to the emotional and cognitive charging of the choice and the moral calculations involved. There may also be a significant difference between human reactions in a thought experiment and real life, however much the same parts of the brain are involved. We know from real examples, and from fiction in film and books, that different people react in different ways to such a choice. They are also affected differently by the longer term affects of any choice. In real life, the moral dilemma continues to have its cognitive and emotional effects, well beyond the moment of choice and in our capacity to imagine and calculate this, this we probably differ greatly from animals.

What have humans added to their evolutionary instinct by developing normative constructs such as *'thou shall not kill?'* These are intended to override emotional responses e.g. to defend a loved one, particularly if associated with a strong, religious based morality. There are many other experiments and too much, real

---

[474] *Primates and Philosophers* p 140 ff. See also Peter Singer, *The expanding Circle*, Clarendon 1981. He is professor of Bioethics and Founding President of the International Association of Bioethics.

life evidence that show how readily humans can kill or harm others, even members of their own in-group. In this, there is a haunting continuity with non-human, violent behaviours and more moral culpability because of our cognitive abilities. There is a *pattern paradox* in the way inherited emotional responses relate to cognitive constructs. We do kill for cognitive, moral values – resource or territory threats, humanitarian intervention and regime change – despite the horrific and unintended negative consequences which we seem so bad at predicting, despite our memory of past horrors. We are capable of killing people we know, went to school with, had dinner parties with, if the ethnic, nationalist card is played with enough conviction and manipulation, as in Bosnia.[475] In Ireland, people killed others who spoke the same language and shared the same economic and social problems, purely on the basis of different religious and political beliefs. We kill and rape as a spontaneous emotional response, despite our best moral codes. Sometimes the cognitive rationale (discontinuous from non-humans) increases the intensity or likelihood of the emotional response (continuous with non-humans) and sometimes the latter is so strong it overrides the former. So, our developed cognitive capacity doesn't always protect us from our inherited responses and is often used to justify the latter. The *veneer*, however understood, is very thin. Cognition can, however, function not only to prevent the bad act, but to produce a rationale to promote the good act. In this case, it increases the intensity or likelihood of the good act as an emotional response. Pushing the above thought experiment even further, we might say that the actions of Jesus represent a contradiction and moral inspiration. He didn't push a friend, a stranger or an ememy 'off the bridge' to save five other lives on the runaway trolley. He threw himself off the bridge to save those who wished him harm. So goes the claim of Christianity, as it looks back on his life and death and tries to understand its meaning.

Singer traces dimensions of the same veneer theory in **De Waal** that the latter dismissed in **Huxley**. **Singer**[476] sees human rationality as marking a clear distinction from non-human 'morality.' The functioning of the relationship between emotional (in-group) and more rational behaviours (to out-groups) must be more complicated than these simple descriptions imply, at least in the case of any one individual. *In-Group* and *Out-Group* behaviours become another *pattern paradox* in this way. De Waal is surely right to say that the roots of human behaviour are to be found in the evolved traits of social animals, but how those evolved patterns developed, varies, not only according to history and personality, but different belief systems which influence rational reflection on

---

[475] I will never forget, on an EU regeneration programme in Mostar, being given vivid personal examples of this.
[476] in his *Expanding Circle*

values and behaviour. As **Singer** says, **Edward Westermarck**[477] rightly distinguishes moral emotions from other emotions. We might ask for the same distinction between purely rational reflections on morality, and inherited and often conflictual beliefs, myths and cultural assumptions. There is certainly more to moral feelings than just raw instinctive emotions, even before they are developed into a cognitive construct. **De Waal** himself believes that moral emotions can be disconnected from immediate situations into a more abstract and disinterested view. The relationship between *interested* and *disinterested* morality is a *pattern paradox* which fluctuates across different kinds of ambiguity, motivation and interpretation. From where does this disinterested or impartial spectator come, if it is not part of our evolved nature? Is there a difference between our failure to put impartial, disinterested morality into practice and our evolved nature? Surely the two are intertwined. In discussing this *pattern paradox*, we dare not create rigid boundaries between evolved nature and personality. The latter includes our varying ability to deal with the *interest-disinterested* or *partial-impartial, pattern paradox* and in most people these boundaries are more porous than the categories imply. De Waal rather rigidly contrasts the morality of the In-group (kinship reciprocity) with moral behaviour to the Out-group, based on a more impartial, disinterested perspective. This may be more true in non-humans than in humans, were there is ample evidence of emotional boundary crossing and inconsistencies in our behaviour within the in-group in relationship to out-groups. Some individuals can demonstrate easy going, empathetic relationships with members of the latter, while others struggle to form reciprocal relationships within the former. Nor are any of these questions new, even though the language used to discuss them has changed through the centuries. Singer quotes the fifth century B.C. Chinese philosopher, **Mozi**, who, in the face of the appalling destruction of war, asked *'what is the way of universal love and mutual benefit – it is to regard other peoples' countries as one's own.'*[478] De Waal believes that the practice of more impartial, moral behaviour to out-groups is fragile - implying echoes of the veneer theory he dismisses elsewhere, but surely that could also be said about our emotional (evolved) reactions to people within our in-groups.

So can we say, more bluntly, that humans have moral systems and apes do not (**Alexander** 1987)? In the course of primate evolution, inter group hostility enhanced group solidarity to the point that 'moral' emotions and even consciousness emerged. In the development of human morality, we moved from inter-personal sensitivity to a consciousness of the greater 'common' good, including the care for the wellbeing of strangers from the other side of our

---

[477] Ibid p 143
[478] Ibid p 145. Cited from W.T Chan *A source book in Chinese Philosophy*, Princeton. 1963. p213

cultural perceptions and boundaries. The paradox is that human morality has some kind of evolutionary relationship to our worst behaviours - when community solidarity is triggered by the external threat of violence and war. As a species, humans have still not outgrown our tendency to treat outsiders differently from insiders, whatever our cognitive morality says. The attempts of international law, not least the Geneva Convention of 1949, have tried to regulate such external relations to protect them against war and its excesses. Despite our extensive knowledge of the suffering caused by war, we do not seem capable of avoiding it, whatever our international controls. In this sense, our emotional evolution has been slow to change. It seems that Jesus's example of breaking boundary conventions has been sustained more by words than actions. Jesus transcended the evolutionary and religious morality of his times, in this regard, demonstrating an empathetic and sympathetic solidarity, not just with the needs of his own racial and other groups, but across a wider human diversity.

Embedded altruism

If a human (or a higher primate) sees their own or another child about to fall in a river, they are likely to react spontaneously, with little cognitive forethought about motive, risks or outcomes. Knowing the right thing to do becomes a very complicated question in the layers of the human brain, as we stand back from immediate, instinctive responses, to make decisions about how society should be organised for the benefit of larger numbers and the protection of individual flourishing within it. As we fashion our moral maturity and control our worst behaviours, we reflect on the causes of instinctive reactions, in an accumulating complexity of intellectual, legal, religious and cultural constructs. As our brain scanning becomes more sophisticated, we see that moral decision making involves different parts of the brain working together, including some extremely 'ancient' pre-human parts. Human morality involves a rational and comparative critique of individual motives and their consequences, however much it is firmly anchored in social emotions, based on evolutionary empathy. The advanced cognitive ordering of these emotions as a distinctively human activity supports, in part, the veneer theorists who regard the human brain as different from all non-human brains in this regard. While earlier developmental psychologists thought that the young child learns her moral lessons through fear of punishment and a desire for praise (adopting and copying parental values to construct a super ego), we now know that very young children are able to distinguish between actions where breaking the rules would harm others, and where that behaviour would only be breaking a social convention. Although the influence of parental and social reward and punishment might be involved in both, they exhibit an evolutionary behaviour in 'naturally' wanting to comfort others in distress.

Reaching out to others is an embedded evolutionary and natural behaviour, but one that needs nourishing and developing, as well as checking and controlling. [479]

We can sustain Darwin's argument that nature favours species adaption for survival without concluding this can only be achieved through aggressive, self-interested behaviour, rather than empathetic, cooperation as well. Therefore, we can say, in Love's Energy that *reaching out in the direction of the other* is part of the *embedded* process by which species adapt within the theory of natural selection. We are naturally selected to be capable of *reaching out in the direction of others*, not only for reproduction, but for levels of co-operation and struggle which benefit our own and our common survival. As humans struggle to enhance their flourishing and survival, they are capable of taking responsibility for their behaviour in relationship to other **others** and to their environment, in ways that other species cannot, at least with the same *range, intensity* and *skills*. Freedom and responsibility, as we saw in LE2, go together. In the *pattern paradox* of *continuity* and *discontinuity*, this connectivity seems to be specially, if not uniquely, present in humans, as they fashion their moral compass, philosophically, socially and technically. We are free to make choices about value. We can choose to *reach out to other others (humans, non-humans and environment)* across stereotypes and boundaries which make the reaching out both courageous and creative. We have used the word Love for this reaching out. While love includes its own emotional and physiological reactions in humans and non-humans, not least for the purposes of reproduction and kinship bonding, it goes far beyond evolutionary 'empathy contagion.' ove takes us into cognition, where choices have to be made freely and thoughtfully within the cultural and political frameworks we have constructed out of historical and comparative experience. Love does not rescue us from moral dilemmas, nor from a pluralism of real conflicts of opinion, attitude and behaviour, as we attempt to balance individual needs with the wider *common* good.[480] Love embeds itself in the process of living within those tensions. It prompts us to find our own wellbeing in the wellbeing of others, despite all the practical difficulties involved. Love ensures a freedom and responsibility which moves beyond all rules and regulations and equips us to act primarily for the benefit of the other, even where that undermines our own self interest. Paradoxically, it is at this point, that Love's energy enhances and nourishes our self –interest, even as it gives it away.

---

[479] Joanne Trollope in an interview with Mark Lawson on 23.03.14 said that civilisation is a very thin veneer over the snakes of Pandora's box.
[480] I am conscious that this phrase is both slippery and controversial – the *common* is often more conflicted and pluralist a concept in itself than the simple word implies.

In reaching out to the other, Jesus' life reminds us that sometimes we might have to break the very, moral regulations which have been invented to protect a rigorous and 'righteous' approach to the moral good of the community. Jesus embedded Love's energy, as he reached out beyond the constraints of organisational norms and Pharisaic, compliance culture, where the complexity or misuse of regulations dis-empowered people's instinct to reach out in the direction of the other, whatever their background. There is a voice that always whispers to us from the extreme left and right of politics - If only we had a tighter regulatory structure, this would prevent the failures in the health service, the economy and individual behaviours. Love places itself in the indeterminacy of the gap between what we hope to achieve and the processes we use to move in that direction, including the unintended negative consequences of our virtuous decisions! Who could argue against the virtuous intent of better health and safety in employment law, or better control of the banks? The former led to about three hundred new employment related acts or amendments passed between 1997 and 2007. Even lawyers found this difficult to understand and manage. Despite this, some employers still behaved irresponsibly and even without this, others behaved more responsibly. I can only speak from my own experience in the public and private sectors. In visiting companies as an industrial chaplain, I observed the radical difference between

➢ Company A - operating mostly with a left brain leadership model so that the regulations were clearly communicated and consistently implemented, but there was rarely evidence of visionary or creative progress beyond the lowest common denominator of behaviours. Motivation and innovation were at low levels and this affected the culture and wellbeing of staff.

➢ Company B – put the diverse needs and experience of employees (and customers) above regulations and processes in the spirit of *'the Sabbath was made for man not man for the Sabbath'*[481] and built leadership and relationships around the values of trust (knowing all the risks), flexibility and encouragement (including rewards). Company success and innovation, adaption and expansion were high, as was staff morale and motivation.

One company of the latter type I visited worked in energy supply. It had grown from nothing to a turnover of many millions, within just 5 years. It regularly broke health and safety rules, but took training and mentoring seriously. The Chief Executive broke the cognitive systems of employment law by giving employees time off, tickets for the ballet, and impromptu holidays as rewards for

---

[481] Mark 2.27

their best ideas and efforts. He invested in trust based responses, including a flexible approach to working environments and hours, producing inspiring and incredible results. When he identified people misusing this trust, he dealt with them through the line management system rather than introducing a generic new regulation that affected everyone, and so reduced the culture of risk and trust. I began to see risk and trust as its own *pattern paradox* during this time. I remember as an Assistant Head of a pioneering Community School in Scotland, we wrestled with this *pattern paradox* regularly on the senior team. We decided to resist union pressures to dictate arrival and leaving times for staff. Most staff were highly motivated to put the needs of children, parents and the community before any employment regulations. We 'managed' the individuals who exploited this on a one to one basis. In the end the teachers' union won in its battle for fixed contractual hours to protect their members' interests, which was of course their job. Slowly, the after school activities that had been *naturally* organised by teachers dwindled. We were living in a new world of public sector negotiation for equity and fairness which seemed to reduce ambition and achievement as far as 'initiative, innovation and inspiration' were concerned. Trust was incrementally edited out of our culture in the name of regulating protections and processes, and what is now called risk management. I sit as a Governor of a local University on the audit committee and watch with horror at the tentacles of the audit industry taking over so many cognitive processes. This is one of the great *pattern paradoxes* of our times – in the name of protecting individuals and organisations from risk, and so increasing the audit processes around those risks, we may be acting against their longer term capacity and motivation for developing responsible, morally mature behaviours.

What I have described as the *reaching out in the direction of the other,* as the ontological and evolutionary basis of all physical reality, involves different kinds of energy transfer and transformation. If love is the motivation and energy required to make connections in this *reaching out*, then this is how nature works and how we evolve our moral behaviours. Seeking alignments between human behaviour and the energy potential for change - present within the ontology of all things -is part of the free exercising of human choices. Because this energy potential for change is part of Love's energy, we are not compelled to act in moral alignment with it, and we may still block its embedded presence within the genes and social instincts that inform our personalities. We remain finite and fallible in our attempts to create a moral compass, built upon alignment with love's energy. In exploring different ways of reaching out, we will find ourselves cohering with much in our evolutionary, empathetic inheritance, if we chose to channel that creatively for the benefit of others, as well as ourselves. Rules and regulations (provided by parents, schools or wider society) have proven their worth as a moral framework for this, and they will be treated by some as their servants and by others as their masters. Following such regulations out of a Kantian duty may help order society, but may not always develop our moral maturity. Whatever social framework we construct (Adam Smith's social institutions and contracts), with all their inadequacies or excesses and changing middle axioms of common sense, humans are capable of ignoring them, honouring, transforming or transcending them, in the development of their varied ways of behaving. We sometimes lack the imagination to see the utilitarian consequences (as **Bentham** and **Mills** might say) of our negative behaviours on others, and yet we sometimes reach out to benefit them, regardless of our immediate self-interest.

It would be wrong to use the Darwinian narrative to explain everything and Spencer's *'survival of the fittest,'* or Tennyson's *'nature red in tooth and claw'* hardly do justice to his understanding of cooperative *reaching out* as well as competition. His image of an *entangled bank* fits our times of global networks and cultural diversity and adaption, even more vividly than his own. Within that entanglement, he gave us a picture of intra-species connectivity across the breadth of time and space, well before the genome project amplified our understanding of its processes. Those who followed his mega-narrative discovered more about our social evolution with its economic and educational processes of natural selection, and their influence on our moral and adaptive development. He also left us with his own view of the issues in this chapter, and his own description of our moral responsibility *'A moral being is one who is capable of comparing his past and future actions or motives and of approving or disapproving of them. We have no reason to suppose that any of the lower animals have this capacity.'* [482]

---

[482] *The Descent of Man and Selection in Relation to Sex* 1871.

CHAPTER 11
Love and Death

In nova fert animus mutatas dicere formas/ corpora [483]

Will humans ever evolve radically new perceptions and relationships within the *pattern paradoxes* I have discussed, so that a new metaphysic of Love's Energy will emerge alongside, or within, new kinds of scientific map mapping? Perhaps not! Perhaps the easier assumptions of science and technology are taking us too far away from this possibility. On the other hand, perhaps this metaphysic is already embedded within the connections of love many have already made and are making in different parts of the world.[484] Because this metaphysic has always been embedded within the processes of physical reality and human perception and the connections that flow from Love's relationships, perhaps it is waiting for us to rediscover its presence and forms, or invent a new linguistic construct to identify and express it. Perhaps, we have to rediscover our trust in the presence of Love to know how potent Love's Energy really is and always has been[485] – within the universe and the diversity of changing human awareness and perceptions. Such a metaphysic might be unrecognisable in terms of traditional religion, let alone contemporary science. The issues Darwin wrestled with, in the *pattern paradox* of order and chance, will certainly remain, as will different forms of Heisenberg's indeterminacy, because Love will contine to be Love, rather than law. Love's metaphysic is not a new ideology or doctrine. Because it remains true to its own nature, it will give no simple answers, even as it *reaches out* in its longing for our continuing autonomy in whatever kind of existence and situations we humans create for ourselves and our world. Nor is there any guarantee that humans will evolve a greater consciousness of Love's energy and embedded presence, whatever our technology and knowledge. Such a guarantee could only be delivered by a metaphysical order that undermines what we know of the universe and the ways of Love's emanating reaching out. Love's purpose remains in the reaching out, but this purpose is no linear teleology. Love remains love even as its purpose is embedded in our response to it. This response will always contain and reflect the reality of indeterminacy, with its different forms of random chance, not to mention its potential for choices that hurt and destroy the autonomy and life of others as well as those who heal and redeem it. Love

---

[483] Opening lines of Ovid's Metamorphoses

[484] Professor Derek Embrey OBE said to me recently 'some of us physicists believe physics is the language of God.'

[485] Professor Embrey claims that the extra energy needed to e xplain continuing expansion of the universe has never been properly explained and showed me the equations that raise this dilemma.

will continue to bear the cost of its kenotic processes, which we perceive as this emanating indeterminacy in the pattern paradoxes of reality.

Nicholai Berdyaev (1874-1948) said *'Only the fact of death puts the question of life in all its meaning.'*[486] The greatest *pattern paradox* of all, and that which contains, or references all others, is that of *life* and *death* – echoed for Christians by the oscillating connections between Good Friday and Easter Sunday. The radical discontinuity between life and death is highlighted by Death's radical finality, whatever constructs we produce about its continuity with life. Darwin's evolutionary language includes both *ascent* and *descent*. The Orthodox Church's theologically normative and symbolically icon of the resurrection is not an *ascent* from a tomb, but a *descent* into the cul de sac of death, symbolized by Hades and its entombment of our old humanity in Adam and Eve. This is the final embedded incarnation into the total reality of life and death in our universe, where there are no escapes or guaranteed futures for any species, however they evolve. By entering this dark cave, the Christ touches that which is locked within it, bringing freedom and transformation from within. This is not another evolutionary variable, but an unveiling of life within death, or the transformation of death's domination over life. In this resurrection icon, life springs out of the dark cave of death in ways that resonate with the light of the new birth of humanity, shining in the dark cave of creation in the nativity icon – where the babe is wrapped in shining white, swaddling clothes that look very much like grave clothes. Life in death, and the life that transforms death, remain a mystery and, at least in this Orthodox tradition, the transformation happens unseen, in the dark of Holy Saturday, not the dawn of Easter morning. Those who witness the results of resurrection on Easter morning, particularly the 'apostle' Mary and other women, are not privy to what actually happened in the dark cave of the tomb, only to the presence of a new form of life that comes out of it. Ever since, despite all the differences in the Gospel accounts[487] and their interpretation, Chistians have looked to resurrection as their central paradigm for the meaning of life and of life in death.

A few months before the Italian, Galileo Galilei[488] looked at the planets of Jupiter through his telescope, an English astronomer and polymath, **Thomas Harriot** (1560 –1621) looked at the moon through his own, very primitive instrument and produced the first ever drawing, based on such a view. It was 26th July 1609. He then went on to draw ever more sophisticated versions. At the same time, in 1610, Shakespeare was writing his last play – the Tempest, capturing something of the

---

[486] I have not been able to find the source for this well known quotation. He was a prolific author of political and religious ideas. A good expression of his philosophy can be found in his *'Spirit and Rea;ity'* of 1946

[487] There is considerable diversity if not disagreement in the Gospel accounts about resurrection appearances which may be one expression of their perceptive reality.

[488] See LE1

spirit of his age. His sources may have included Erasmus and Ovid's Metamorphoses – and its various themes[489] of transformation – *'I intend to speak of forms changed into new entities.'* The themes of transformation have been a continuing dimension of Love's Energy and now, in this last chapter, we face perhaps the greatest of them all, We join with those, through the ages, who have wondered about death as opening up a *'form changed into new entities.'* In the Tempest, Prospero, with his arts *'rapt in secret studies,'* and *'whose library was dukedom enough'* conjured Ariel, a spirit of nature, to create the storm, a symbol of death and new opportunity on a distant shore. Perhaps the idea of Prospero was based on the famous mathematician, astronomer, astrologer and adviser to Elizabeth, the Welshman **John Dee** (1527-1609) who dealt in codes and secret calculations, very much in the way Newton would later deal in alchemy and physics.[490] In the Tempest, the travellers have arrived at a transition point – a shipwreck that isolates them between the past and the future and Prospero says to Miranda, his daughter, *'What seest thou else in the dark backward and abysm of time? If thou remember'st aught ere thou camest here, how thou camest here thou mayst.'* On this newly discovered island, the spirits play their games to reconcile different political factions and then Prospero says *'These our actors, As I foretold you, were all spirits and Are melted into air, into thin air: And, like the baseless fabric of this vision, The cloud-capp'd towers, the gorgeous palaces, The solemn temples, the great globe itself, Ye all which it inherit, shall dissolve And, like this insubstantial pageant faded, Leave not a rack behind. We are such stuff As dreams are made on, and our little life Is rounded with a sleep.'*[491] Each age paints its own pictures of the *rack* we inherit and leave behind, and the finitude of all existence, even that of our sun and its solar system, let alone the teeming and evolving species of this *great globe. Our little lives* are often *rounded* by more painful reality than the *stuff that dreams are made of.* Shakespeare knew something of the political and personal coastline of this reality, and helped us navigate through its wonders (as indeed does science itself) and face some of its horrors. It is a coastline where the contours of literature, culture, theology and science interweave in the transition points and *pattern paradoxes* of time and space, and we are left wondering, with Prospero, as with Einstein, Heisenberg,

---

[489] from human to inanimate object, constellations, from animals to humans etc. Book One begins with the creation. If Darwin read this I wonder what he made of its vivid descriptions of transmutations!

[490] His *Brytannicae reipublicae synopsis* (1570), mapped the achievements and potential of the Elizabethan world and his *General and rare memorials pertayning to the Perfect Arte of Navigation* 1576, advised Elizabeth 1 to build a strong navy and imperial expansion. Dee made a formal claim to North America on the back of a map drawn in 1577–80 and in his *Title Royal* 1580, even proposed that King Arthur and Madog ab Owain Gwynedd had discovered America, no doubt to show England's claim to that 'New World' was stronger than Spain's. Numbers were the key to all knowledge and gave humans the potential for spiritual power. They were the basis of a religion that could unify the Catholic and Protestant divisions and recapture the purer theology of the ancients.

[491] Act 1V Scene 1. Capital letters indicate where a new line should begin.

Darwin and some of the others we have considered, what else there is to see and discover in the *dark abysm of time.*

As we have seen, the story of science has progressed via the intuitive speculations of great individuals which then inspired the determination and courage to pursue a *probability* or *possibility* to its next stage. We know how pre-cognitive intuitions are responsible for much in our conscious thinking. However much we understand of how the brain functions behind, and in, these intuitions, our existing knowledge invites us to remain open minded about the larger context of science itself. Love is clearly about far more than science, but we do well to listen to the latter's best findings, as well as the instincts and feelings of the heart. In claiming that the nature of God is Love, I hope the reader will understand that I view this not as a reduction of the other attributes of God, in the belief systems of different faith communities, but as an expansion and summary. This is surely the genius and distinctiveness of Christianity – that in the Love of God expressed and emanating in Christ we see the best way to be human and our best glimpse of God. I believe that this nature of God transcends the dilemmas of our *pattern paradoxes,* even as it is embedded within them. God's Love, of course, also transcends our multi-facetted experiences of love, in and with others, and our own attempts to live up to the possibilities and callings of love. As we *reach out* across different *pattern paradoxes,* we learn more of Love's calling and ambiguities, in our dissolving experiences of its *presence* and its *absence.* Clearly my use of Love, as an energy having its own ontology, lifts it beyond the *inherited rack* of our daily struggles and failures in love, but it includes those experiences as reference and access points in the fabric of its embedded forms, particularly as interpreted in the humanity of Jesus of Nazareth. We have looked at how these *forms* become new *entities* in the continuing presence of Love's energy, and we have seen how the early mystics saw them as emanations of ek-static and kenotic Love itself. If anything of what I have said is true, then Love must be *ontological* in ways that do not contradict the best of science *epistemologically,* now or in the future. But it must be far more than this. In LE1, we looked at the *pattern paradox* of *some-thing* and *no-thing,* with all its ontological and epistemological challenges and unknowns. If death is a return to nothingness as many believe, others suspect and most of us fear, then is there a *pattern paradox* in that state of being *no-thing* which has its own kind of *some-thing* within the Energy of Love? We certainly know, within many human experiences of particular *some-things,* the haunting presence of *no-thing,* in all its emotional ambiguity and diversity. We may long for a particular some-thing, only to discover its meaning and value disappears into *no-thing* when we acquire or possess it. Our lives may seem full of *some-thing* and then it unpredictably runs away like the sand of time in our hands, and we are left in the pain of the *pattern paradox* of *connectivity* and *separation*. The *pattern paradox* of *life* and *death* reflects the mystery of many of the others, including *light* and *dark, certainty* and *uncertainty, knowing* and *not knowing.* In Love, says the Christian faith, we see how life and death overlap, and speak to each other if not merge, even when they are most separated in physical reality. In cosmology and evolution, we have seen how life comes from death, as embedded Love becomes the source of transmutation and transformation with all the *indeterminacy* of its emerging and outcomes.

In *'Proof of Heaven,'*[492] the neuro surgeon Dr Eben Alexander tells his own story of encountering Love as part of a Near Death Experience (NDE). *And if I had to boil it down further, to just one word, it would be, simply Love. Love is, without doubt the basis of everything...Not much of a scientific insight. Well I beg to differ. I'm back from that place and nothing could convince me that this is not only the single most important emotional truth in the universe, but also the single most important scientific truth as well.*[493] Of course, we know something of the neurological causes of NDEs and Eben Alexander knew them too. Recent experiments on rats shows heightened brain arousal activity after death, following cardiac arrest.[494] Alexander also quotes Einstein and Heisenberg as well! We are reminded frequently in his book that he remains committed to science. *To say that there is still a chasm between our current scientific understanding of the universe and the truth as I saw it is a considerable understatement. I still love physics and cosmology, still love studying our vast and wonderful universe. Only I now have a greatly enlarged conception of what 'vast' and 'wonderful' really mean.*[495] He certainly believes in a world of connected vibrations of energy, where the observer cannot be separated out of the process and calls for the inclusion of 'consciousness' within any theory of everything. *Before my experience out beyond, I was generally aware of all these modern scientific ideas, but they were distant and remote. In the world I lived and moved in -the world of ...operating tables and patients who did well or not depending partially on whether I operated on them successfully – these facts of subatomic physics were rarefied and removed. They might be true, but they didn't concern my daily reality. But when I left my physical body behind I experienced these facts directly.*[496] He puts together two interesting and compelling words here. Firstly that of *'experience,'* which remains of particular significance, after the different

---

[492] Piatkus 2012
[493] Ibid page 71
[494] 'Within 30 seconds after suffering a cardiac arrest, all the animals displayed a short-lived surge of widespread, highly synchronized brain activity. We were surprised by the high levels of activity,' Dr George Mashour, one of the US researchers from the University of Michigan in Ann Arbor. August 2013
[495] Ibid page 82
[496] Ibid page 151

work of Heisenberg and Eddington, referred to in LE1, not to mention the ancient mystical theologians referred to in LE2. He then adds something quite surprising for a man so immersed in science - *'these facts directly.'* Experience *and* Facts! The context is also important. This allegedly took place when he left his *'physical body behind.'* So we have an *experience* of *facts* that seemed to him to be a *'direct'* experience. The *facts* he refers to are those other scientists had talked about, at least those he had studied. Therefore we are dealing with his *perception* of what other scientists had meant. His *experience*, he claims, confirms them directly.

However strange a choice this may seem for the final pages of this book on Love's energy in cosmology and evolution, I choose it because, whatever we think of Eben Alexander's work, he is struggling to engage with a powerful (NDE) experience of Love, *via his knowledge of science.* This is no hippy trip for him, but a life changing experiential, or experimental 'vision,' during a life changing illness, caused by bacterial meningitis.[497] Of course, history is full of many different kinds of vision, not all of them of Love. He describes his 'memory' of what he experienced in considerable and vivid detail, but sums it up as an experience of Love. *I continued moving forward and found myself entering an immense void, completely dark, infinite in size, yet also infinitely comforting. Pitch black, as it was, it was also brimming over with light; a light that seemed to come from a brilliant orb that I now sensed near me. An orb that was living and almost solid..This being was so close that there seemed to be no distance at all between God and myself. Yet at the same time, I could sense the infinite vastness of the Creator, could see how completely minuscule I was by comparison.*[498] Yes, he is interpreting this NDE experience of an *immense void..brimming over with a light that seemed to come from a brilliant orb'* in relation to the idea of God. I find his next sentence to be of particular interest. *'This being was so close that there seemed to be no **distance** at all between God and myself.'* Let us, for one moment omit the identification of God with this light, in case that could be seen as projection. Let us pause instead on the main observation – *that there seemed to be no distance at all between 'it' and myself.'* This is surely of interest. Perhaps this observation of *'no distance'* is also a projection from his knowledge of science onto this NDE. It implies a level of connectivity between A and B that is remarkable. I join with him in recognising how much more there is to be discovered of Love and Love's energy, not only in science, but in our daily living. This is no theoretical matter. Far too many people (and not just the elderly and bereaved) live with an experience of isolation, or lost purpose, or feeling unloved

---

[497] I also acknowledge another connection and perhaps reason for including this – my grandson Theo contracted bacterial meningitis in his first weeks of life and its effects will be with him throughout his life.
[498] Ibid page 47

or overwhelmed with fears and anxiety. This too is part of our human condition. Love matters to them a great deal, whatever their experience of its different meanings in its presence or absence. Finding our own ways of loving others matters to all of us, even though the practical opportunities to do so vary a great deal across the globe, or within a single life. The human condition is full of the stories of Love's oscillating pathways from A to B. Too often those pathways seem blocked, hidden, torturous, forgotten, ignored, hard or undervalued, but they are present, waiting to be discovered and explored in their unpredictability, indeterminacy and possibility.

In Love's Energy is there a place called heaven?

What can we say about this difficult question in the light of Love's Energy? All religions have given their own kind of answer. Islamic teaching seems to have infected many suicide bombers with the belief that their 'martyrdom' in killing others will merit the 'fleshy' delights of Paradise. Other religions have taught a circular, never ending kind of loop, where each life lived provides opportunities to escape the particularities and limits of our Karma. In Judaism, Christianity and Islam the word 'heaven' is often used as a metaphor for the after life. This is confusing and problematic. Firstly, we know from the Genesis texts that God created the 'heavens and the earth.' Heaven, in this sense, was the visible dome above the earth, the place of the 'lights' with all their symbolic meaning. Science has taught us that the *space* of the heavens with all its different solar systems and galaxies is indeed greater than we can imagine, but part of a universe that 'makes itself.' It cannot be the *space* of the after life, the life we *pass*[499] into after we die. The implication of the language of 'passing' is that 'we' have moved from one state into another. Einstein taught us that space and time are interrelated. This means that where there is any kind of space, there is an effect on time and vice versa. There is no space or time in an 'after life,' otherwise it would still be part of this cosmological or biological reality. So, if there is no time-space, or space-time, an 'after life,' has to be outside time and certainly outside the space mostly called 'heaven' in the Bible. In the Lord's Prayer, recorded as the words of Jesus in the New Testament, we pray *'Our Father which art in heaven.'* I assume this doesn't refer to a place in the cosmos, but a 'theological' space. Jesus was teaching people to pray from within their first century cosmological and religious assumptions about the 'location' of God, or he was meaning something different. Jesus said that the Kingdom of God is *'within us,'* and at the same time pointed to a coming Kingdom where the values and presence of God would be unveiled more completely. Being *'within'* us and *'outside'* us, constitutes the *pattern paradox* of ekstasis. That which is within us must be outside us in some way, to connect with other *others* and something more. We saw, in LE2, that the nature

---

[499] Passing or passing on is now a common phrase for death

of Love is ekstatic. Does that help us understand the nature of the Kingdom of God as being within, but always reaching out in its own kind of connectivity and potentia? Is this *pattern paradox* of connectivity helpful, as we face the relationship between Love as embedded ek-statically in living things and Love in death?

What did Jesus mean – that the Kingdom and heaven were the same 'thing?' If so, how in his mind, did this relate to an after life? There is no description in the New Testament of the geography or 'interior design' of heaven, and Jesus warns us about speculating. In John 14.2 we find, in different translations, 'In my Father's house (*mansion*) there are many *rooms*, or *dwelling places*, or *abodes* or *more than enough room* for you.. I go to prepare a place for you..that where I am you may be also.' I have read many interpretations of rooms or mansions as meaning e.g. different degrees of spiritual state, different religions, the Jerusalem Temple as *'my Father's house,'* Jesus' suffering and death, Jesus as the Way, the Truth and the Life, participation in the Spirit, or Jesus' post resurrection life and 'Sacred Tabernacles in which the Real Presence of Christ dwells in the Holy Eucharist.' [500] I also think of Robert Hook looking down those early and primitive microscopes at a section of cork – seeing for the first time in human history what reminded him of tiny monks 'cells,' and drawing the image in his Micrographica. It was only much later that we realised all living things are made up of these 'dwelling places' and then, much later again, that we discovered that these cells contained membranes, protons, chromosomes and then DNA.[501] We are learning that the secret of life cannot be explained by the coding of DNA alone, but probably only in the intricate and interrelated communication and connection of the energy in trillions of cells in human bodies. We still do not know exactly how this is achieved. In the Greek of John's Gospel we find μονή (monē) as meaning a staying, abiding, dwelling or abode in the οἰκία (oikia) of the Father which means a domestic house or household, inhabited edifice, or dwelling. I see these two words together as reinforcing, not so much a physical place, with a certain large number of rooms, but the sense of 'being with,' or 'abiding in,' as a deeper kind of relationship. In the pain of loss and bereavement, some people like to visualise a description of a place, and others a more abstract sense of a 'being with,' or a 'a dwelling with,' or within God. In truth, it is hard for us who live in rooms, if not mansions, and certainly in space-time, to cross the boundary limits of human understanding or imagination in our different beliefs.

The ancients commonly used the word 'eternity' to describe the after life. Clearly this meant much 'more' than an *'impossible to imagine'* duration of time.

[500] Catholic Doors Ministry
[501] See earlier discussion on the work of Wilkins, Franklin, Crick and Watson.

Eternity can't be a measure of a length of time, just as heaven can't be a description of a place, in some kind of space. It must transcend time and space. So, Christians and thinking Muslims, who take science seriously, have to reject any picture of Paradise that is about place, or time, or activities in that 'place.' Nor will it be of any help to take the contribution of Heisenberg's indeterminacy principle here, tempting though it might be. That, too, only operates within our understanding of the relationship between particles of energy matter from within the workings of quantum in our physical cosmos of space time. However useful it is to our spiritual understanding of connectivity in those relationships, it cannot apply to anything outside of space time, material reality. I have been saying, in this Trilogy, that Love reaches outside of itself in our direction, and is also embedded in the energy matter freedoms and connectivity of our kind of cosmos and perhaps of any kind if others exist. Love is the source and nature of the energy that creates the *possibility* or *propensity* of those energy-matter freedoms to come into existence and 'make themselves.' Love, in this ekstatic state in the created Otherness of the cosmos, with its diversity of other *others* and in the humanity of Jesus, is all we can 'know' of the nature of Love. Is it possible then that Love is the source of a life after this life and that this life in Love, after this life, has both a radical continuity and a radical discontinuity with what we know as 'life?' It must have a radical discontinuity because of the science of space time quoted, but does it have a radical continuity, because it is true to itself in its reaching out to create the *possibility* or *propensity* of an energy that enables life to make itself and is always, somehow conserved? Is it possible that Love is the connection between these two radical polarities, because that is its nature? If Love's Energy can be embedded within space time, without contradicting its own nature, can it continue in a relationship with us after our existence in that space time comes to an end? At that point, even though we can never understand the nature of the connectivity, is it possible that the relationship with Love is not extinguished?

So, does the Energy of Love move through and beyond what all species know as the maximum entropy of death? Most religions hold to a belief or hope that this is the case. It cannot be *known*, but, as we have seen, the relationship between *ontology* and *epistemology* is porous rather than fixed, and there are many unseen frequencies of connectedness between them. I do not believe we can *know* as science *knows,* but perhaps there is another kind of knowledge that science will take more seriously in the future. It is within this knowledge that Christians, and people of many other faiths, have believed that Love is stronger than death, or flows through it and beyond it. Christians believe that Christ who is the embodiment of God's love lives, not as we live now, but beyond the limits of that living, because the Love that was in Christ was fully ekstatic. It stood outside our experience of 'living,' and was also deeply embedded in it – so much so that it embraced what we know of death as its *final* incarnation. Love embeds itself kenotically into embedded reality to the degree that it transforms that reality, including the relationship between life and death. It opens up what was the *closed* system of death into the *open* system of another kind of risen life. If we can find, in our living, a way of being in the flow of Love's energy, or as Christians might say in Christ, then, perhaps, we will experience a communion or connectedness with an energy that stands outside the limits of a closed, life and death system. In this sense, death is the final *stasis* and Love's energy opens it up, or transforms it, into an *ekstasis*, a standing outside of itself. The focal point of Christian belief is the resurrection, not the idea of eternity or an after life, although as stasis moves to ekstasis there may be a connection. In Christ's death, Love brings life and so demonstrates that life continues to long for life and can stand outside, as well as through and in death.

The New Testament witness is built around two apparently contradictory eschatological beliefs. Firstly, that there is a continuity between eternal life as a state of being, *after* this life, and eternal life as something knowable or accessible *within* this present life. Secondly, there is a discontinuity between those two things and from the perspective of our knowledge within the present *now* of space time, the future is unknowable and indeterminate, and therefore radically different from our understanding of it. What stays the same is Love. If there is what the Bible calls a final judgement to our lives, it will surely be the judgement of Love which continues beyond space and time in what we might call eternity. If God is our judge, then it will not be as in a human court where a final punishment decision is made *to fit the crime.* If this were the case, it would deny the death and resurrection, and the life and ways of Jesus as the expression of the love of God. Jesus told us to forgive unto seventy times seven, which is a number that evokes an eternity of love's forgiveness. Dr Barry Morgan, the Archbishop of Wales,[502] said that when Peter betrayed Jesus at the latter's trial, Jesus didn't say anything, but only looked at him. We cannot know, but in that look we might assume there was the judgement of Love and love's capacity to forgive in ways that changed lives. Nelson Mandela, on the very same day as that sermon – December 15th 2013 - was buried in the village where he had been born. The whole world remembered how his forgiveness[503] of his Apartheid persecutors[504] inspired people and leaders across the world, changed a whole nation and saved thousands of white people from violence and civil war, even if

---

[502] preaching in St Mary's Swansea on December 15th 2013

[503] On his journey from hatred and violence to the capacity to forgive, perhaps the suffering of his years in prison forged a transformation into love.

[504] Which somehow came out of his years of suffering in Prison and the transformation of his own beliefs

many black people continued to be killed and different forms of apartheid waited to be solved. Our failure of the values and ways of Love in our lives, and our failure to forgive and love others, as God continues to reach out towards us with an unconditional love, is our own judgement. In the life of Jesus, we see the judgement of Love. So, we can imagine the Love that will not let us go, but yet, after death, still respects the freedom and autonomy of our created nature for all eternity, because of Love. This kind of spiritual map leads me to believe that Love and Love's synergy will not let us go, whatever our rejection of its presence in our continuing state of Love's freedom. In this sense, the un-conditionality of Love is present in death and in eternity, as well as in life and our mortality. We are judged by love, and all of us have failed that love, in one way or another. But we are also saved by this same love, because its forgiveness of our failures is so much greater than we can imagine, or have ever shown to others. As we have seen, Love cannot be untrue to itself and its own ways, and therefore cannot force any kind of future on anyone, or predetermine it – although Love is free of space time constraints, even as it is embedded within them. Love, as we have seen, reaches out to make possible the conditions for the Other to exist within its own freedom and potentiality. I believe this continues beyond and despite the coastline of death. Connectedness continues because of the nature of love and its ekstatic reaching out across all boundaries and barriers, even that boundary between what we know as life, and what we know and imagine of death. Contemplating such a possibility moves us to *'wonder, love and praise'* as **Charles Wesley's** (1707-1788) famous hymn puts it.

Finish, then, Thy new creation;
Pure and spotless let us be.
Let us see Thy great salvation
Perfectly restored in Thee;
Changed from glory into glory,
Till in heaven we take our place,
Till we cast our crowns before Thee,
Lost in wonder, love, and praise.[505]

I hope, with Eben Alexander and countless others through the ages, that underneath our subjective, connecting and engaged relationship with Love's Energy, Love will embrace us beyond the limits of physical reality,[506] as it does from within those limits. I believe that Love is always the great NOW of God which transcends space-time and, therefore, is both eschatological past and

---

[505] Final verse of Charles Wesley's most famous hymn written in 1747. Interestingly it is commonly chosen for both Weddings and Funerals – those two great services of the offering of love, which book wedge are greatest experiences of Love.

[506] Guru Nanak, (1469-1539), the founder of Sikhism, called this sense of beyond the limits of time and space 'Anannt.'

eschatological future in its present presence, propensity and potential. This is the great I AM THAT WHICH I AM[507] of the God we have dared to name as Love. This is the great divine relativity of BEFORE ABRAHAM WAS I AM[508] with which Jesus identified something of himself, and out of which, his Love flowed, reaching out in our direction, redemptively and creativity. This Love is always NOW because it embraces and energises, as well as transcends all our human time and space measures of relativity and relationship. As I gaze up at the stars and space, or look down through a microscope, I have to acknowledge that the language we use to describe our faith in God, or Love, is itself limited by being a construct of language. This boundary limit may affect all that we believe as well as know, but, within its own terms, we glimpse the possibility of a relationship, however relative, with the boundary crossing energy of Love itself. This Love is always NOW because it abolishes the *difference* and the *distance* between every A and B, at the deepest possible level of energy-matter connectedness, and yet enables A and B to be freely what they most truly are, in their dynamic relationship and potential. I hope that this same NOW of Love will call us home to its source, even as it has joined us in our travelling over the past 13.7 billion years. I hope that there, beyond space and time, Love's energy will harmonise all distance and difference between the As and Bs of our separation with it, and with each other. I hope that deep within the uncertainty and indeterminacy of this, as well as the separation, there will be an instantaneous movement-moment of connectedness, well beyond our imagining, in the *where* and *when* of NOW. I hope that in the *'then'* and *'there'* of that NOW, the waves of Love's light and life will be seen - beyond the present limits of our seeing - to flow through all things, so their potential and their telos are finally, beautifully, and completely fulfilled, within the un-entangled bank of the self-giving and ecstatic energy of God's Love.

The Welsh born priest and poet **George Herbert** (1593-1633) did not write to be published. At the end of his life, he sent his poems to Nicholas Ferrar (1592-1637) of the Little Gidding community, simply suggesting that if they could *'turn to the advantage of any dejected poor soul'* then they could be published; if not, they should be burnt. He wrote out of and for his Love of God in an internal, spiritual life that engaged with ordinary things. He wrote in between the times of Shakespeare and the extremes of religious and political life that were about to rip England apart. In the following poem,[509] he did not speak of God, only of Love. When I was studying the Metaphysical Poets at school,[510] this poem was one of the first to really engage me, and embed itself in my life and faith - after

---

[507] Exodus 3.14

[508] John 8.5

[509] In the collection entitled *The Temple* of 1633

[510] And later in Helen Gardner's lectures at Oxford.

all, poetry is a form of language through which we *reach out* to all that is already *embedded,* but waiting to be discovered in our species creativity.[511] So, it is with this poem that I choose to end this trilogy of Love's Energy, and entrust the rest to the work and thoughts of other *others* within their own indeterminacy!

Love (III)

Love bade me welcome, yet my soul drew back,
Guiltie of dust and sinne.
But quick-ey'd Love, observing me grow slack
From my first entrance in,
Drew nearer to me, sweetly questioning
If I lack'd anything.

A guest, I answer'd, worthy to be here;
Love said, You shall be he.
I the unkinde, ungrateful? Ah my deare,
I cannot look on thee.
Love took my hand, and smiling did reply,
Who made the eyes but I?

Truth Lord, but I have marr'd them; let my shame
Go where it doth deserve.
And know you not, sayes Love, who bore the blame?
My deare, then I will serve.
You must sit down, sayes Love, and taste my meat;
So I did sit and eat.

---

[511] See *The Poetic Species – A conversation between Edward Wilson and Robert Hass.* Bellevue Literary Press. 2014

The Love's Energy Equation

Offered as a bit of fun to those who like the thought of equations, but who, like me, were never good at them; offered also, as a way of reaching out, however inadequately, to those who are good at them, and spend their professional lives working with them, to help understand physical reality.

$$LE/\,ke => O/o^n + f(em+st)$$

KEY

LE	=	Love's Energy
k	=	Kenotic
e	=	Ekstasis
O	=	the Otherness of the universe
o^n	=	the plurality and diversity of other others
f	=	Freedom
em	=	energy matter
st	=	space time
=>		Material implication

04.09.14
The author,
drawn at a conference
on John Berger
by Chris Glynn,
Senior Lecturer, Illustration
Cardiff School of Art & Design
Cardiff Metropolitan University

INDEX

www.ingramcontent.com/pod-product-compliance
Lightning Source LLC
Chambersburg PA
CBHW081433170526
45166CB00008B/2189